T0348583

VOLUME FORTY FOUR

ADVANCES IN
INSECT PHYSIOLOGY

VOLUME FORTY FOUR

Advances in
INSECT PHYSIOLOGY
Target Receptors in the Control of Insect Pests: Part I

Edited by

EPHRAIM COHEN
*Robert H. Smith Faculty of Agriculture,
Food and Environment,
The Hebrew University of Jerusalem,
Rehovot, Israel*

AMSTERDAM • BOSTON • HEIDELBERG • LONDON
NEW YORK • OXFORD • PARIS • SAN DIEGO
SAN FRANCISCO • SINGAPORE • SYDNEY • TOKYO
Academic Press is an imprint of Elsevier

Academic Press is an imprint of Elsevier
The Boulevard, Langford Lane, Kidlington, Oxford, OX51GB, UK
32 Jamestown Road, London NW1 7BY, UK
Radarweg 29, PO Box 211, 1000 AE Amsterdam, The Netherlands
225 Wyman Street, Waltham, MA 02451, USA
525 B Street, Suite 1900, San Diego, CA 92101-4495, USA

First edition 2013

Notice
No responsibility is assumed by the publisher for any injury and/or damage to persons or
property as a matter of products liability, negligence or otherwise, or from any use or
operation of any methods, products, instructions or ideas contained in the material herein.
Because of rapid advances in the medical sciences, in particular, independent verification of
diagnoses and drug dosages should be made

ISBN: 978-0-12-394389-7
ISSN: 0065-2806

For information on all Academic Press publications
visit our website at store.elsevier.com

Printed and bound by CPI Group (UK) Ltd, Croydon, CR0 4YY
Transferred to digital print 2012

CONTENTS

CONTRIBUTORS

Ephraim Cohen
Robert H. Smith Faculty of Agriculture, Food and Environment, The Hebrew University of Jerusalem, Rehovot, Israel

Chaoxian Geng
Dow AgroSciences, Discovery Research, Indianapolis, Indiana 46268, USA

Peter Lümmen
Bayer CropScience AG, Monheim, Germany

Yoshihisa Ozoe
Department of Life Science and Biotechnology, Faculty of Life and Environmental Science, Shimane University, Matsue, Shimane, Japan

Thomas C. Sparks
Dow AgroSciences, Discovery Research, Indianapolis, Indiana 46268, USA

Motohiro Tomizawa
Graduate School of Medical Sciences, Nagoya City University, Nagoya, Japan

Gerald B. Watson
Dow AgroSciences, Discovery Research, Indianapolis, Indiana 46268, USA

PREFACE

This special volume (part I) consists of five chapters that comprehensively review basic and practical aspects of several major target receptors that are involved in crucial physiological mechanisms in insects. Disruption of the normal function and control of such receptors via natural products and synthetic ligands is the underlying basis for insect pest control or insect pest management.

Sustained homeostasis of water and electrolytes, which is of extreme importance for insects to survive environmental challenges, is clearly a target for interference. Such maintenance is critical for insects to subsist in subzero, xeric or salty habitats, or for those ingesting large volumes of fluids such as blood or plant sap. The nAChR, which forms a postsynaptic membrane-spanning ligand-gated ion channel, is target for interference with agonistic ligands such as nicotines, neonicotinoids and spinosyns. The ligand-gated calcium channels, which are selective to calcium conductance, are important *inter alia* in gene regulation, membrane transport processes, release of neurotransmitters and muscle contraction. The ryanodine receptors, forming the intracellular calcium channels, are the molecular target of insecticides such as the phthalic acid diamides and anthranilamides. Inhibitory synaptic transmission mediated by GABA-gated and glutamate-gated chloride channels interacts with insecticidal receptor ligands such as fipronil and macrocyclic lactones, respectively.

In general, insecticidal bioactive compounds were discovered accidentally or by screening programs. The following research was aimed at revealing their mode of action, while concomitantly at designing selective and efficient pest control agents. Large-scale screening of synthetic compounds or natural products has been successful in identifying chemical structures that were commercially developed and subsequently applied to control a wide range of damaging pests in agriculture, forestry and public health. The intensive studies on the mechanisms of action of insecticidal compounds have generated seminal information regarding agonistic or antagonistic interactions with their respective receptors. Crystallography of receptor proteins, *in silico* docking simulations, recombinant and site-directed mutagenesis studies, and photoaffinity labelling experiments, combined with QSAR analyses has been harnessed to design active new compound leads. Most of the above target systems have already been exploited as the basis of commercialized pest control agents. Nevertheless, despite extensive research based on highly active lead compounds, some systems have so far failed to generate commercial products.

Additional well-characterized receptors and their physiological agonists, which play crucial roles in insect development, metabolism, neurotransmission, behaviour and reproduction, will be published in a following part II volume. This special volume will include the following topics: voltage-gated sodium channels targeted by DDT, pyrethroids (as well as by scorpion and spider toxins); the octopamine receptor that is the site of action of insecticidal formamidines; the ABC transporters in the context of insecticide resistance and of controlling insect pests; the juvenile hormone (JH) receptor(s) and molecular mechanisms of JH action; acetylcholinesterases as target enzymes; and receptors targeted by neuropeptides. I do hope that the two AIP volumes will provide comprehensive and broad accounts on major target systems with pivotal roles in insect physiology and, consequently, utilization in insect pest management.

Unfortunately, only a limited number of target receptors, in particular, those related to the nervous system, have been adequately characterized and utilized for commercial development of insecticides. Conceivably, a plethora of target receptors exist within the nervous system and in other tissues as well, that have not yet been tapped. It is hopeful that advances in genomic, molecular biology and proteomic research on these receptors and their corresponding agonist or antagonist ligands will provide essential information and experimental tools to develop novel and selective pest control agents.

EPHRAIM COHEN, Editor

CHAPTER ONE

Water Homeostasis and Osmoregulation as Targets in the Control of Insect Pests

Ephraim Cohen
Robert H. Smith Faculty of Agriculture, Food and Environment, The Hebrew University of Jerusalem, Rehovot, Israel

Contents

Abstract

Insects, which are characterized by a comparative large surface area versus a relative small size, are generally challenged by osmotic stresses and, in particular, when exposed to hot, dry, salty and freezing environments. Furthermore, insects feeding on large volumes of fluids such as blood, phloem-sap or nectar confront a different type of osmotic challenges. The digestive tract and associated Malpighian tubules are the major organs active in alleviating such challenges. Various passive and active transepithelial transport facilitators, such as pumps, ion exchangers, co-transporters and ion and water channels

Advances in Insect Physiology, Volume 44
ISBN 978-0-12-394389-7
http://dx.doi.org/10.1016/B978-0-12-394389-7.00001-6

1

(aquaporins), are under the hormonal control by biogenic amines and various neuro-peptides. Interaction of the neurohormones with their respective receptors induces cas-cade pathways that trigger diuretic or antidiuretic responses. Since water homeostasis and osmoregulation are crucial for the survival of insects, the elaborate physiological mechanisms that have evolved in maintaining balance of water and electrolytes are po-tential targets for interference. Synthesis of metabolically and environmentally biostable neuropeptidergic analogs in combination with receptor assay-guided structure–activity studies could be instrumental in discovering highly potent molecular leads that are vital in generating efficient and selective insect control agents.

LIST OF ABBREVIATIONS

AQP aquaporin
ADF antidiuretic factor
ADH antidiuretic hormone
CAP cardioacceleratorty protein
CHO Chinese hamster ovary
CNS central nervous system
CRF corticotrophin-releasing factor
CT calcitonin
DH diuretic hormone
DNJ deoxynojirimycin
DRIP Drosophila integral protein
GPCR G protein-coupled receptor
IP$_3$ inositol (1,4,5) triphosphate
ITP ion transport protein
LEA late embryogenesis abundant
MIT major intrinsic protein
MT Malpighian tubule
NO nitrogen oxide
NOS NO synthase
V-H$^+$-ATPase vacuole-type H$^+$- ATPase

1. INTRODUCTION

Homeostasis is the dynamic tendency of a biological system to main-tain constant internal equilibrium and relative stability. Water is a universal solvent that enables transportation of nutrients, metabolites, hormones and bodily waste within organisms. Water is also the common environment in which cellular and extracellular metabolic activities take place. Water ho-meostasis delineates active physiological processes that sustain and preserve an internal steady state of water within cells (cytoplasmic fluid) and two other compartments (extracellular spaces and circulating fluids) of an organism. Water homeostasis is maintained by regulation of the osmotic pressure (osmoregulation) of the organism's fluids and electrolytes within certain physiological range limits and tolerable boundaries.

Insects, and in particular terrestrial species, are under continuous osmotic stress as they tend to lose water due to their small size and therefore comparative small volume versus a relatively large surface area. Water is lost due to evaporation across the cuticular surface area, respiratory transpiration though gas exchange, and as a result of anal excretion of urine and faeces. Insects gain water by direct ingestion of fluids, feeding, generation of metabolic water or in particular cases by absorption of atmospheric water vapour with the help of hyperosmotic or hygroscopic solutions excreted from mouth and anus (Machin, 1983). Insects confront severe osmotic stresses in dry, hot or salty environments as well as under freezing conditions, whereas in contrast, sanguinivorous arthropods (e.g. ticks, mosquitoes, some flies), nectar (e.g. bees, lepidopteran adults) or plant-sap feeders (e.g. aphids, whiteflies) must void excess of ingested fluids to maintain tolerable water balance. Post-feeding diuresis in nectar-feeding lepidopterans and postprandial diuresis in lepidopterans and mosquitoes have been documented (Bushman et al., 1989; Dores et al., 1979; Nicolson, 1980). Insects are characterized inter alia by their outstanding adaptations to survive such environmental conditions. Insects (and other terrestrial arthropods) have therefore developed behavioural and physiological/biochemical strategies to overcome unfavourable environmental conditions. Behavioural avoidance includes migration and aggregation, escape to favourable microhabitats and seeking moist microclimate by borrowing and tunnelling in niches and into substrates. Nocturnal activity and seasonal dormancies (aestivation, overwintering, diapause), which are characterized by hypometabolism, are additional strategies used to avoid stressful climate conditions by minimizing water loss (Benoit, 2010; Chown et al., 2011).

Physiological and biochemical strategies in keeping the water balance are involved in the manipulation of spiracles' opening, composition of cuticular lipids, generation of antifreeze proteins and low-molecular-weight cryoprotectants. Of paramount importance is the dynamic regulation of fluids and solutes movement across the alimentary canal and attached Malpighian tubules (MTs) and in association with their anatomical structure and conserved plasma membrane-associated transport proteins. The proteins include apical vacuole-type H^+-ATPases, basolateral Na^+-K^+-ATPases and inward rectified K^+ channels, chloride channels, ion exchangers and co-transporters, organic solute transporters and water channels (aquaporins, AQPs). Their physiological functions are hormonally regulated via second messengers such as cAMP, cGMP, nitrogen oxide (NO) and calcium (Dow and Romero, 2010) .

2. MORPHOLOGICAL ASPECTS OF WATER BALANCE

Cuticular transpiration depends on the lipid/hydrocarbon composition of the epicuticular wax layer (Hadley, 1994), and the respiratory transpiration is largely controlled by opening and closing of the spiracles. Extensive loss of water occurs during flight activity as a large volume of air is ventilating through the respiratory system. Yet, at the same time, generation of metabolic water takes place.

The alimentary canal and associated MTs (Fig. 1.1A) are the major organs in handling water and electrolyte balance in insects. Their functional multifaceted cellular pumps, ion transporters and various ion and water channels control the pH, osmolarity and composition of the hemolymph and determine the final components of the secreted urine. In addition, these organs secrete and eliminate nitrogenous waste and toxic components. The digestive tract consists of three main parts: foregut and attached coeca, midgut and the hindgut (ileum, rectum and rectal pads). Morphological

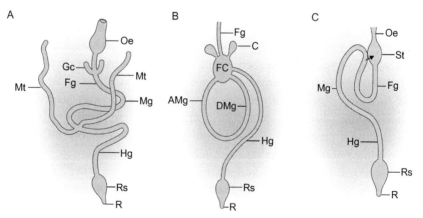

Figure 1.1 Schemes of insect digestive tracts: A, typical insect; B, whitefly; C, aphid. AMg, ascending midgut; C, coecum; DMg, descending midgut; Fc, filter chamber; Fg, foregut; Gc, gastric coecum; Hg, hindgut; Mg, midgut; Mt, Malpighian tubule; Oe, oesophagus; R, rectum; Rc, rectal sac; St, stomach. Anatomical gut structures associated with water-shunting in whiteflies and aphids are depicted in B and C, respectively. Cohen, E and OMICS Publishing Group (ISSN-2161-0983) are credited for this figure. (See Color Plate 1 at the back of the book.)

modifications in plant-sap feeders like filter chambers (in whiteflies and cicadas) (Fig. 1.1B) or close proximity of foregut and midgut (in aphids) (Fig. 1.1C) enable water bypass. The MTs, which are blind tubes connected to the digestive tract between midgut and hindgut, float in the surrounding hemolymph, and the attached accessory musculature in many insects enable a better contact with the bathing body fluid. As MTs are not innervated, their essential physiological roles are controlled by neurohormones released into the hemolymph.

MTs are an efficient excretory system that maintains water and salt balance by producing primary urine and its subsequent elimination into the gut. Recently, the developmental, physiological, genomic and molecular aspects of MTs have been comprehensively reviewed (Beyenbach et al., 2010; Dow, 2009). Structurally, the MTs consist of one layer of apical brush-bordered epithelial cells that separates its lumen from the surrounding hemolymph. Generally, MTs have two functional cell types: the large and abundant principal cells with large and dense microvilli and the intercalated smaller, star-shaped and scattered stellate cells with sparse and smaller microvilli (Fig. 1.2) that are devoid of mitochondria. It is noteworthy that some insect species, such as the tsetse fly, *Glossina morsitans* (Cabrero et al., 2004), and the house cricket, *Acheta domesticus* (Hazelton et al., 1988), lack stellate cells, whereas collembolans, thysanurans and certain hemipterans, such as aphids and whiteflies, lack the MT organ system (Schowalter, 2006). While, in general, the distal regions of the MTs have diuretic functions and secrete electrolytes and organic solutes in addition to water, the proximal ends reabsorb water and solutes. The hindgut, and in particular the rectum, is

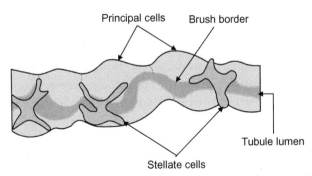

Figure 1.2 The scheme depicts the two different cell types of nearly all Malpighian tubules. Principal cells are abundant, large and densely brush-bordered. Stellate cells are smaller, star-shaped, scattered among principal cells and have fewer and smaller apical microvilli. (See Color Plate 2 at the back of the book.)

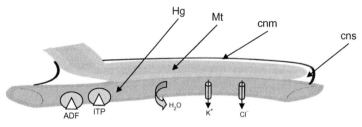

Cryptonephridial complex

Figure 1.3 A generalized scheme of a cryptonephridial complex. Hg, hindgut; Mt, Malpighian tubule attached to the hindgut; cnm, cryptonephridial membrane; cns, cryptonephridial space. The putative receptors for ADF (antidiuretic factor) and ITP (ion transport peptide) neurohormones are depicted in circles.The cylinders represent channels for K^+ and Cl^- ions. (See Color Plate 3 at the back of the book.)

the primary site of water conservation by reabsorption and determines the ionic composition of the urine by selectively regulating ion reuptake. In certain coleopterans and in larvae of most lepidopteran species, the MTs are not free-floating in the hemocoel but form a cryptonephridial complex in which the distal ends are attached to the surface of the rectum and enclosed within a perinephric space by a perinephric membrane (Fig. 1.3).

3. PHYSIOLOGICAL ASPECTS OF ION AND WATER BALANCE

Para- and transcellular water transport across MTs was attributed to principal and stellate cells (Beyenbach et al., 2010). Eight major membrane-associated components are involved in the functional osmoregulatory system of MTs (Beyenbach and Piermarini, 2011; Dow and Romero, 2010) (Fig. 1.4). (1) The vacuolar-type H^+-ATPase (V-H^+-ATPase), which is abundant and densely expressed in the apical brush border membranes of the principal cells, pumps protons across plasma membranes and generates an electrochemical gradient that drives the active transepithelial cation (Na^+, K^+) transport (Wieczorek et al., 2009). This membrane-bound V-H^+-ATPase is composed of a complex assembly of 15 different protein subunits, encoded in *Drosophila melanogaster* by 33 genes (Allan et al., 2005). (2) The basolateral inward rectified K^+ channels, which transport potassium ions across principal cells, are encoded by three genes in *D. melanogaster* MTs (Döring et al., 2002). Inhibition of fluid secretion in *Drosophila* and in the yellow mealworm (*Tenebrio molitor*) by glibenclamide,

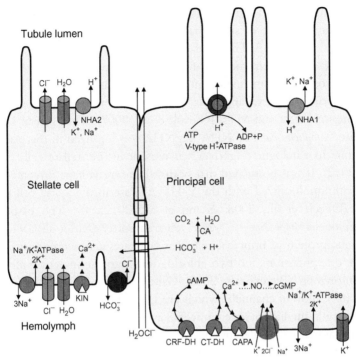

Figure 1.4 A generalized schematic diagram summarizing various current models for ion and water transport as well as receptor signalling pathways in principal and stellate cells of Malpighian tubules. Cylinders in basal and apical domains of the stellate cell represent transcellular chloride and water channels (aquaporins), whereas the transmembrane basolateral-rectified potassium ion channel is depicted in the principal cell. In addition, the paracellular pathway for water molecules and chloride ions is included. The apical ion exchangers NHA1 and NHA2 are represented in the principal cell and in the stellate cell, respectively. The V-type H^+-ATPase that transports protons across apical cell membranes is illustrated in the principal cell. Basolateral G-protein-coupled receptors (GPCRs) for kinins (KIN), corticotrophin-releasing factor-like DHs (CRF-DH), cardioacceleratory peptide (CAP_{2b}), and calcitonin-like DHs (CT-DH) are depicted. Intracellular second messengers such as cAMP and Ca^{2+}, which are triggered by CRF-DH and CT-DH and by KIN, correspondingly, are presented. Generation of cAMP by interaction of serotonin with its GPCR is not included in the diagram. The cascade pathway generating cGMP following interaction of CAP_{2b} with its receptor starts with Ca^{2+} activating NOS, and the produced NO interact with guanylate cyclase (GC) to form cGMP. The K^+, Na^+, $2Cl^-$ co-transporter (red circle) in the principal cell and the Na^+/K^+-ATPase (purple circle) in both cell types are represented. The bicarbonate anion (HCO_3^-) produced in principal cell by carbonic anhydrase (CA) is transported via septate junctions to the stellate cell where the basal bicarbonate/chloride ions exchanger functions. (See Color Plate 4 at the back of the book.)

a known sulphonylurea blocker of inwardly rectifying K^+ channels, indicates the involvement of such channels in diuresis (Evans et al., 2005; Wiehart et al., 2003a). (3) The basolateral Na^+-K^+-ATPase of principal cells is a highly conserved enzyme that functions to actively transport Na^+ and K^+ ions across the plasma membrane in opposite directions. (4) The apical NHA antiporter that exchanges Na^+ or K^+ for protons is expressed in the gastrointestinal tract and MTs (Rheault et al., 2007). The African malaria mosquito *Anopheles gambiae* NHA (AgNHA1) is localized in the principal cells while its paralogue (AgNHA2) is present in the stellate cells (Xiang et al., 2012). It was postulated that both NHAs are voltage driven as they are co-immunolocalized with the V-H^+-ATPase in the apical cell membranes (Okech et al., 2008; Xiang et al., 2012). (5) The basolateral bumetanide-sensitive Na^+-K^+-$2Cl^-$ co-transporter (Xu et al., 1994) is a membrane protein of principal cells that facilitates active transport of one sodium, one potassium and two chloride ions into the insect tubule from the hemolymph (Beyenbach, 2003; Gámez et al., 2012; Gillen et al., 2006). (6) Chloride channels, which are located in the apical membrane of stellate cells, mediate passive transcellular secretion of chloride. Two types of such channels with different conductance rates and kinetics were identified in the yellow fever mosquito *Aedes aegypti* MTs (O'Connor and Beyenbach, 2001). There is a debate concerning transepithelial transport of chloride ions as, in addition, a passive paracellular pathway was demonstrated as a major route for its translocation in this insect species (Yu and Beyenbach, 2004). (7) The bicarbonate–chloride (HCO_3^--Cl^-) anion exchanger, located in the basolateral membrane of stellate cells (of *Aedes*), extrudes intracellular bicarbonate ions in exchange of extracellular chloride anions and may also play a role in transepithelial fluid secretion (Piermarini et al., 2010). As the anion exchanger is not present in principal cells, it was suggested that bicarbonate anions generated by carbonic anhydrase activity in these cells are transported to the stellate cells via gap junctions to mediate chloride influx (Beyenbach and Piermarini, 2011). Additional chloride anion exchangers (sulphate-chloride, oxalate-chloride and formate-chloride exchangers) are highly expressed in gut and MTs of dipteran species (*D. melanogaster* and *An. gambiae*) (Hirata et al., 2012). Still, their physiological relevance in maintaining ionic homeostasis in relation to osmoregulation requires further studies. (8) The AQPs are transmembrane channels that facilitate bidirectional conductance of water and largely small neutral solutes (see details in AQPs and osmoregulation in this chapter). Other ions, sugars, amino acids and organic solutes cross the cell membranes by passive diffusion or active transport, and water follows the osmotic gradients or transported via AQPs.

Comprehensive reviews on systems and mechanisms implicated in water and ion homeostasis in insects have been published over the years (Coast et al., 2002; Coast, 2007; Schooley et al., 2005). The diagram (Fig. 1.4) that summarizes the cellular components of the MT osmoregulatory system is largely based on studies with few model insects such as *D. melanogaster* and blood feeders such as the kissing bug *Rhodnius prolixus* and the mosquitoes *A. aegypti* and *An. gambiae*. Limitations and inadequacy of such a diagram to provide a comprehensive analysis of the relative osmoregulatory functional importance of cellular components of MTs should be kept in mind.

There is a close relationship between water conservation and nitrogenous waste generated by the metabolism of proteins and nucleic acids. Since elimination of toxic ammonia and urea requires large volumes of water, terrestrial insects excrete nitrogen metabolism waste as a less toxic uric acid (guanine in mites, ticks and spiders) which is insoluble in water and can be also accumulated and stored in fat body cells.

4. COPING WITH EXTREME ENVIRONMENTAL CONDITIONS

Insects develop behavioural, biochemical and physiological adaptation strategies to cope with osmotic stress imposed by severe dehydration due to high temperatures and drought, as well as exposure to sub-zero temperatures (Benoit, 2010; Block, 1996; Duman et al., 1991; Lee, 2010; Lee and Costanzo, 1998; Storey, 1997; Storey and Storey, 1988, 2012). Lack of water is imposed by desiccating high temperatures and equally so by sub-zero conditions as liquid water is transformed into ice. Mechanisms that induce cold hardiness are largely divided into two types: (a) *freeze tolerance* that is characterized by the ability to produce ice-nucleating proteins and the restriction of ice formation to extracellular spaces, by producing antifreeze proteins and glycoproteins as well as by generating low-molecular-weight cryoprotectants (Duman et al., 1991) and (b) *freeze avoidance* that denotes avoiding of ice formation in both extracellular and intracellular compartments. Freeze avoidance, in which bodily liquids remain fluid due to the supercooling effect, is accomplished by minimizing or eliminating ice nucleation sources, by generating antifreeze and stabilizing proteins and by enhanced levels of low-molecular-weight cryoprotectants. An extreme but rare strategy named vitrification is utilized by larvae of the Alaskan beetle *Cucujus clavipes puniceus* (Sformo et al., 2010). Larvae exposed to extremely deep supercooling conditions (below −58 °C) are vitrified (as water is converted into a solid amorphous glassy state) and can survive and avoid freezing even at −100 °C. Both freeze-tolerant and freeze-avoidant insects utilize antifreeze proteins, glycoproteins and chaperones as well as diverse low-molecular-weight polyols (glycerol, sorbitol), disaccharides (trehalose, sucrose) and free amino acids (proline) that alone or jointly reduce ice formation and stabilize proteins and cellular membranes (Nieminen et al., 2012; Rinehart et al., 2007; Zhang et al., 2011).

Genes encoding several molecular chaperones such as heat shock proteins were upregulated under osmotic changes induced by freezing, desiccation or rehydration. Drought acclimation, which induced the synthesis of the heat shock protein Hsp70 in a soil collembolan species, is most likely involved in cross-tolerance to cold conditions (Bayley et al., 2001). Cold tolerance adaptation of the firebug *Pyrrhocoris apterus* was associated with upregulation of *Hsp70* (Koštál and Tollarova-Borovanská, 2009), and the heat shock protein genes *smHsp*, *Hsp70* and *Hsp90* are upregulated in larvae of the Antarctic midge, *Belgica antarctica* in response to changes in the hydration state (Lopez-Martinez et al., 2009). Enhanced transcript expression of *Hsp70* and *Hsp23* was induced in pupae of the flesh fly *Sarcophaga crassipalpis* by desiccation (Hayward et al., 2004; Tammariello et al., 1999), while freezing, dehydration and rehydration induce the expression of *Hsp70* in three mosquito species (*A. aegypti*; *An. gambiae*; and the common house mosquito, *Culex pipiens*) (Benoit et al., 2010). Silencing of *Hsp70* and *Hsp90* genes in *A. aegypti* using RNAi knockdown methodology demonstrated the functional role of these chaperones in dehydration tolerance. The increase of chaperone proteins (Hsp110, Hsp70, Hsp40, Hsp40, Grp78 and αB-crystallin) during late fall and winter in larvae of the goldenrod gall fly, *Eurosta solidaginis*, suggests their cold tolerance role as protectants and stabilizers of cellular macromolecules (Zhang et al., 2011). Different types of proteins, the late embryogenesis abundant proteins (LEAs), stabilize cellular proteins and membranes during desiccation (Hand et al., 2011), and the expression of LEA transcripts is increased in dehydration-stressed larvae of the sleeping chironomid, *Polypedilum vanderplanki* (Kikawada et al., 2006). Levels of antioxidant enzymes (catalase, superoxide dismutase), which are increased during dehydration, apparently cope with damaging reactive oxygen species, including oxygen radicals, that are formed under desiccation-stress conditions (Benoit, 2010).

Accumulation of low-molecular-weight cryoprotectants (glycerol, trehalose) was correlated with cold resistance to environmental or laboratory-induced cold conditions. Glycerol as a common cryoprotectant derived from degradation of stored glycogen in insects (Churchill and Storey, 1989) is non-toxic at high concentrations and confers cold hardiness in insects adapted to survive in sub-zero temperatures. Glycogen reserves accumulated in the fall in last instar larvae of the freeze-tolerant *E. solidaginis* and in the freeze-avoidant goldenrod gall moth, *Epiblema scudderiana*, are converted to glycerol (Joanisse and Storey, 1995). Accumulation of high glycerol levels was detected in diapausing larvae of the rice stem borer, *Chilo suppressalis* (Ishiguru et al., 2007); overwintering *Diplolepis* gall wasps on roses (Williams et al., 2002); larvae of the midge, *B. antarctica* (Michaud et al., 2008); the flesh fly, *Sarcophaga bullata* (Yoder et al., 2006) and over-wintering third instar larvae of the gall fly, *E. solidaginis* (Morrissey and Baust, 1976).

Trehalose, which stabilizes the structure and activity of macromolecules, is accumulated in many organisms in response to stress (Sampredo and Uribe, 2004; Singer and Lindquist, 1998). A build-up of this non-reducing disaccharide was demonstrated in overwintering diapausing larvae of the codling moth, *Cydia pomonella* (Khani et al., 2007), *E. solidaginis* larvae (Morrissey and Baust, 1976), an arctic springtail species (Clark et al., 2009) and in freeze-tolerant coleopteran species (Fields et al., 1998). Trehalose was implicated in facilitating cross desiccation tolerance of the midge, *B. antarctica*, to both cold and heat conditions (Benoit et al., 2009). Desiccation tolerance in the Antarctic springtail, *Cryptopygus antarcrticus*, was associated with the accumulation of organic osmolytes such as glycerol, glucose and trehalose (Elnitsky et al., 2008), and the trehalose accumulation in the Arctic collembolan, *Onychiurus arcticus*, due to breakdown of glycogen following exposure to sub-zero temperatures was reported by Worland et al. (1998). Accumulation of cryoprotecitve sugars (trehalose, glucose) and polyols (sorbitol, manitol, threitol, erythritol) reaches a sum level of 0.9 M during winter and consequently increases the osmolarity of the supercooled body fluids in the freeze-avoidant spruce bark beetle, *Ips typographus* (Koštál et al., 2011). Larvae of the cryptobiotic sleeping chiron-omid (*P. vanderplanki*) accumulate trehalose under desiccation conditions (Kikawada et al., 2005; Watanabe et al., 2002), and a trehalose transporter gene is induced by desiccation and regulates hemolymph trehalose levels (Cornette et al., 2010; Kanamori et al., 2010). Apparently, the cryoprotec-tive accumulation of nonessential amino acids contributes to cold hardiness of the freeze-tolerant ectoparasitic fly, *Lipoptena cervi* (Nieminen et al., 2012).

Larvae of the anhydrobiotic chironomid *P. vanderplanki* resist almost complete dehydration only under a tolerable slow desiccation period (Kikawada et al., 2005; Nakahara et al., 2008). Such slow dehydration requirement, which is essential for activating physiological and biochemical mechanisms vital for the insect's survival, is met naturally by the water-retaining capability of the tubular nest built by the larvae (Kikawada et al., 2005). Using differential scanning calorimetry measurement and Fourier-transform infrared analyses, Sakurai et al. (2008) provided evidence that vitrification is required for *P. vanderplanki* larvae to attain a successful anhydrobiotic state. It was suggested that trehalose, which is accumulated during desiccation (Kikawada et al., 2005) and spreads out over the entire body of desiccated larvae, replaces water-bound molecules, and as sugar glass, immobilizes and protects intracellular macromolecules and cell membranes (Sakurai et al., 2008).

In summary, various studies have demonstrated inducible build-up of protective proteins (heat shock proteins, LEAs, antioxidant enzymes) as well as low-molecular-weight compounds (largely glycerol and trehalose) during desiccation and freezing conditions. Such responses help prevent cellular damage inflicted by protein denaturation, protein aggregation, and harmful reactive oxygen species that impair membrane lipids.

5. AQUAPORINS AND OSMOREGULATION

AQPs are ubiquitous and highly conserved transmembrane channels that facilitate bidirectional passive flux of largely water molecules (water-specific AQPs). Uncharged solutes mainly glycerol and low-molecular-weight polyols, urea, hydrogen peroxide, dissolved gases (CO_2, NO, NH_3) and certain metalloids might also be transported across similar such channels. The specific structural architecture and characteristics of AQP are largely based on analyses of high-resolution atomic structures of mammalian and bacterial protein crystals (Gonen et al., 2004; Harries et al., 2004; Ho et al., 2009; Horsefield et al., 2008; Savage et al., 2003). Basically, the channels are organized *in situ* as homotetramers, and each individual monomer is composed of six tilted transmembrane α-helices connected by five loops and the C- and N-terminals are cytoplasmic. One of the cytoplasmic and one of the extracellular loops, which fold from opposite directions, converge half way into the lipid bilayer, and their conserved asparagine–proline–alanine (NPA) amino

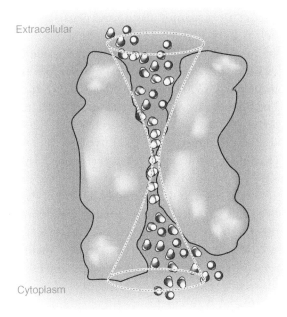

Figure 1.5 "Hourglass" model of AQP (side view). Circles represent water molecules inside the narrow channel and in the extracellular and cytoplasmic vestibules. Cohen, E. and OMICS Publishing Group (ISSN-2161-0983) are credited for this figure. (See Color Plate 5 at the back of the book.)

acid triplet motif creates a pore the size of a water molecule in water-specific AQPs. The pore and the narrow channel are flanked by conically shaped external and cytoplasmic vestibules suggesting an "hourglass" model (Fig. 1.5). Water molecules traverse the channels as a single file due to reorientation caused by disruption of the hydrogen bonds between their atoms and adjacent hydrogens and formation of hydrogen bonds with amido groups of the NPA's asparagine residues that are extended into the pore. Passage of protons and hydronium ions (H_3O^+) is precluded by the reorientation layout of water molecules as well as by another barrier/filter at the extracellular site of the channel. It is composed of highly conserved aromatic and arginine amino acid residues (ar/R) located around the NPA motifs. In addition to protons, the positive charges at the ar/R constriction region act as an electrostatic barrier that repels charged molecules and ions from approaching the pore. An additional barrier is formed by arginine and histidine residues at the cytoplasmic and extracellular regions that fend off protons and charged molecules from traversing the channel. In comparison with the water-selective AQPs, the relatively more hydrophobic canals with a larger pore size, termed aquaglyceroporins, facilitate accommodation and permeation of larger neutral molecules such as glycerol (Fu et al., 2000; Nollert et al., 2001; Savage et al., 2010). Due to an accessible cysteine residue located near the NPA motif, most AQPs are sensitive to mercurial compounds (Preston et al., 1993; Savage and Stroud, 2007). Freeze-fracture electron microscopic studies revealed orthogonal array configuration of AQPs in native tissues of certain vertebrate (Rash et al., 1998; Nicchia et al., 2009; Verbavatz et al., 1997) and insect (Beuron et al., 1995) species.

The overall physiological roles of AQPs in excretion and osmoregulation in insect and other arthropods species have recently been reviewed (Campbell et al., 2008; Cohen, 2012; Spring et al., 2009). Part of insect adaptation to survive under extreme environmental conditions such as sub-zero temperatures and drought or to subsist on excessive liquid source of food such as blood or plant saps (phloem and xylem) involves functional AQPs. The channel proteins were detected in organs and tissues with physiological functions related to bidirectional movement of water, glycerol and other small neutral solutes. AQPs in digestive tracts and associated MTs, and to a lesser degree the salivary and silk glands, play important roles largely in water homeostasis and cryptobiosis. The MTs of D. melanogaster as a genetically tractable model insect were obvious candidates to study channel proteins involved in water transport and water homeostasis. Seven putative AQP homologues were identified in the Drosophila genome, one of which is the previously described highly water-selective AQP called DRIP (for Drosophila integral protein) (Dow et al., 1995), that was detected in stellate cells of adult MTs (Kaufmann et al., 2005). DRIP mRNA expression in the Xenopus oocytes swelling assay and in the yeast secretory vesicle membranes displayed high rates of water permeability (Kaufmann et al., 2005). DRIP, which is also expressed in other organs related to water conductance such as the gut regions and the posterior spiracles of the embryonic stages, is indication of its important role in osmoregulation (Kaufmann et al., 2005). A different function related to fecundity was recently demonstrated for DRIP expressed in certain corazonin-producing brain neurons of Drosophila (Bergland et al., 2012).

Based on public data, Campbell et al. (2008) constructed a phylogenetic tree out of 39 full-length AQPs in insects, whereas Kambara et al. (2009) depicted 72 AQP sequences in 17 insect species divided into 4 major clusters. However, despite the relatively large numbers of available insect AQP gene sequences, only a very small part of them was functionally characterized.

5.1. Aquaporins in blood-feeding arthropods

Sanguinivorous insects and ticks, and insects feeding on plant sap or nectar are osmotically stressed with large volumes of ingested fluids that must be rapidly and effectively voided. AQPs in such arthropods are largely and abundantly expressed in organs that maintain water balance such as digestive tracts, MTs and salivary glands. Ticks, for example, endure enormous osmotically stress following blood meals that increase their initial body weight by up to 100-fold (Campbell et al., 2010). The salivary glands in ticks play a major dual role in water management by helping absorb atmospheric water vapour via secretion of hygroscopic droplets (Bowman and Sauer, 2004; Gaede and Knülle, 1997) in off-host periods between meals, and by returning ingested blood–meal water ($\sim$75%) back to the host via the saliva (Kaufman and Phillips, 1973). The cloned and sequenced water-selective AQP genes from the brown dog tick, *Rhipicephalus sanguineus* (*RsAQP1*) (Ball et al., 2009), and the sheep tick, *Ixodes ricinus* (*IrAQP1*) (Campbell et al., 2010), were abundantly expressed not only in the salivary glands but also in tissues involved in massive water transport such as gut, MTs and rectal sacs.

A water–selective AQP (AgAQP1) that is sensitive to $HgCl_2$ and is highly expressed in tissues such as MTs, alimentary canal, and ovaries was detected and characterized in the malaria vector mosquito *An. gambiae* (Liu et al., 2011). It is noteworthy that the channel protein was immunolocalized in the stellate cells known to be involved in water permeation. AgAQP1 expression is higher in adults in comparison to immatures, and higher in female than in males. Its expression in female mosquitoes is elevated at high relative humidity, and RNA interference studies revealed that compared to controls, insects survived longer periods of severe desiccation when *AgAQP1* is silenced (Liu et al., 2011). Unlike *An. gambiae*, the water-selective AQP of the yellow fever mosquito, *A. aegypti* (AeaAQP), is associated with the respiratory system (essentially the tracheoles) of the MTs (Pietrantonio et al., 2000). The channel protein was immunolocalized within the tracheoles, and by using the *Xenopus* swelling system, it was demonstrated that water permeation is inhibited by mercury ions that apparently engaged the Cys^{79} residue near the NPA-conserved motif (Duchesne et al., 2003). The rapid flow of water during diuresis following blood meal and after adult emergence is firmly related to the physiology of the MT-associated tracheoles at times of great need for oxygen supply. A recent bioinformatic study by Drake et al. (2010) identified six putative AQPs in the *A. aegypti* genome, and based on RNAi-mediated gene silencing experiments, three of the AQPs, which are intensely expressed in female MTs, are the main water channels that possibly play a functional role in diuresis. Furthermore, it was even speculated that AeaAQP may also serve as an osmosensor of MTs (Pietrantonio et al., 2000).

The triatomid bug *R. prolixus* has been used as a model insect to study water homeostasis and its regulation. An AQP (Rp-MIT) cDNA from MTs was cloned and sequenced, and its water permeability was demonstrated in the *Xenopus* oocytes expression assay (Echevarría et al., 2001; Martini et al., 2004). *Rp-MIT* transcripts were upregulated in MT after a blood meal and following exposure to the diuretic hormone (DH) serotonin or to cAMP (Martini et al., 2004).

5.2. Aquaporins in plant sap-feeding insects

Xylem and phloem plant saps are nutritional source of hemipteran insects characterized by high volumes of ingested fluids that must be voided to alleviate osmotic pressures. Phloem feeders are challenged with an additional stress associated with the high sucrose content of the ingested phloem. The alimentary canals of certain hemipterans such as leafhoppers, froghoppers, psyllids and whiteflies have a characteristic water-conducting anatomical complex called the "filter chamber" (Fig. 1.1B) that bypasses part of the digestive tract and functions to shunt rapidly and directly large amounts of fluids from the foregut to the midgut (Cicero et al., 2009; Hubert et al., 1989). Aphids lack the elaborated filter chamber complex, and water bypass is facilitated by a close proximity between distal and proximal regions of the gut (Douglas, 2003) (Fig. 1.1C). Studies focusing on the filter chamber of the green leafhopper, *C. viridis*, revealed the first water-specific AQP (AQPcic) in this xylem-sap feeder (Guillam et al., 1992). Later, AQP-like proteins were also detected in several other hemipteran species (Le Cahérec et al., 1997; Shakesby et al., 2009). The cDNA encoding a 26-kDa protein was cloned from a *C. viridis* filter-chamber cDNA library, and AQPcic protein expressed in *Xenopus* oocytes assay instigated a 15-fold increase in water flux comparable to that of the human AQP1 (Le Cahérec et al., 1996). A scanning transmission electron microscopy study of the *C. viridis* filter-chamber membranes revealed dense packing of oriented AQP homotetramers (Beuron et al., 1995; Hubert et al., 1989). Recently, a water-specific and mercury-sensitive AQP (BtAQP1) was cloned from and characterized in the sweet potato whitefly, *Bemisia tabaci* (Mathew et al., 2011). The channel protein is highly expressed in second instar nymphs and is primarily located in the filter chamber and in the anterior ileum region of the hindgut of adult whiteflies.

Osmoregulation in aphids feeding on high sucrose content of phloem-sap is partly maintained via transglycosylation by which the sugar is enzymatically cleaved along with subsequent oligomerization of glucose and the consequent reduction in the osmotic pressure in the gut lumen (Douglas, 2003, 2006). In addition, the aforementioned anatomical construct of aphids' alimentary tract (Fig. 1.1C) may account for the recycling of water and the dilution of the high sucrose content in the gut lumen. A water-specific AQP (ApAQP1) was identified in the pea aphid, *Acyrthosiphon pisum*, and since it was immunolocalized in the stomach and the adjacent loop of the distal intestine, its probable function relates to uphold water shunting (Shakesby et al., 2009). Silencing of ApAQP1 by oral administration of dsRNA resulted in a significant increase in the osmotic pressure of the aphid's hemolymph (Shakesby et al., 2009). A second *A. pisum* AQP (ApAQP2), which is predominantly expressed in fat body cells and in bacteriocytes enclosing the symbiotic bacterium *Buchnera*, has been recently discovered (Wallace et al., 2012). In addition to water molecules, the mercury-insensitive ApAQP2 channel permeates polyols such as glycerol, sorbitol and mannitol as well as other uncharged compounds such as formamide and urea. It is noteworthy that only a very few aquaglyceroporins have been so far detected in insects. The novel multifunctional aquaglyceroporin of the pea aphid apparently plays physiologically important roles in water balance and in carbon nutrition that involves the insect symbionts (Wallace et al., 2012).

5.3. Cold hardiness and drought resistance as related to aquaporins

The mechanism by which anhydrobiosis is induced in larvae of the sleeping chironomid, *P. vanderplanki*, is strongly dependent on transmembrane water channels (Kikawada et al., 2008). Two cDNAs encoding mercury-sensitive, water-selective AQPs (*PvAQP1* and *Pv AQP2*) with different tissue distribution and physiological roles were isolated from *P. vanderplanki* larvae. PvAQP2 is restricted to the fat body and apparently controls water homeostasis in this tissue under normal condition, whereas its counterpart, PvAQP1, which is expressed in tissues such as the epidermis, fat body, midgut and muscle, is involved in water removal during the onset of dehydration-induced periods.

A single mercury-sensitive, water-selective AQP (BaAQP1) was cloned and sequenced from a related chironomid species, the Antarctic midge,

B. antarctica (Goto et al., 2011). BaAQP1 is expressed in a variety of larval tissues such as salivary glands, foregut, midgut and MTs but is weakly expressed in the fat body. Despite sequence similarity to PvAQP1, which is dehydration-inducible, BaAQP1 is refractory to dehydration or rehydration induction and appears to be constitutively expressed (Goto et al., 2011). Additional AQPs, homologous to *Drosophila* DRIP and mammalian AQP2 and AQP3, were immunolocalized in all tissues of *B. antarctica*, except the MTs (Yi et al., 2011). The role of such proteins in bestowing cold hardiness and preventing desiccation is indicated by the increased levels of an AQP2-like protein in larvae in response to dehydration, rehydration and freezing, and by enhanced levels of DRIP-, AQP2- and AQP3-like proteins in adults during dehydration (Yi et al., 2011).

A water-selective AQP (EsAQP1) was detected in larvae of *E. solidaginis* that survives sub-zero climatic conditions (Philip et al., 2011). Although EsAQP1 was detected in overwintering larvae in tissues such as the guts, MTs and salivary glands, it is highly abundant in the nervous system and practically absent in the fat body. It is rather surprising that albeit seasonal adaptation to freezing temperatures and acquisition of cold hardiness by *E. solidaginis* larvae are linked to accumulation of glycerol as a cryoprotectant (Morrissey and Baust, 1976), no aquaglyceroporins have been detected so far in this or any other dipteran species (Drake et al., 2010). Glycerol movement across cell membranes is most likely independent on glycerol-facilitating transmembrane channels.

Unlike hematophagous and plant sap-feeding insects, most terrestrial insects exposed to arid conditions must conserve water and consequently produce dry faeces under xeric ambiences. Lepidopteran larvae are a good example indicating the paramount physiological importance of maintaining water homeostasis. Water balance in larval tissues of the silkmoth *Bombyx mori* is mediated by two AQPs that are present in water-transporting epithelia such as midgut, hindgut, MTs and silk glands (Kataoka et al., 2009a). The major water-specific AQP (AQP-Bom1) is abundant in the apical surfaces of the colon and rectal epithelia but is also detected in midgut, MTs and silk glands. The expression of this AQP in silk glands in feeding larvae decreased in the spinning stage, a phenomenon that coincide with the switching off of the $V-H^+$-ATPase (Azuma and Ohta, 1998). Both membraneous systems function in water and ion flux and help in stabilizing the liquid silk in the glandular lumen (Miyake and Azuma, 2008). The minor second AQP isoform (AQP-Bom2) is characterized as an aquaglyceroporin able to permeate glycerol in addition to water molecules (Kataoka et al., 2009a). The physiological role of AQP-Bom2, which is expressed in MTs and the posterior region of the midgut, is unclear but could be involved in urea transportation. Two similar AQP isoforms (AQP-Gra1 and the aquaglyceroporin AQP-Gra2) were reported in the oriental fruit moth *Grapholita molesta* (Kataoka et al., 2009b). Recently and in addition to AQP-Bom1, a functional water-specific and mercury-sensitive AQP (AQP-Bom3) was reported to be prevalently expressed in the hindgut of *B. mori* larvae (Azuma et al., 2012). Unlike the apical location of AQP-Bom1, AQP-Bom3 is expressed at the basal plasma membranes of the hindgut cells. The different spatial location of both water-specific AQPs within the same epithelial cell suggests a fine-tuning of water recycling and reabsorption via the cryptonephridial complex (Fig. 1.3) of lepidopteran larvae (Azuma et al., 2012).

Water-specific functional AQPs were detected in several other insect species such as the Formosan subterranean termite, *Coptotermes formosanus* (Kambara et al., 2009); the Korean firefly, *Pyrocoelia rufa* (Lee et al., 2001) and a DRIP-like AQP in the house cricket, *A. domesticus* (Spring et al., 2007).

5.4. Osmoregulation in phloem-feeding insects

Phloem-feeding hemipterans such as aphids, whiteflies, psyllids and planthoppers are osmotically challenged by high fluid intake as well as by high sugar content in the phloem-sap. As phloem feeders such as whiteflies and many aphids lack MTs, morphological and biochemical adjustments were developed to facilitate osmoregulation. By-passing excess fluids in the alimentary canal either by a special anatomical structure like the filter chamber in whiteflies (Fig. 1.1B) or by a specific anatomical folding of the digestive tract in aphids (Fig. 1.1C) was mentioned before. Also, the osmoregulatory role played by water-specific AQPs in the gut tissues of hemipteran species was previously discussed. Recent evidence suggests that aphids may ingest low osmolarity xylem sap to regulate high osmotic pressure contributed by the high sucrose levels in phloem sap (Pompon et al., 2011). The following section will address the particular issue of avoiding desiccation due to challenging osmotic pressures inflicted by the high sucrose content in phloem saps.

5.4.1 Transglucosylation

The disaccharide sucrose, which is the principal sugar in phloem saps that can range in concentrations from 0.2 to 1.0 M and beyond, presents a serious dehydration challenge as the osmotic pressure in the gut is higher than that of the hemolymph (Douglas, 2006). Sucrose, which is not absorbed by midgut cells, is hydrolyzed by α-glucosidase to yield fructose and glucose. Reduction in gut osmotic pressure is obtained by absorption of most fructose molecules, while part of the undigested glucose is trans-glucosylated (primarily at high dietary sucrose levels), forming (particularly in aphids) a variety of oligosaccharides (Ashford et al., 2000; Rhodes et al., 1997; Woodring et al., 2006). As a result of transglycosylation and dilution by water flow from distal intestine regions, the excreted honeydew of phloem-sucking insects is osmotically similar to the hemolymph (Douglas, 2003; Wilkinson et al., 1997). Experiments with antibiotic-treated aphids demonstrated that the production of oligosaccharides is related to ingestion of dietary high sucrose levels and indicated that the associated symbiotic microflora are not connected with such production (Wilkinson et al., 1997).

The enzyme α-glucosidase (EC 3.2.1.20, α-D-glucoside glucohydrolase) catalyzes the cleavage of α-glucosyl residues from the non-reducing end of sucrose. The process of transglucosylation occurs as this enzyme also transfers a glucose residue to an acceptor hydroxyl group in another glucose mono-mer or a glucose oligomer (instead of water molecules) forming an α-glycosidic bond. The role played by α-glucosidase in osmoregulation is clearly indicated by the increase in hemolymph osmotic pressure in the pea aphid following inhibition of the enzyme (Karley et al., 2005). The pu-rified enzyme was immunolocalized in the posterior midgut of the pea aphid (Cristofoletti et al., 2003), and cDNA encoding a membrane-bound gut-specific α-glucosidase was cloned and its transcripts were expressed in the aphid gut (Price et al., 2007).

Despite the similar source of ingested plant fluid, whiteflies differ from aphids as trehalulose (D-glucose-1-1-α–D-fructose) (Bates et al., 1990) is a major disaccharide sugar detected in the honey dew (>30% of total carbo-hydrates), and it was postulated to be involved in osmoregulation and in egesting excess sugars (Byrne et al., 2003; Cristofoletti et al., 2003). Trehalulose, which is the isomerization product of sucrose, is carried out by trehalulose synthase that is apparently associated with membranes in the filter chamber and hindgut of *Bemisia argentifolii* (Salvucci, 2000). In addition to trehalulose, tri (bemisiose), tetra (maltosucrose) and penta (diglucomelezitose), oligosaccharides were detected in *Bemisia* honeydew (Hendrix and Wei, 1994; Hendrix et al., 1992; Wei et al., 1997).

6. REGULATION OF WATER HOMEOSTASIS

6.1. Regulation of AQPs

Like any other membrane bound protein, generation of functional AQPs undergoes dynamic sequential cellular events starting from generation of mRNA transcripts, their translation, post-translation modification, recruitment as vesicular cargo, transport to appropriate plasma membrane domains, and finally precise docking on and fusion with cell membranes. In addition, *in situ* flux of water and solutes can be inhibited or modulated by gating the entrances of functional AQPs, and the channel protein might be subjected to internalization in endosomal compartments for subsequent recycling or degradation. Obviously, regulation of the above biochemical and biophysical processes can be critical for general water homeostasis, resistance to sub-zero ambient temperatures and for tolerance to desiccation or overhydration. Much of the studies on function and regulation AQPs have been conducted with vertebrate and plant systems (Agre, 2006; Benga, 2009; Gomes et al., 2009; Gonen and Walz, 2006; Shao et al., 2008; Törnroth-Horsefield et al., 2010; Verkman, 2011) and as information relevant to insects is rather scattered and meagre, any extrapolation, though plausible, must be viewed with caution.

Experiments using temperature-sensitive *Drosophila* mutants indicated that 20-hydroxyecdysone regulates the expression of DRIP in stellate cells via possible interaction with the ecdysone receptor isoform, EcR-B2 (Gautam and Tapadia, 2010).

Some evidence, mostly indirect, indicates the involvement of phosphorylation in stimulating, mobilizing and targeting of insect AQPs. An increase in diuresis of *R. prolixus* following a blood meal is accompanied by upregulation of a DRIP-like water channel in the MTs (Martini et al., 2004), and it was suggested that such stimulation is triggered by serotonin and mediated by the cAMP-dependent pathway. As transcript levels of the chironomid *PvAQP1* are increased 6 h after the onset of desiccation, it was hypothesized that regulation of the AQP trafficking or gating involves protein kinase C (PKC)-mediated phosphorylation of serine residues (Kikawada et al., 2008). Possible phosphorylation sites at the C-terminus of *Bemisia* BtAQP1 were speculated to target this channel protein to its plasma membrane domain (Mathew et al., 2011). Activation of adenylyl cyclase, by which cAMP levels are increased, was implicated in the induced salivation in ticks by the dopaminergic system (Bowman and Sauer, 2004; Kaufman and Phillips, 1973). It was further suggested that translocation of the brown dog tick RsAQP1 relates to dopamine-dependent phosphorylation that is mediated via PKC activity (Ball et al., 2009).

6.2. Diuretic and antidiuretic factors

Basically, diuretic and antidiuretic activities that are closely associated with the alimentary canal and MTs determine water and electrolytes homeostasis via strict neurohormonal regulation. Physiologically, this regulation establishes the volume and ion composition of the hemolymph and removal of nitrogenous waste and toxins. Various model organisms, such as the osmotically challenged hematophagous insects (*R. prolixus*, *A. aegypti*, *An. gambiae*), the genetically amenable *D. melanogaster*, phloem–sap feeders (*A. pisum*), lepidopteran species (*Manduca sexta*, *B. mori*, *Heliothis virescens*), coleopteran species (*Tribolium castaneum*, *T. molitor*) and large orthopteran insects (*Locusta migratoria*, *Schistocerca gregaria*, *A. domesticus*) provided essential information vis-a-vis the above endocrine control of water and solutes homeostasis.

Various biogenic amines and neuropeptides and their respective recep-tors are involved in the chemical signalling of diuresis/antidiuresis as well as in embryonal and postembryonal development, feeding, metabolism, be-haviour (mating, oviposition, migration), pheromone biosynthesis and myotropic activity. They are highly specific (some acting at extremely low doses in the pM or even fM range), and in blood feeders such as *Aedes*, for example, they coordinate activities such as behaviour, digestion, and oogenesis that are tightly related to or affected by diuresis (Kersch and Pietrantonio, 2011). Available genomic database of a number of insect species has greatly facilitated studies of genes encoding neuropeptides and their respective receptors.

Diuresis can be assessed by changes in total body weight and haemolymph volume (Bushman and Nelson, 1990; Kataoka et al., 1989; Proux et al., 1982; Rafaeli et al., 1987). *In vitro* methods like the Ramsay assay (Coast, 1988; Ramsay, 1954) using single isolated MTs from a variety of insect species has widespread use in measuring diuresis. The inverted rec-tum or ileum assays of large insects such as locusts for measuring reabsorption rates have been employed (Goh and Phillips, 1978; Goldbard et al., 1970; Proux et al., 1984). The above *in vitro* assays, albeit highly useful, have been insufficient to establish the exact *in vivo* roles played by neurohormones in diuresis or antidiuresis (Coast, 1988).

Diuretic and antidiuretic neurohormones as multiple and versatile extracellular signalling molecules can be classified into seven major groups: biogenic amines (serotonin, tyramine), corticotrophin-releasing factors (CRFs)-like diuretic peptides, leucokinins, cardioacceleratory peptides (CAP), calcitonin (CT)-like diuretic peptides, antidiuretic factors (ADFs) and ion transport peptides (ITPs). The size of the neuropeptides varies and ranges from short amino acid sequence (6–16) kinin and CAPA peptides and RYamides, intermediate CT-like DHs (31), and relatively long CRF-like DHs (30–60) or ITP (72) (Table 1.1). Nearly all neuropeptides involved in diuresis have an amidated carboxyl terminus that is essential for their biological activity (Roberts et al., 1997). Interaction of membrane-associated receptors (mostly G-protein-coupled receptors, GPCRs) with neurohormones generates intracellular second messengers such as cyclic nucleotides (cAMP, cGMP), NO, inositol (1,4,5) triphosphate (IP$_3$) and Ca^{2+} (Coast, 2007; Schooley et al., 2005). The induced signal transduction triggers transepithelial transport of ions (H$^+$, Na$^+$, K$^+$, Cl$^-$) and the consequent manipulation of diuresis, antidiuresis and the homeostasis of electrolytes.

Neurohormones are synthesized in neurosecretory cells in the brain or the ventral nerve cord and are secreted into the hemolymph from neurohemal sites (such as the corpora cardiaca) and from perivisceral organs and endings of the abdominal nerve cord (Coast, 2007; Te Brugge et al., 2005). Biogenesis of neuropeptides is an integrated and regulated cascade of temporal and spatial episodes that includes transcriptional, post-transcriptional and translational events, intracellular translocation and post-translational modifications (proteolytic cleavage of prepropeptide precursors, amidation) and release into the hemolymph from neurohemal sites or nerve endings (Masler et al., 1993). The distributed signalling diuretic and antidiuretic neurohormones bind to their respective receptors (mainly GPCRs but some are tyrosine kinase receptors) that are expressed largely in the digestive tracts and in the MTs (Li et al., 2008).

Since neurohormones regulate a wide array of essential physiological processes that in addition to water homeostasis include—behaviour, feeding, reproduction, metabolism, metamorphosis, myotropism, diapause induction and moulting—they may serve as potential applicative avenue for future pest control agents (Keeley and Hayes, 1987; Masler et al., 1993; Nachman et al., 1993; Scherkenbeck and Zdobinsky, 2009).

Table 1.1 Selected peptides acting as diuretic or antidiuretic neurohormones in various insect species

Neuropeptide	Insect	Peptide sequence	References
CRF-like DHs			
Manse-DP II	*M. sexta*	SFSVNPADVILQHRYMEKVAQNNRNFLNRV-NH$_2$	Blackburn et al. (1991)
Zoone-DH	*Z. nevadensis*	TGAVPSLSIVNPLDVLRQRLLLEIARRRMR QSQDQIQANREMLQTI-NH$_2$	Baldwin et al. (2001)
Drome-DH$_{44}$	*D. melanogaster*	NKPSLSIVMPLDVLRQRLLLEIARRQMKE NTRQVELNRAILKNV-NH$_2$	Cabrero et al. (2002)
Rhopr-DH	*R. prolixus*	MQRPQGPSLSVANPIEVRLRSRLLLEIARR RMKEQDASRVSKNRQYLQQI-NH$_2$	Te Brugge et al. (2011)
Tenmo-DH$_{37}$	*T. molitor*	SPTISITAPIDVLRKTWEQERARKQMVK NREFLNSLN-OH	Furuya et al. (1995)
Kinins			
Leucokinin I	*L. maderae*	NPAFNSWG-NH$_2$	Holman et al. (1986)
Drosokinin	*D. melanogaster*	NSVVLGKKQRFHSWG-NH$_2$	Terhzaz et al. (1999)
Aedeskinin I	*A. aegypti*	NSKYVSKQKFWSWG-NH$_2$	Veenstra (1994)
Achetakinin I	*A. domesticus*	SGADFYPWG-NH$_2$	Holman et al. (1991)
Helicokinin I	*H. virescens*	YFSPWG-NH$_2$	Blackburn et al. (1995)
Locustakinin	*L. migratoria*	AFSSWG-NH$_2$	Schoofs et al. (1992)
CAPA peptides[a]			
Drome-CAP$_{2b}$	*D. melanogaster*	GANMGLYAFPRV-NH$_2$	Kean et al. (2002)
Anoga-CAP$_{2b}$	*An. gambiae*	GPTVGLFAFPRV-NH$_2$	Pollock et al. (2004)

Table 1.1 Selected peptides acting as diuretic or antidiuretic neurohormones in various insect species—cont'd

Neuropeptide	Insect	Peptide sequence	References
Manse-CAP$_{2b}$	M. sexta	DGVLNLYPFPRV-NH$_2$	Huesmann et al. (1995)
Nezvi-CAP$_{2b}$	N. viridula	DQLFPFPRV-NH$_2$	Predel et al. (2006)
Locmi-CAP$_{2b}$	L. migratoria	AAGLFQFPRV-NH$_2$	Clynen et al. (2003)
Peram-CAP$_{2b}$	P. americana	GASGLIPVMRN-NH$_2$	Predel et al. (1995)
Musdo-CAP$_{2b}$	M. domestica	AGGTSGLYAFPRV-NH$_2$	Nachman et al. (2005)
Leuma-CAP$_{2b}$	L. maderae	GSSGLIPFGRT-NH$_2$	Predel et al. (2000)
Calcitonin (CT)-lke DHs			
Dippu-DH$_{31}$	D. punctata	GLDLGLSRGFSGSQAAKHLMGLAAANY–AGGP-NH$_2$	Furuya et al. (2000)
Drome-DH$_{31}$	D. melanogaster	TVDFGLARGYSGTQEAKHRMGLAAANF–AGGP-NH$_2$	Coast et al. (2001)
Anoga-DH$_{31}$	An. gambiae	TVDFGLSRGYSGAQEAKHRMAMAVANF-AGGP-NH$_2$	Coast et al. (2005)
ADF			
Tenmo-ADFa	T. molitor	VVNTPGHAVSYHVH-OH	Eigenheer et al. (2002)
Tenmo-ADFb	T. molitor	YDDGSYKPHIYGF-OH	Eigenheer et al. (2003)
ITP			
Schhgr-ITP	S. gregaria	SFFDIQCKVYDKSIFARLDRICEDCYNLFREPQLHSLCR SDCFKSPYFKGCLQALLLIDEEEKFNQMVEIL–NH$_2$	Audsley et al. (1992a)
RYamides			
RYamide-1	D. melanogaster	PVFFVASRY-NH$_2$	Collin et al. (2011)

^{a}CAP$_{2b}$ is assigned to the CAPA peptides according to the revised nomenclature by Coast and Schooley (2011) and names of the insect species are abbreviated. ADF, antidiuretic factor; CAPA, cardioacceleratory peptides; CRF, corticotropin-releasing factor; DH, diuretic hormone; ITP, ion transport peptide.

6.2.1 Biogenic amines

Tyramine and serotonin are two biogenic monoamines that play a role in diuresis of a limited number of insect species. Tyramine, the product of tyrosine decarboxylation in *Drosophila* MTs, is a neuroactive factor with diverse physiological and behavioural functions including *inter alia* stimulation of diuresis (Lange, 2009) via the chloride shunt conductance at a 1-nM threshold level (Blumenthal, 2003). Urine production rate was reduced by the alkaloid yohimbine acting as an antagonist to tyramine-induced signaling via the tyramine receptor expressed in the MTs (Blumenthal, 2003). As immunohistochemical staining indicated that tyramine is synthesized in principal cells, its action as a paracrine hormone on the neighbouring stellate cells, where chloride conductance occurs, was suggested (Lange, 2009).

5-Hydroxytryptamine (serotonin)   4-Hydroxyphenethylamine (tyramine)

Serotonin stimulated fluid secretion by isolated MTs of various insects including the mosquito *A. aegypti* (Clark and Dradley, 1998; Veenstra, 1988), the tobacco hornworm *M. sexta* (Skaer et al., 2002), the citrus swallowtail butterfly, *Papilio demodocus* (Nicolson and Millar, 1983), the locust, *L. migratoria* (Morgan and Mordue, 1984) and the house cricket *A. domesticus* (Coast, 2011). However, most studies on serotonin as a DH with a major regulatory role were conducted with the blood-gorging bug, *R. prolixus* as a model insect (Coast, 2009; Lange et al., 1989; Maddrell et al., 1969, 1991; Orchard, 2006). Serotonin was immunolocalized in neurons of the mesothoracic ganglionic mass and the axons that reach into neurohemal sites along the ventral abdominal nerves (Lange et al., 1989; Maddrell, 1963, 1966; Orchard, 1989). The sanguinivorous insect ingests blood ∼10 times its initial mass and the diuretic activity of the MTs rapidly increases ∼1000-fold to manage osmotic pressure and boost up mobility by which predation risk is reduced (Coast, 2009; Martini et al., 2007). Serotonin is released (triggered by stretch receptors in the abdomen) into the circulation within a few minutes following the onset of blood feeding. It should be mentioned that serotonin acts on diverse tissues and synchronizes additional physiological activities associated with the voluminous blood meal intake such as cuticle plasticization, salivation, enhanced gut motility and increased contraction rates of the dorsal vessel and the anterior midgut (Orchard, 2006). Immediately following a blood meal, the neurohormone stimulates simultaneously fluid absorption from the anterior midgut, fluid secretion from the upper regions of the MTs and selective K^+ reuptake from the proximal parts of the MTs (Coast, 2009). These three activities are coordinated by serotonin via cAMP in the targeted tissue sites to ensure osmoregulation, removal of water and excess NaCl present in vertebrate blood, and to reduce the depletion of essential potassium ions. Serotonin titer drops when blood meal is completed resulting in reduced diuresis in the anterior midgut and upper MTs. However, the proximal MT regions, which are apparently sensitive to lower serotonin levels, continue to reabsorb KCl. Serotonin as a diuretic neurohormone is apparently an adaptation of a blood-feeding insect as it lacks any effect on plant-feeding hemipterans (Coast et al., 2010). In addition to serotonin, diuresis and electrolyte homeostasis in *Rhodnius* are regulated in conjunction with other neurohormones (to be discussed in the following sections).

6.2.2 Peptidergic signaling molecules

A plethora of highly powerful peptidergic DHs and antidiuretic hormones are released into the circulatory systems of insects and target MTs or digestive tract (largely the hindgut) tissues. The temporal and coordinated discharge of various neuropeptides guarantees the fine-tuning of fluid and electrolyte homeostasis that is crucial for insect survival. The available information regarding relevant genes, posttranslational modifications (enzymatic cleavage, amidation), GPCR receptors and ligand binding as well as the biological signaling by the triggered generation of second messengers will be summarized below.

Diuretic and antidiuretic neuropeptide hormones are divided into four major classes: CRF-like diuretic peptides, leucokinins, CAPs and CT-like diuretic peptides.

CRF-like DHs—CRF-like DH peptides (30–54 amino acid residues; 25 peptide sequences) were identified in a variety of orthopteran, dictyopteran, isopteran, hemipteran, dipteran, lepidopteran and coleopteran species (Te Brugge et al., 2011, Table 1.1). The neuropeptides, which bear structural homologies with and are functionally related to vertebrate CRF-related DHs (sauvagine, urotensin; Lovejoy and Balment, 1999), stimulate transepithelial fluid secretion by isolated MTs in various insects via a cAMP-signaling pathway (Audsley et al., 1995; Cabrero et al., 2002; Patel et al.,

1995; Tobe et al., 2005; Wiehart et al., 2003b). The CRF-like DH peptides share similar amino acid sequences, and except for the free acid C-terminal of *T. molitor* (Tenmo-DH_{37}) and the almost identical *T. castaneum* (Trica-DH_{37}) sequences, other peptides are C-terminally amidated (Te Brugge et al., 2011). It was suggested that the N-terminus of the peptide is essential for receptor activation while the rest sequence plays a role in receptor binding (Schooley et al., 2005). In general, a single peptide molecule is produced, yet as a result of alternate gene splicing, some insects such as *M. sexta*, *T. molitor* and *T. castaneum* have two CRF-like DH isoforms, while *B. mori* has three isoforms (Coast and Schooley, 2011; Roller et al., 2008).

The first neuropeptide (41 amino acid residues) was isolated from heads of *M. sexta* pharate adults (Manse-DH) (Kataoka et al., 1989), and later, a shorter peptide (Manse-DPII; 30 amino acid residues) was identified (Blackburn et al., 1991). There is a single genomic copy of Manse-DH, and its encoded precursor peptide consists of dibasic amino acid processing sites for endopeptidase proteolytic cleavage, and a putative signal sequence for secretion (Digan et al., 1992). It was proposed that Manse-DH is degraded in the MTs of *M. sexta* larvae by the combined action of proteolytic activity and oxidative metabolism (Li et al., 1997). Later, CRF-like peptides in other insect species were reported. Identical peptide sequences were isolated and identified in the house fly *Musca domestica* and the stable fly *Stomoxys calcitrans* (Clottens et al., 1994), and an almost identical sequence of the above dipterans was described for *D. melanogaster* (Drome-DH_{44}) neuropeptide (Cabrero et al., 2002). Two CRF-like DH peptides, isolated from *T. molitor* (Tenmo-DH37 and Tenmo-DH47) (Furuya et al., 1995, 1998), are highly potent neurohormones stimulating diuresis in isolated MTs at EC_{50} values of 12 and 27 nM, respectively (Wiehart et al., 2002), and two similar DHs (DH37 and DH47) were isolated from the closely related *T. castaneum* species (Li et al., 2008). Other CRF-like DHs from a variety of insect species include the diuteric hormone Dippu-DH_{46} isolated from the Pacific beetle cockroach, *Diploptera punctata* (Furuya et al., 2000); Zoone-DH was reported from the dark Pacific Coast rottenwood termite, *Zootermopsis nevadensis* (Baldwin et al., 2001) and a diuretic peptide (Peram DH) was obtained from the American cockroach *Periplaneta americana* (Kay et al., 1992).

A CRF-like factor with diuretic activity previously found in *Rhodnius* (Te Brugge et al., 2002) was later identified by using database sequences of the *Rhodnius* genome project (Te Brugge et al., 2011). The Rhopr DH_{31} affects the MTs and anterior midgut and was immunolocalized in the medial neurosecretory cells of the brain that send neuronal projections to various neurohemal release sites. Non-native CRF-like DH (Zoone-DH) (Donini et al., 2008) and the native Rhopr DH along with the previously mentioned serotonin stimulate water absorption from the anterior midgut and secretion by the upper MT of *Rhodnius* via the cAMP signalling pathway (Te Brugge et al., 2011). A CRF-like DH in *A. aegypti* is a natriuretic peptide acting by managing the surplus NaCl in vertebrate blood meal via a cAMP-dependent mechanism (Beyenbach, 2003).

Cross-species assays using CRF-like DHs demonstrated that despite their sequence dissimilarity in comparison to native neuropeptides, non-native peptides elicit biological response in the respective heterologous receptors at extremely low levels. Zoone-DH stimulates diuretic secretion by upper MTs and water absorption from the anterior midgut of *Rhodnius* but with no effect on potassium ion reuptake from its proximal MT regions (Te Brugge et al., 2011), and Zoone-DH also stimulated secretion by isolated MTs of the predaceous spined soldier bug, *Podisus maculiventris* (Coast et al., 2011). Synthetic Manse-DH and Tenmo-DH$_{37}$ stimulated *in vitro* fluid secretion by isolated MTs of the tobacco budworm, *H. virescens*, and the dung beetle, *Onthophagus gazelle*, respectively (Holtzhausen and Nicolson, 2007; Keeley et al., 1992).

Two functional GPCRs encoded by *CG8422* (Johnson et al., 2004) and *CG12370* (Hector et al., 2009) genes for the *Drosophila* CRF-like diuretic neurohormome, Drome-DH$_{44}$ were described. Both receptors have different sensitivities to Drome-DH$_{44}$ but are associated with elevated cAMP levels as well as recruitment of β-arrestin-GFP reporter which regulates GPCR-mediated signal transduction.

Kinins (leucokinins)—Kinins are short (6–15 amino acid residues) peptidergic hormones with both myotropic and diuretic activity in orthopteran, dictyopteran, hemipteran, dipteran and lepidopteran species (Coast et al., 2002; Scherkenbeck and Zdobinsky, 2009; Schooley et al., 2005, Table 1.1). They are characterized by an evolutionary conserved C-terminal pentapeptide motif (FX$_1$X$_2$WG-amide) core (Taneja-Bageshwar et al., 2009), sufficient to exert full myotropic and diuretic activities with extremely low EC$_{50}$ values of 10^{-10} to 10^{-11}M (Coast et al., 1990). The biological activity of the neuropeptide is lost when the terminal amide is replaced by the negatively charged acid group or when the core aromatic residues Phe or Trp are replaced by alanine (Taneja-Bageshwar et al., 2006).

The first two extremely potent kinins (L-I and L-II) isolated from head extracts of the Madeira cockroach, *Leucophaea maderae*, have myotropic effects in this insect *in vitro* hindgut assays (Holman et al., 1986). Likewise, hindgut contractions were induced by kinins isolated from other insect species such as *A. domesticus* (Holman et al., 1991), *L. migratoria* (Schoofs et al., 1992), the unbanded saltmarsh mosquito, *Culex salinarius* (Hayes et al., 1994) and *A. aegypti* (Veenstra et al., 1997). Since kinins have both myotropic and diuretic activities, the hindgut bioassay has served as a useful tool in isolation and characterization of diuretic neuropeptides from various insect species. Synthetic *Leucophaea* kinins (leucokinins) affect fluid secretion in isolated MTs of *A. aegypti* (Hayes et al., 1989) and *D. melanogaster* (O'Donnell et al., 1996), and the myotropical structurally similar achetakinins (of *A. domesticus*) exhibit diuretic activity in isolated *A. domesticus* MTs (Coast et al., 1990). Although kinins are regarded as ubiquitously diuretics, leucokinin I does not effect fluid secretion by *R. prolixus* upper MTs (Donini et al., 2008) and the *T. castaneum* genome lacks both kinin and its cognate receptor genes (Li et al., 2008).

The single *Drosophila* kinin (Drosokinin) is encoded by a single gene (*pp*) located at 70E3–70F4 of chromosome 3L (Terhzaz et al., 1999) and is expressed in the *pars intercerebralis* of adults and in the larval central nervous system (CNS) (Radford et al., 2002). Unlike *Drosophila*, however, three leucokinins, which were isolated and identified from the mosquitoes, *A. aegypti* and *An. gambiae* (Radford et al., 2004; Veenstra, 1994), stimulated *in vivo* production of urine (Cady and Hagedorn, 1999a). These leucopeptides are encoded by a single cDNA in the abdominal ganglia, and the translated proleucokinin is cleaved to yield one copy of each of the three *Aedes* kinin isoforms (Veenstra et al., 1997). Also, three diuretic peptides designated Helicokinin I, II and III were isolated from the abdominal nerve cord of the adult corn earworm *Heliothis zea* (Blackburn et al., 1995). The extremely potent neuropeptides stimulated diuresis in the heterologous *H. Virescens* (Seinsche et al., 2000) and *M. sexra* isolated MTs (Blackburn et al., 1995) at levels below 10^{-11}M. All three leucokinins injected into fifth instar *Heliothis* larvae caused reduction in weight gain and even significant mortality when exposed to Helicokinin I (Seinsche et al., 2000).

Interaction of Drosokinin with its GPCR induces an increase of intracellular Ca^{2+} as a second messenger in stellate cells with consequent stimulation of ion secretion and diuresis by the chloride shunt conductance pathway (O'Donnell et al., 1998; Radford et al., 2002; Terhzaz et al., 1999). The diuretic activity, which is mediated via increased intracellular calcium level, is dependent on both *norpA*-encoded phospholipase Cβ (PLCβ) and *itpr*-encoded IP$_3$ receptor in MT epithelium (Pollock et al.,

2003). Binding of Drosokinin to its receptor mobilizes IP_3 to interact with its receptor (a ligand-gated Ca^{2+} channel). Such interaction triggers the calcium signaling cascade that facilitates the chloride shunt conductance pathway. *A. aegypti* kinin peptides released from abdominal ganglia interact with a single-GPCR kinin receptor (*Ae* KR) expressed in cultured Chinese hamster ovary (CHO)-K1 cells (Pietrantonio et al., 2005), and an RNAi silencing study established the functional *AeKR* role in fast post-feeding liquid secretions in female mosquitoes (Kersch and Pietrantonio, 2011). *Ae* KR and the cognate *An. gambiae* GPCR, which were immunolocalized in the stellate cells of female MTs, regulate transepithelial chloride transport (Lu et al., 2011a; Radford et al., 2004). Like in *Drosophila*, chloride conductance and diuresis are stimulated in *Aedes* via increase of IP_3 levels that in turn trigger the release of Ca^{2+} from IP_3-sensitive intracellular calcium stores (Cady and Hagedorn, 1999b). Although kinin receptors are localized in stellate cells of mosquitoes and the fruit fly, it appears that chloride transport pathway is transepithelial via *Drosophila* stellate cells but paracellular via intercellular junctions in *Aedes* principal cells (Lu et al., 2011a). As *Acheta* MTs lack stellate cells, its principal cells are apparently involved in the chloride conductance pathway.

CAPs—CAPs, which were identified in a number of insects, exert myotropic effects largely on the heart muscle, and depending on the insect species, act as either diuretic or antidiuretic peptidergic hormones (Predel and Wegener, 2006). They are largely expressed in the abdominal nerve cord ganglia and are usually characterized by a consensus C-terminal $FPRV-NH_2$ sequence (Table 1.1) and by leucine at position 7 from the C-terminus (Predel and Wegener, 2006). To avoid ambiguities, a new nomenclature for CAPs was suggested (Coast and Schooley, 2011) by which CAPA and periviscerokinin names are replaced by CAP_{2b} family having the above conserved C-terminal motif. The first CAP_{2b} neuropeptide that was isolated from ventral abdominal nerve cord of pharate adults *M. sexta* increased heart contractions and stimulated fluid secretion by MTs in both *Manduca* and *Drosophila* (Davies et al., 1995; Huesmann et al., 1995). CAP_{2b} peptides of *M. sexta* and *M. domestica* stimulated diuresis by the housefly MTs (Nachman et al., 2005), and CAP_{2b} neuropeptides isolated from the stable fly *S. calcitrans* stimulated fluid secretion by its MTs at nM levels (Nachman et al., 2006). In contrast to the dipteran species, *Manduca* CAP_{2b} at low doses yielded antidiuretic responses in MTs of *R. prolixus* (Quinlan et al., 1997) and *T. molitor* (Wiehart et al., 2002). Also CAPA peptides, which were isolated and identified from the Southern green stink bug

Nezara viridula (Predel et al., 2006, 2008), demonstrated antidiuretic activity (Coast et al., 2010) and most likely act *in vivo* to terminate cAMP-stimulated diuresis.

It has been established that unlike CRF-like DHs, CAP_{2b} neuropeptides act via cGMP as a second messenger (Broderick et al., 2003; Davies et al., 1995, 1997; Dow et al., 1994; Pollock et al., 2004; Quinlan et al., 1997). Studies with *Drosophila* have demonstrated that stimulation of MT fluid secretion is mediated by the complex NO-cGMP-signalling pathway cascade (Davies, 2000; Davies et al., 1995, 1997). The gene *capability* (*capa*) of *Drosophila* encodes three neuropeptides, two of which belong to the CAP_{2b} family, according to their amino acid sequences (Kean et al., 2002). Both peptides, which are diuretics via a cascade pathway that includes stimulation of NO synthase (NOS), enhanced NO production that subsequently activates soluble guanylate cyclase, that down the road, resulted in elevating cGMP and intracellular Ca^{2+} levels in principal cells of *Drosophila* MTs (Kean et al., 2002). NOS, which is Ca^{2+}/calmodulin dependent (Regulski and Tully, 1995), is expressed in *Drosophila* MTs (Dow et al., 1994) and was immunodetected in tubules of other dipteran species (*A aegypti*; *An. gambiae* and the tsetse fly, *Glossina moristans*) (Pollock et al., 2004). CAPA peptides encoded by *D. melanogaster* (CAPA-1) and *An. gambiae* (two putative CAPA peptides) stimulated NOS activity and cGMP level in all these insect species (Pollock et al., 2004). Although NOS was detected in *S. gregaria* MTs, it is apparently not involved in orthopteran diuresis via the CAPA-mediated NO/cGMP signaling pathway (Pollock et al., 2004). cGMP activates protein kinases which in turn stimulate ion transport and fluid secretion by interacting with apical $V-H^+$-ATPases and perhaps with ion channels. CAP_{2b} like the previously mentioned Drosokinin elevates IP_3 levels by stimulating PLCβ and interaction of IP_3 with its receptor is essential for the calcium signaling cascade in both principal and stellate cells (Pollock et al., 2003). cGMP stimulates fluid secretion by *Drosophila* MTs, and apparently, the NO-cGMP signaling pathway plays a role in fluid secretion by MTs as NOS is expressed in *Drosophila* MTs (Dow et al., 1994) and is stimulated by CAP_{2b} in this tissue (Davies et al., 1997). NOS was immunodetected in MTs of certain dipterans (*D. melanogaster*, *A. aegypti*, *An. stepheni* and *G. moristans*) and in an orthoptran species (*S. gregaria*) (Pollock et al., 2004).

Stimulation of fluid secretion by serotonin and the CRF-like DH peptide in *Rhodnius* is negated by a native $CAPA_{2b}$ (RhoprCAPA-2) neurohormone, which apparently plays a role in regulating antidiuretic activity in MTs and across the anterior midgut (Ianowsky et al., 2010; Paluzzi and

Orchard, 2010; Paluzzi et al., 2008). The antidiuretic function of the neuropeptide in MTs is independent of the NO pathway (Quinlan et al., 1997), and its antidiuretic effect in the anterior midgut is seemingly not mediated by cGMP or calcium (Ianowsky et al., 2010). Unlike *Drosophila*, where cAMP and cGMP stimulate diuresis, exogenously applied cGMP inhibits urine secretion by *Rhodnius* upper MT's and is antagonistic to cAMP by an apparent activation of a cAMP-degrading phosphodiesterase (Quinlan and O'Donnell, 1998). It appears that the antidiuretic effect of CAP_{2b} is a common feature of the blood-feeding *Rhodnius* and the plant feeding hemipterans such as *N. viridula* and the green stink bug *Acrosternum hilare* (Coast et al., 2010). It is also noteworthy that in addition to the antidiuretic activity of their native CAPA peptides, insect kinins have no activity in both *Rhodnius* and the stink bugs.

Two *R. prolixus* genes, RhoprCAPA-α and Rhopr-CAPA-β, are expressed in the CNS, and each encodes a prepropeptide of 157 amino acids that yields three CAPA peptides. Two identical CAP_{2b} (containing the FPRV-amide motif) peptides of each gene (Rhopr-CAPA-α2 and Rhopr-CAPA-β2) (Paluzzi and Orchard, 2010) activate the *Rhodnius* GPCR (capa-r1) that is highly expressed in MTs and in the anterior midgut (Paluzzi et al., 2010). Recent experiments by (Paluzzi et al., 2012) suggest that Rhopr-CAPA-α2 does not inhibit $V-H^+$-ATPase in *Rhodnius* MTs. but rather functions by reducing the rapid stimulatory effect of the diuretic hormones with concomitant inhibition of the associated natrioresis. The *D. melanogaster* and *An. gambiae* (Ang-Capa-R) GPCRs cloned and expressed in CHO cells were activated by two respective CAP_{2b} (*Drosophila* capa-1; *Drosophila* capa-2 and Ang-capa-1; Ang-capa-2) neuropeptides (Iversen et al., 2002; Olsen et al., 2007). The capa receptor gene (*CG14575*) of *Drosophila*, which binds to the capa neuropeptides with a subsequent induction of the calcium signaling pathway, is highly expressed in the MT principal cells of larvae and adults (Terhzaz et al., 2012). The induced diuresis by and the elevation of Ca^{2+} levels in principal cells, in which the capa receptor was silenced, were reduced, whereas resistance of insects to desiccation was increased in adults of transgenic capa receptor RNAi lines (Terhzaz et al., 2012).

CT-like DHs—A number of CT-like DHs, either sequenced from tissue extracts or predicted from available genomes, are comprised of 31 amino acid residues and display high degree of sequence similarity, and in certain insects even sequence identity (Te Brugge et al., 2005; Zandawala, 2012, Table 1.1). The first molecule (Dippu-DH_{31}) of this peptidergic family was extracted and identified from the brain and corpora cardiaca of *D. punctata* (Furuya et al., 2000). This neuropeptide stimulated fluid secretion by MTs of *D. punctata, L. migratoria* (Furuya et al., 2000) and the large milkweed bug, *Oncopletus fasciatus* (Te Brugge and Orchard, 2008). Unlike the CRF-like DH (Dippu-DH$_{46}$) isolated from the same tissues and that activates fluid secretion by a cAMP-dependent mechanism, this second messenger is not involved in the diuretic activity in *D. punctata* (Tobe et al., 2005). In contrast, the *Drosophila* neuropeptide (Drome-DH$_{31}$) stimulated fluid secretion by MTs by activating the V-H$^+$-ATPase system in the apical membrane of principal cells via cAMP (Coast et al., 2001). *Drosophila* transcriptome analysis revealed that a CT-like gene (CG17415) is enriched 17-fold in the MTs (Wang et al., 2004), and Drome-DH$_{31}$ GPCR, which is selectively expressed in MT principal cells, requires co-expression of CT receptor-like receptor (CLR) accessory proteins for its activity (Johnson et al., 2005). The mosquito natriuretic hormone that controls cAMP-mediated sodium-rich urine by *A. aegypti* MTs (Beyenbach and Petzel, 1987; Petzel et al., 1987) was later identified by Coast et al. (2005) as a CT-like DH. A synthetic *An. gambiae* (Anoga-DH$_{31}$), which is identical in sequence to the *Aedes* CT-like peptide (Aedae-DH$_{31}$), displayed diuretic and natriuretic activities that are stimulated by cAMP production (Coast et al., 2005). Dippu-DH$_{31}$ was immunolocalized in the CNS and in tissues like hindgut, salivary glands and neurohemal release sites in fifth instar *R. prolixus* (Te Brugge et al., 2005). Recently, Zandawala et al. (2011) determined the complete cDNA sequences of three *Rhopr-DH$_{31}$* splice variants and demonstrated the different expression level of the transcripts in the CNS of unfed *Rhodnius* fifth instar nymphs.

6.2.3 Other neuropeptides

ADFs—The first endogenous antidiuretic factor (FopADF) that inhibits fluid secretion by MTs was purified from hemolymph and abdomen extracts of the red wood ant *Formica polyctena* (Laenen et al., 2001). At the same time, a potent ADF from head extracts of the Colorado potato beetle, *Leptinotarsa decemlineata* was reported by Lavigne et al. (2001). Later, two additional ADFs (Tenmo-ADFa and Tenmo-ADFb) were isolated from head extracts of *T. molitor* pupae and sequenced (Eigenheer et al., 2002, 2003). The peptides, which were immunolocalized in the protocerebrum, are extremely potent molecules that act via cGMP to inhibit fluid secretion by isolated *Tenebrio* MTs (EC$_{50}$ value of ~ 10 fM with ADFa) (Eigenheer et al., 2002). Tenmo-ADFa inhibited the rate of fluid secretion by isolated *A. aegypti* MTs via elevated intracellular cGMP concentrations (Massaro et al., 2004), but unlike the antidiuretic activity of the CAP$_{2b}$-mediated cGMP/NO pathway, NO production is not involved in the antidiuretic mechanism in the beetle (Eigenheer et al., 2002). Based on the experimental results, it was suggested that the $Na^+/K^+/2Cl^-$ co-transporter and the Na^+/H^+ exchangers are also possible targets for Tenmo-ADFa in *Aedes* MTs (Massaro et al., 2004). Five genes with similar gene structures of Tenmo-ADFb antidiuretic peptide and CRF-like DH gene (Li et al., 2008) were identified within the *T. castaneum* genome sequence. As the ADF genes were not detected in sequenced genomes of other insects, it is conceivable that they play a specific crucial role in water conservation in desiccation-resistant beetle species such as *Tenebrio* and *Tribolium*.

ITP—ITPs are antidiuretic peptides that stimulate transport of ions (Cl^-, Na^+, K^+) and fluid reabsorption via cAMP in ilea and recta of orthopterans (Audsley et al., 1992b; Jeffs and Phillips, 1996; Meredith et al., 1996; Phillips and Audsley, 1995) with sequence similarity to the family of peptidergic crustacean hyperglycaemic hormones (Montagné et al., 2010). The first IPT (Schge-ITP) was purified from corpora cardiaca of the locust *S. gregaria* (Audsley et al., 1992a), and ITP-related neuropeptides were immunolocalized in the corpora cardiaca of several other orthopteran species (Macins et al., 1999). A cDNA with transcripts restricted to the brain and corpora cardiaca encoded a putative ITP prohormone that yields an active amidated peptide of 72-amino acid residues (Meredith et al., 1996; see Table 1.1 for peptide sequence). Synthetic ITP prepared according to the deduced propeptide cDNA demonstrated similar biological activity as the native peptide, thus confirming the complete amino acid sequence (King et al., 1999). Apparently, IPT is not unique to orthopterans because a cDNA encoding an ITP putative peptide was isolated from brains and thoracic ganglia of *B. mori* larvae (Endo et al., 2000). Conserved genes encoding ITP and a related non-amidated, inactive ITP-like peptide (ITPL) by alternative splicing were identified in *Manduca*, *Bombyx* and *Aedes* (Dai et al., 2007), and in *T. castaneum* (Begum et al., 2009). Furthermore, similar conserved genomic DNA structures and peptide precursor sequences were also detected in *A. mellifera*, *An. gambiae* and *D. melanogaster* (Dai et al., 2007). The alternatively spliced forms of *D. melanogaster itp* gene (CG13586) encodes two similar ITPL and a shorter (73 amino acids) ITP neuropeptides that were immunolocalized in distinct neurons of the CNS and the peripheral nervous system (Dircksen et al., 2008). As the hindgut is the apparent target for antidiuresis, it was hypothesized that the ITPL non-amidated isoforms play a regulatory role as competitive inhibitors to the antidiuretic ITPs at still unknown hindgut receptors (Dircksen, 2009).

Orcokinins (OKs) and RYamides—OKs, which are a new family of neu-ropeptides identified in arthropod species, control various physiological functions (Christie, 2008; Christie et al., 2010; Yamanaka et al., 2011). Re-cently, it has been established that OKB neuropeptide is expressed in the brain and the anterior midgut of *R. prolixus*, suggesting its regulatory role in diuresis (Sterkel et al., 2012). RYamides are a recently discovered class of neuropeptides detected by far in all insects whose genome was sequenced (Collin et al., 2011; Hauser et al., 2010). RYamides are characterized by a unique FFXXRRY-NH$_2$ C-terminal sequence motif, and their receptors in *Drosophila* and *Tribolium* were identified (Collin et al., 2011). Since the gene coding for RYamide in *Drosophila* (see amino acid sequence in Table 1.1) is appreciably expressed in the hindgut of adults, a role played by this neuropeptide in water reabsorption was postulated (Collin et al., 2011).

7. TARGET SITES FOR INTERFERENCE

Osmoregulation and water homeostasis in insects are natural and prom-ising targets for interference in order to discover selective and environmentally compatible pest control agents. Common *in vitro* water conductance assays (isolated MTs and hindguts, *Xenopus* oocytes swelling system), transfected cells, high-throughput yeast-based phenotype method using transformed cells, as well as screening systems based on *in silico* simulations have been available to probe potent agonist and antagonist compounds as candidate lead compounds. Nevertheless, despite extensive research, there are no commer-cially effective and selective compounds acting as blockers or modulators that target water channels, neurohormone synthesis and processing or neurohor-mone receptors.

7.1. Aquaporins

The quaternary ammonium compound, tetraethylammonium (TEA) chlo-ride (Fig. 1.6), a known blocker of the voltage-activated potassium channel (MacKinnon and Yellen, 1990), inhibits water conductance in several mam-malian AQPs both via native channels and through channels expressed in the *Xenopus* oocytes system (Brooks et al., 2000; Detmers et al., 2006; Küppers et al., 2008; Müller et al., 2008; Yool et al., 2002). Most likely, the inhibition involves an interaction with Tyr[186] residue (in bovine AQP1) in the external loop E domain (Detmers et al., 2006). The corresponding Tyr[185] residue in the *Anopheles* (AgAQP1) channel protein apparently interacts with TEA as its replacement by Phe eliminates the sensitivity to the inhibitor (Liu et al., 2011). However, studies with mammalian AQP1 (Søgaard and Zeuthen, 2008; Yool et al., 2002) and with *Bemisia*, BtAQP (Mathew et al., 2011) claimed that water transport was not affected by TEA. A number of sulfon-amide antiepileptic drugs were subjected to *in silico* docking studies and to standard *in vitro Xenopus* assays for water flux inhibition using mammalian AQP4 (Haddoub et al., 2009; Huber et al., 2007, 2009). Acetazolamide (Fig. 1.6) was the most active compound; yet like TEA, other studies with AQP4 failed to demonstrate inhibition of water permeation by a large series of antiepileptic drugs (Yang et al., 2008). Acetazolamide had no effect on fluid secretion by the upper MTs of *Rhodnius*, but bumetanide (Fig. 1.6), which acts on bumetanide-sensitive $Na^+/K^+/2Cl^-$ co-transporter (Ianowski and O'Donnell, 2001; Ianowski et al., 2004), inhibited secretion rate by 80% at 50 μM (Gámez et al., 2012). By and large, all compounds show rather weak inhibitory effects, and even the most potent 4-aminopyrimidine carboxamide analog of the bumetanide series (AqBO13, Fig. 1.6) displayed only a mild effect ($IC_{50} \sim 20$ μM) on water permeation via mammalian AQP-1 and AQP-4 (Migliat et al., 2009). *In silico* docking experiments of AqBO13 to rat AQP-4 suggest a direct interaction with a channel protein by occluding the cytoplasmic water-pore entrance (Migliat et al., 2009).

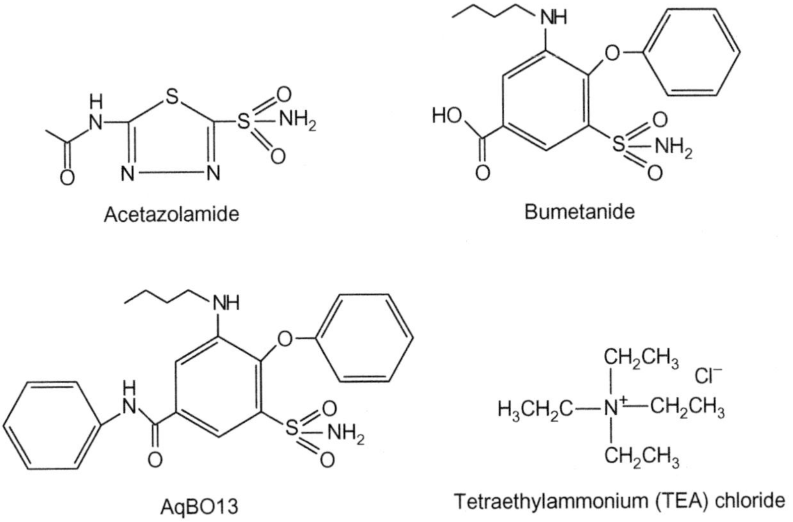

Figure 1.6 Chemical structure of aquaporin inhibitors.

7.2. Peptidergic neurohormones

Diuretic and antidiuretic neuropeptides and their respective receptors may serve as invaluable experimental tools for providing leads to highly selective and environmentally acceptable insect control agents (Scherkenbeck and Zdobinsky, 2009). The great advantage lies in their high specificity, biological activity at extremely low dosage and short amino acid sequence that is amenable to chemical modifications yielding effective and stable peptidomimetics. There are, however, many hurdles in the above approach related to instability of compounds in the environment, short half-life due to fast metabolic degradation, difficult in accessibility that involve penetration and pharmacokinetics as well as anticipated problems in application because of poor solubility in organic and aqueous solutions. Other applicative approaches such as recombinant baculoviruses and transgenic plants expressing signaling neuropeptides were also suggested (Broeck et al., 1997).

The short kinin peptides (6–15 amino acid residues) with the bioactive amidated C-terminal pentapeptide core motif ($F-X^1-X^2-W-G-NH_2$) have been palpable candidates for chemical synthesis of peptidomimetic molecules (Taneja-Bageshwar et al., 2009). It has been confirmed that the C-terminal amide and the β-turn conformation by residues 1–4 of the core peptide region is essential for bioactivity (Roberts et al., 1977; Taneja-Bageshwar et al., 2006). Biostable analogs depend on protecting the vulnerable hydrolysis-sensitive sites of the neuropeptide molecule (Nachman et al., 2002; Taneja-Bageshwar et al., 2009). The primary hydrolysis site is protected when the bulky α-amino isobutyric acid (Aib), which sterically mimics a critical β-turn configuration essential for bioactivity, is incorporated into the third position of the core motif (Taneja-Bageshwar et al., 2009). Biostability was further enhanced by the introduction of a second bulky moiety close to the secondary cleavage site located outside the core region. Recombinant Southern cattle tick (*Boophilus microplus*) (Holmes et al., 2003) and mosquito (*A. aegypti*) (Pietrantonio et al., 2005) receptors expressed in CHO-K1 cells facilitated examination of kinin analog using the intracellular calcium bioluminescence plate assay (Lu et al., 2011b). A superagonist analog, in which Aib is incorporated into the kinin active core motif, was eightfold more active in the isolated *A. domesticus* MT fluid secretion assay than the most active native Achetakinin (Nachman et al., 1997). Several of the hydrolytically stable agonist peptides were similar and even surpassed the activity of the controls. One kinin analog (K-Aib-1 = [Aib]FF[Aib] $WGNH_2$) was five- and eightfold more potent than the control agonist

(FFFSWGNH$_2$) on the tick and mosquito receptors, respectively (Taneja–Bageshwar et al., 2009). This analog is twofold more active than the native *Aedes* kinin-II on the mosquito receptor and displayed considerable diuretic activity when assayed with isolated *Aedes* MTs. Aib-containing peptidase-resistant insect kinin analog was as potent as Muscakinin in *in vivo M. domestica* diuretic assay (Nachman et al., 2002). The most potent analog (NSKYVSKQ [Aib]FF[Aib]WG–NH$_2$) (Fig. 1.7C) and the native *Aedes* kinin demonstrated similar stimulation of fluid secretion rate by isolated *Aedes* MTs. The same biostable kinin analog added to an *A. pisum* artificial diet significantly reduced honeydew formation and was lethal at extremely low concentration (LC$_{50}$ at

A. Cyclo[Tyr-Phe-Ser-Pro-Trp-Gly]
    Cyclohelicokinin I

B. Keto-Tyr-Phe-Ser-Pro-Trp-Gly-CONH$_2$
    Helicokinin analog

C. Asp-Ser-Lys-Tyr-Val-Ser-Lys-Gln-[Aib]-Phe-Phe-[Aib]-Trp-
    Gly-CONH$_2$

    Aedes kinin I analog

D.

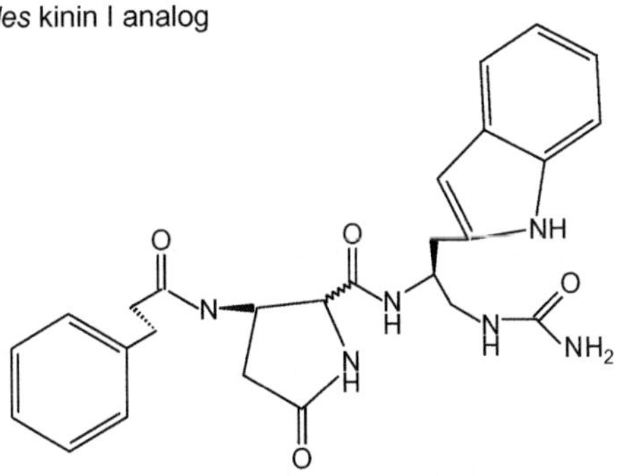

Analog of kinin pentapeptide core motif

Figure 1.7 Structures of peptidergic insect kinin analogs. (A) A cyclic analog of helicokinin I. (B) Helicokinin analog in which the C-terminal tyrosine is replaced with keto tyrosine [β-(4′-hydroxy-2′-benzoyl) alanine] residue (Scherkenbeck et al., 2009). (C) *Aedes* kinin biostable analog in which two α-amino isobutyric acid (Aib) residues were incorporated to reduce hydrolysis at primary and secondary peptidase sites (Taneja-Bageshwar et al., 2009). (D) A biostable kinin analog in which the X$_1$X$_2$ amino acids in the kinin core motif (FX$_1$X$_2$WG-amide) are replaced by the β-turn mimetic 4-aminopyroglutamic acid, and the Phe residue is replaced with hydrocinnamic acid (Nachman et al., 2012).

63 pmol/µl diet) (Smagghe et al., 2010). Using measurements of fluid secretion by isolated MTs in conjunction with the receptor calcium bioluminescence assay, a meticulous structure–activity relationship (SAR) study with Helicokinin peptidomimetics (see Fig. 1.7B for a typical linear bioactive analog) was carried out by Scherkenbeck et al. (2009). Comparison of SAR data from MT versus receptor-expressed system assays shows rather large disparities and inconsistencies among the peptide analogs. For example, the truncated pentapeptide stimulated fluid secretion but its receptor activation was reduced by more than two orders of magnitude compared to the three Helicokinins. The cyclic peptide (cyclo[YFSWG]), which was active in stimulating fluid secretion, was inactive in the receptor assay. A study with end-to-end cyclic Helicokinin analogs (Fig. 1.7A) found them practically inactive using the receptor activating assay in a membrane-mimicking environment composed of sodium dodecyl sulphate or dodecylphosphocholine micelles (Zdobinsky et al., 2009).

Recently, novel non-peptide agonist analogs of kinin, in which the native C-terminal pentapeptide core motif (F-X_1-X_2-W-G-NH_2) was drastically modified, displayed considerable diuretic activity in isolated *A. domesticus* MTs fluid secretion assay (Nachman et al., 2012) (Fig. 1.7D). The X_1 and X_2 residues were replaced by a 4-aminopyroglutamate type VI β-turn mimetic, and resistance to proteolytic inactivation was achieved by eliminating the peptide bond between tryptophan and glycine and by replacing the phenylalanine by hydrocinnamyl residue (Fig. 1.7D).

7.3. Transglucosylation and α-glucosidase

α-Glucosidase-mediated transglucosylation has been considered as a target for disrupting osmoregulation in phloem-feeding insects (Price et al., 2007). α-Glucosidase inhibitors from plant and bacterial sources, such as pseudo aminosugars, carbasugars, iminosugars and conduritol derivatives (de Melo et al., 2006), interfere with enzyme activity, oligomerization and osmoregulation in insects. The pseudotetrasaccharide acarbose (Fig. 1.8), an effective α-glucosidase inhibitor which has a medical use as an anti-diabetic drug, is a potent pea aphid enzyme inhibitor with an IC_{50} value of 6×10^{-8} M (Karley et al., 2005). The level of oligosaccharides was reduced in the excreted honeydew whereas the hemolymph osmotic pressure was increased in *A. pisum* feeding on sucrose-containing artificial diet supplemented with acarbose, and the mortality rate of aphids exposed to the inhibitor indicated dysfunctional osmoregulation (Karley et al., 2005).

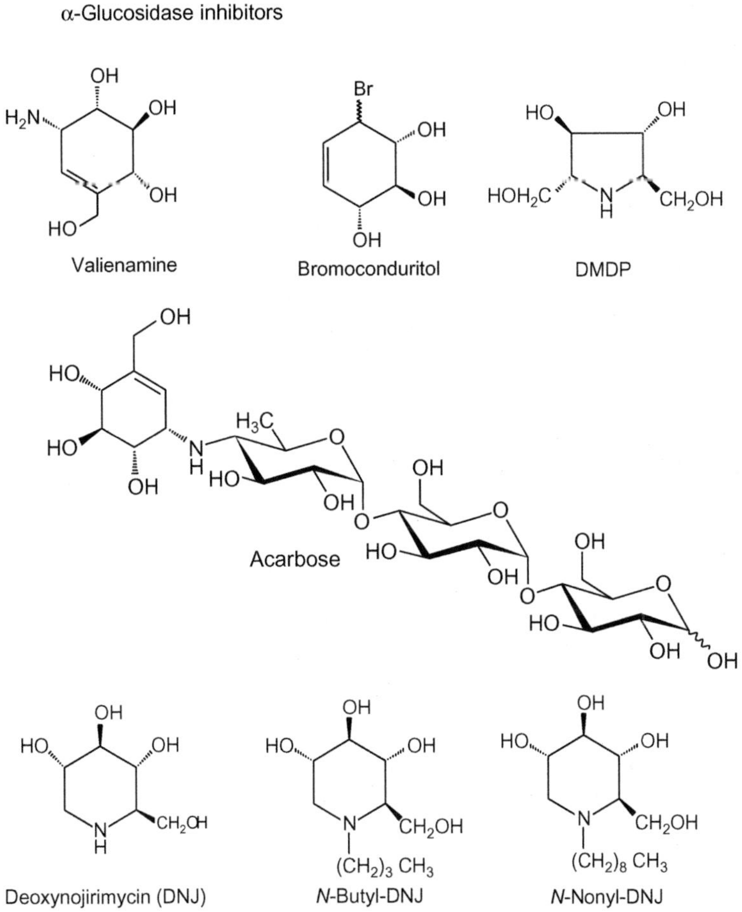

α-Glucosidase inhibitors

Valienamine   Bromoconduritol   DMDP

Acarbose

Deoxynojirimycin (DNJ)   N-Butyl-DNJ   N-Nonyl-DNJ

Figure 1.8 Chemical structure of α-glucosidase inhibitors.

The halocyclitol compound bromoconduritol (Fig. 1.8), a specific irreversible α-glucosidase inhibitor (Datema et al., 1982), inhibited the activity of this enzyme in addition to trehalulose synthase in *Bemisia* extracts (Salvucci, 2000). Exposure of whiteflies to the dietary bromoconduritol considerably decreased the amount of excreted sugars without changing their relative distribution. Other compounds like the pseudoaminosugar valienamine were reported as strong reversible inhibitors of the Asian honeybee (*Apis cerana*) α-glucosidase (Zhang et al., 2007), and the iminosugar DMDP (2R, 5R-dihydroxymethyl-3R, 4R-dihydroxypyrrolidine) (Fig. 1.8) was demonstrated as an efficient inhibitor in a number of terrestrial insects (Scofield et al., 1995), aquatic insects and certain crustaceans (Jüttner and Wessel, 2003).

Iminosugars, which are metabolically inert polyhydroxylated alkaloids in which the endocyclic oxygen of sugars is replaced by the positively charged nitrogen atom, are widespread in plants and microorganisms. The configuration of iminosugar compounds, which mimics the transition state of enzyme-catalyzed glycoside bond hydrolysis, is the basis of their potent biological activity (Asano et al., 2001). 1-Deoxynojirimycin (DNJ) (Fig. 1.8) was the first reported glucose mimetic iminosugar. It was isolated from several *Bacillus* and *Streptomyces* strains and from leaves of mulberry trees, and has been recognized as a potent α-glucosidase inhibitor (Asano, 2003; de Melo et al., 2006). Mulberry latex, which contains high levels of various iminosugars including DNJ and toxic to generalist herbivore caterpillars, is non-toxic to the specialist *B. mori* (Asano et al., 2001; Hirama et al., 2007; Konno et al., 2006). Inhibition experiments using DNJ demonstrated that the Eri silkmoth, *Samia recini*, midgut sucrase is sensitive to the iminosugar at μM levels, while the *B. mori* enzyme is insensitive to concentrations higher than 1 mM (Hirama et al., 2007). This enzymatic response may be part of the adaptive mechanism of silkworms, evolved to circumvent the toxic effects of latex-containing inhibitory alkaloids in mulberry trees.

DNJ and two N-alkylated DNJ analogs (N-butyl DNJ and N-nonyl DNJ) (Fig. 1.8) have diverse inhibitory effects on the adult *B. tabaci* α-glucosidase activity. DNJ is the most potent iminosugar with IC_{50} value of 2 μM, trailed by NB-DNJ and NN-DNJ with IC_{50} values of 10 and 100 μM, respectively (Cohen, E., unpublished). In comparison with the whitefly, inhibition of the pea aphid enzyme by DNJ is lower ($IC_{50} \sim 30 \mu$M) (Karley et al., 2005). When whitefly adults were exposed to dietary iminosugars, the rate of mortality showed that NN-DNJ instigated the most toxic effects (EC_{50} of 5 mM), while DNJ was least toxic (EC_{50} value of 50 mM) (Cohen, E., unpublished). The difference between the *in vitro* and *in vivo* experiments may be attributed to the relatively more hydrophobic NN-DNJ with better access to the membrane-bound α-glucosidase compartment. Similarly, toxicity to cultured bovine kidney cultured cells was enhanced with increased chain length of N-alkylated DNJ (Mellor et al., 2002). Insect mortality, nevertheless, might not be solely due to disrupted osmoregulation, as the iminosugars can possibly act on essential non-digestive α-glucosidases.

8. CONCLUDING NOTES

A large body of evidence has established the elaborate physiological mechanisms operating in insects to sustain water homeostasis and to manage extreme osmotic challenges. Although the basic components of the cellular machinery that functions to keep proper balance of water and electrolytes are similar, the specific temporal and spatial regulation of diuresis and antidiuresis, triggered by neurohormones, varies among insect species. For example, CAP_{2b} that is diuretic in dipterans is antidiuretic in hemipterans; serotonin, which is a major neurohormone in the blood-sucking *Rhodnius*, has no effect on plant-feeding hemipterans; ADH genes detected in *T. castaneum* have not been discovered in genomes of other insects, whereas genes for a diuretic kinin and its receptor are lacking in this xeric beetle. It appears that insects have evolved different hormonal strategies to regulate diuresis and antidiuresis. Largely, such modified regulations are associated with the particular insect genome, nutritional sources and habitats.

Peptidergic diuretic and antidiuretic neurohormones as well as their respective receptors have been recognized as exceptionally potent targets for disrupting water balance in insects. Activity at extremely low doses and amenability to synthetic modifications due to short amino acid sequences (particularly kinins) underlie the remarkable advantage of neuropeprtides. Many metabolically biostable linear and cyclic analogs were synthesized, part of which were equally potent and even surpass the activity of native neuro-hormones. Unfortunately and disappointedly, despite decades of intensive research, not even a single compound has been commercialized to date as an insect control agent. That is also the case with other insect neurohormones such as pheromone biosynthesis-activating neuropeptide, adipokinetic hor-mone, allatotropin or the eclosion hormone.

In addition to the obvious course of disrupting water balance by chem-ical measures, other routes have been considered as future approaches to control medically- and agriculturally important insect pests. Such measures, albeit trendy, include transgenic baculoviruses and plants expressing neuro-peptides or knock down of essential genes encoding neuropeptides and their receptors by exposing insects to respective dsRNAs either directly or indi-rectly via transgenic plants.

ACKNOWLEDGEMENTS

I thank Prof. Uri Gerson from the Hebrew University of Jerusalem and Dr. Hans Merzendorfer from the University of Osnabrueck for carefully reading the chapter and for their useful comments. I greatly acknowledge the creative illustrations prepared by Mr. Yossi Maoz.

REFERENCES

Agre, P., 2006. The aquaporin water channels. Proc. Am. Thorac. Soc. 3, 5–13.
Allan, A.K., Du, J., Davies, S.A., Dow, J.A.T., 2005. Genome-wide survey of V-ATPase genes in *Drosophila* reveals a conserved renal phenotype for lethal alleles. Physiol Geno-mics 22, 128–138.

Asano, N., 2003. Naturally occurring iminosugars and related compounds: structure, distribution, and biological activity. Curr. Top. Med. Chem. 3, 471–484.

Asano, N., Yamashita, T., Yasuda, K., Ikeda, K., Kizu, H., Kameda, Y., Kato, A., Mash, R.J., Lee, H.S., Ryu, K.S., 2001. Polyhydroxylated alkaloids isolated from mulberry trees (*Morus alba* L.) and silkworms (*Bombyx mori* L.). J. Agric. Food Chem. 49, 4208–4213.

Ashford, D.A., Smith, W.A., Douglas, A.E., 2000. Living on high sugar diet: the fate of sucrose ingested by a phloem-feeding insect, the pea aphid *Acyrthosiphon pisum*. J. Insect Physiol. 46, 335–341.

Audsley, N., McIntosh, C., Phillips, J.E., 1992a. Isolation of a neuropeptide from locust corpus cardiacum which influences ileal transport. J. Exp. Biol. 173, 261–274.

Audsley, N., McIntosh, C., Phillips, J.E., 1992b. Actions of ion-transport peptide from locust corpus cardiacum on several hindgut transport processes. J. Exp. Biol. 173, 275–288.

Audsley, N., Kay, I., Hayes, T.K., Coast, G.M., 1995. Cross reactivity studies of CRF-related peptides on insect Malpighian tubules. Comp. Biochem. Physiol. A 110, 87–93.

Azuma, M., Ohta, Y., 1998. Changes in H^+-translocating vacuolar-type ATPase in the anterior silk gland cell of *Bombyx mori* during metamorphosis. J. Exp. Biol. 201, 479–486.

Azuma, M., Nagae, T., Maruyama, M., Kataoka, N., Miyake, S., 2012. Two water-specific aquaporins at the apical and basal plasma membranes of insect epithelia: molecular basis for water recycling through the cryptonephric rectal complex of lepidopteran larvae. J. Insect Physiol. 58, 523–533.

Baldwin, D.C., Schegg, K.M., Furuya, K., Lehmberg, E., Schooley, D.A., 2001. Isolation and identification of a diuretic hormone from *Zootermopsis nevadensis*. Peptides 2, 147–153.

Ball, A., Campbell, E.M., Jacob, J., Hoppler, S., Bowman, A.S., 2009. Identification, functional characterization and expression pattern of a water-specific aquaporin in the brown dog tick, *Rhipicephalus sanguineus*. Insect Biochem. Mol. Biol. 39, 105–112.

Bates, R.B., Byrne, D.N., Kane, V.V., Miller, W.B., Taylor, S.R., 1990. N.m.r. characterization of trehalulose from excrement of the sweet potato whitefly, *Bemisia tabaci*. Carbohydr. Res. 201, 342–345.

Bayley, M., Petersen, S.O., Knigge, T., Köhler, H.-R., Holmstrup, M., 2001. Drought acclimation confers cold tolerance in the soil collembolan *Folsomia candida*. J. Insect Physiol. 47, 1197–1204.

Begum, K., Li, B., Beeman, R.W., Park, Y., 2009. Function of ion transport peptide-like in the red flour beetle *Tribolium castaneum*. Insect Biochem. Mol. Biol. 39, 717–725.

Benga, C., 2009. Water channel proteins (later called aquaporins) and relatives: past, present, and future. IUBMB Life 61, 112–133.

Benoit, J.B., 2010. Water management by dormant insects: comparison between dehydration resistance during summer aestivation and winter diapause. Prog. Mol. Subcell. Biol. 49, 209–229.

Benoit, J.B., Lopez-Martinez, G., Elnitski, M.A., Lee Jr., R.E., Denlinger, D.L., 2009. Dehydration-induced cross tolerance of *Belgica antarctica* larvae to cold and heat is facilitated by trehalose accumulation. Comp. Biochem. Physiol. A 152, 518–523.

Benoit, J.B., Lopez-Martinez, G., Phillips, Z.P., Patrick, K.V., Denlinger, D.L., 2010. Heat shock proteins contribute to mosquito dehydration tolerance. J. Insect Physiol. 56, 151–156.

Bergland, A.O., Chae, H.-S., Kin, Y.-J., Tatar, M., 2012. Fine-scale mapping of natural variation in fly fecundity identifies neuronal domain of expression and function of an aquaporin. PLoS Genet. 8, e1002631.

Beuron, F., Le Cahe'rec, F., Guillam, M.-T., Cavalier, A., Garret, A., Tassan, J.-P., Delamarche, C., Schultz, P., Mallouh, V., Rolland, J.-P., Hubert, J.-F., Gouranton, J., Thomas, D., 1995. Structural analysis of a MIP family protein from the digestive track of *Cicadella viridis*. J. Biol. Chem. 270, 17414–17422.

Beyenbach, K.W., 2003. Transport mechanisms of diuresis in Malpighian tubules of insects. J. Exp. Biol. 206, 3845–3856.

Beyenbach, L.W., Petzel, D.H., 1987. Diuresis in mosquitoes: role of a natriuretic factor. News Physiol. Sci. 2, 171–175.

Beyenbach, K.W., Piermarini, P.M., 2011. Transcellular and paracellular pathways of trans-epithelial fluid secretion in Malpighian (renal) tubules of the yellow fever mosquito *Aedes aegypti*. Acta Physiol. 202, 387–407.

Beyenbach, K.W., Skaer, H., Dow, J.A.T., 2010. The developmental, molecular, and transport biology of Malpighian tubules. Annu. Rev. Entomol. 55, 351–374.

Blackburn, M.B., Kingan, T.G., Bodnar, W., Shabanowitz, J., Hunt, D.F., Kempe, T., Wagner, R.M., Raina, A.K., Schnee, M.E., Ma, M.C., 1991. Isolation and identification of a new diuretic peptide from the tobacco hornworm, *Manduca sexta*. Biochem. Biophys. Res. Commun. 181, 927–932.

Blackburn, M.B., Wagner, R.M., Shabanowitz, J., Kochanskym, J.P., Hunt, D.F., Raina, A.K., 1995. The isolation and identification of three diuretic kinins from the abdominal ventral nerve cord of adult *Heliothis zea*. J. Insect Physiol. 41, 723–730.

Block, W., 1996. Cold or drought—the lesser of two evils for terrestrial arthropods. Eur. J. Entomol. 93, 325–339.

Blumenthal, E.M., 2003. Regulation of chloride permeability by endogenously produced tyramine in the *Drosophila* Malpighian tubule. Am. J. Physiol. Cell Physiol. 284, C718–C728.

Bowman, A.S., Sauer, J.R., 2004. Tick salivary glands: function, physiology and future. Parasitology 129, S67–S81.

Broderick, K.E., MacPherson, M.R., Regulski, M., Tully, T., Dow, J.A.T., Davies, S.A., 2003. Interactions between epithelial nitric oxide signaling and phosphodiesterase activity in *Drosophila*. Am. J. Physiol. Cell Physiol. 285, C1207–C1218.

Broeck, J.V., Schoofs, L., de Loof, A., 1997. Insect neuropeptides and their receptors. New leads for medical and agricultural application. Trends Endocrinol. Metab. 8, 321–326.

Brooks, H.L., Regan, J.W., Yool, A.J., 2000. Inhibition of aquaporin-1 water permeability by tetraethylammonium: involvement of loop E pore region. Mol. Pharmacol. 57, 1021–1026.

Bushman, D.W., Nelson, J.O., 1990. The role of the frontal ganglion and corpora cardiaca/corpora allata complex in post-feeding weight loss in adult *Heliothis zea*. Physiol. Entomol. 15, 269–274.

Bushman, D.W., Raina, A.K., Nelson, J.O., 1989. Post-eclosion diuresis in adult *Heliothis zea*. Physiol. Entomol. 14, 391–396.

Byrne, D.N., Hendrix, D.L., Williams III, L.H., 2003. Presence of trehalulose and other oligosaccharides in hemipteran honeydew, particularly Aleyrodidae. Physiol. Entomol. 28, 144–149.

Cabrero, P., Radford, J.C., Broderick, K.E., Costes, L., Veenstra, J.A., Spana, E.P., Davies, S.A., Dow, J.A.T., 2002. The *Dh* gene of *Drosophila melanogaster* encodes a diuretic peptide that acts through cAMP. J. Exp. Biol. 205, 3799–3807.

Cabrero, P., Pollock, V.P., Davies, S.A., Dow, J.A.T., 2004. A conserved domain of alkaline phosphatase expression in the Malpighian tubules of dipteran insects. J. Exp. Biol. 207, 3299–3305.

Cady, C., Hagedorn, H.H., 1999a. The effect of putative diuretic factors on in vivo urine production in the mosquito, *Aedes aegypti*. J. Insect Physiol. 45, 317–325.

Cady, C., Hagedorn, H.H., 1999b. Effects of putative diuretic factors on intracellular second messenger levels in the Malpighian tubules of *Aedes aegypti*. J. Insect Physiol. 45, 327–337.

Campbell, E.M., Ball, A., Hoppler, S., Bowman, A.S., 2008. Invertebrate aquaporins: a review. J. Comp. Physiol. B 178, 935–955.

Campbell, E.M., Burdin, M., Hoppler, S., Bowman, A.S., 2010. Role of an aquaporin in the sheep tick *Ixodes ricinus*: assessment as a potential control target. Int. J. Parasitol. 40, 15–23.

Chown, S.L., Sørensen, J.G., Terblanche, J.S., 2011. Water loss in insects: an environmental change perspective. J. Insect Physiol. 57, 1070–1084.

Christie, A.E., 2008. Identification of putative crustacean neuropeptides using *in silico* analyses of publicly expressed sequence tags. Gen. Comp. Endocrinol. 157, 174–185.

Christie, A.E., Durkin, C.N., Hartline, P., Ohno, P., Lenz, P.H., 2010. Bioinformative analyses of the publicly accessible crustacean expressed sequence tags (ESTs) reveal numerous neuropeptide-encoding precursor proteins, including ones from members of several little studied taxa. Gen. Comp. Endocrinol. 167, 164–178.

Churchill, T.A., Storey, K.B., 1989. Seasonal variation in the temperature-stimulated inter-conversion of glycogen and glycerol pools in a freeze avoiding moth larva. Cryo-Letters 10, 127–136.

Cicero, J.M., Brown, J.K., Roberts, P.D., Stansly, P.A., 2009. The digestive system of *Diaphorina citri* and *Bactericera cockerelli* (Hemiptera: Psyllidae. Ann. Entomol. Soc. Am. 102, 650–665.

Clark, T.M., Dradley, T.J., 1998. Malpighian tubules of larval *Aedes aegypti* are hormonally stimulated by 5-hydroxytryptamine in response to increased salinity. Arch. Insect Biochem. Physiol. 34, 123–141.

Clark, M.S., Thorne, M.A.S., Purać, J., Burns, G., Hillyard, G., Popović, Z.D., Gruber-Lajšić, G., Worland, M.R., 2009. Surviving the cold: molecular analysis of insect cryoprotective dehydration in the arctic springtail *Megaphorura arctica* (Tullberg). BMC Genom. 10, 328.

Clottens, F.L., Holman, G.M., Coast, G.M., Totty, N.F., Hayes, T.K., Kay, I., Mallet, A.I., Wright, M.S., Chung, J.-S., Truong, O., Bull, D.L., 1994. Isolation and characterization of a diuretic peptide common to the house fly and stable fly. Peptides 15, 971–979.

Clynen, E., Huybrechts, J., De Loof, A., Schoofs, L., 2003. Mass spectrometric analysis of the perisympathetic organs in locusts: identification of novel periviscerokinins. Biochem. Biophys. Res. Commun. 300, 422–428.

Coast, G.M., 1988. Fluid secretion by single isolated Malpighian tubules of the house cricket, *Acheta domesticus*, and their response to diuretic hormone. Physiol. Entomol. 13, 381–391.

Coast, G., 2007. The endocrine control of salt balance in insects. Gen. Comp. Endocrinol. 152, 332–338.

Coast, G.M., 2009. Neuroendocrine control of ionic homeostasis in blood-sucking insects. J. Exp. Biol. 212, 378–386.

Coast, G., 2011. Serotonin has kinin-like activity in stimulating secretion by Malpighian tubules of the house cricket *Acheta domesticus*. Peptides 32, 500–508.

Coast, G.M., Schooley, D.A., 2011. Toward a consensus nomenclature for insect neuropep-tides and peptide hormones. Peptides 32, 620–631.

Coast, G.M., Holman, G.M., Nachman, R.J., 1990. The diuretic activity of a series of cephalomyotropic neuropeptides, the achetakinins, on isolated Malpighian tubules of the house cricket, *Acheta domesticus*. J. Insect Physiol. 36, 481–488.

Coast, G.M., Webster, S.G., Schegg, K.M., Tobe, S.S., Schooley, D.A., 2001. The *Drosophila melanogaster* homologue of an inset calcitonin-like diuretic peptide stimulates V-ATPase activity in fruit fly Malpighian tubules. J. Exp. Biol. 204, 1795–1804.

Coast, G.M., Orchard, I., Phillips, J.E., Schooley, D.A., 2002. Insect diuretic and antidiuretic hormones. Adv. Insect Physiol. 29, 279–409.

Coast, G.M., Garside, C.S., Webster, S.G., Schegg, K.M., Schooley, D.A., 2005. Mosquito natriuretic peptide identified as a calcitonin-like diuretic hormone in *Anopheles gambiae* (Giles). J. Exp. Biol. 208, 3281–3291.

Coast, G.M., Te Brugge, V.A., Nachman, R.J., Lopez, J., Aldrich, J.R., Lange, A., Orchard, I., 2010. Neurohormones implicated in the control of Malpighian tubule secretion in plant sucking heteropterans: the stink bugs *Acrosternum hilare* and *Nezara viridula*. Peptides 31, 468–473.

Coast, G.M., Nachman, R.J., Lopez, J., 2011. The control of Malpighian tubule secretion in a predaceous hemipteran insect, the spined soldier bug *Podisus maculiventris* (Heteroptera, Pentatomidae). Peptides 32, 493–499.

Cohen, E. (2012). Roles of aquaporins in osmoregulation, desiccation and cold hardiness in insects. Entomol Ornithol Herpetol http://dx.doi.org/10.4172/2161-0983 S1-001

Collin, C., Hauser, F., Krogh-Meyer, P., Hansen, K.K., Gonzalez de Valdivia, E., Williamson, M., Grimmelikhuijzen, C.J.P., 2011. Identification of the *Drosophila* and *Tribolium* receptors for the recently discovered insect RYamide neuropeptides. Biochem. Biophys. Res. Commun. 412, 578–583.

Cornette, R., Kanamori, Y., Watanabe, M., Nakahara, Y., Gusev, O., Mitsumasu, K., Kadono-Okuda, K., Shimomura, M., Mita, K., Kikawada, T., Okuda, T., 2010. Identification of anhydrobiosis-related genes from an expressed sequence tag database in the cryptobiotic midge *Polypedilum vanderplanki* (Diptera; Chironomidae). J. Biol. Chem. 285, 35889–35899.

Cristofoletti, P.T., Ribeiro, A.F., Deraison, C., Rahbe, Y., Terra, W.R., 2003. Midgut adaptation and digestive enzyme distribution in a phloem feeding insect, the pea aphid *Acyrthosiphon pisum*. J. Insect Physiol. 49, 11–24.

Dai, L., Zintan, D., Adams, M.E., 2007. Strategic expression of ion transport peptide gene products in central and peripheral neurons in insects. J. Comp. Neurol. 500, 353–367.

Datema, R., Romero, P.A., Legler, G., Schwarz, R.T., 1982. Inhibition of formation of complex oligosaccharides by the glucosidase inhibitor bromoconduritol. Proc. Natl. Acad. Sci. U.S.A. 79, 6787–6791.

Davies, S.-A., 2000. Nitric oxide signaling in insects. Insect Biochem. Mol. Biol. 30, 1123–1138.

Davies, S.A., Huesmann, G.R., Maddrell, S.H., O'Donnell, M.J., Skaer, N.J., Dow, J.A., Tublitz, N.J., 1995. CAP2b, a cardioacceleratory peptide, is present in *Drosophila* and stimulates tubule fluid secretion via cGMP. Am. J. Physiol. 269, R1321–R1326.

Davies, S.A., Stewart, E.J., Huesmann, G.R., Skaer, N.J.V., Maddrell, S.H.P., Tublitz, N.J., Dow, J.A.T., 1997. Neuropeptide stimulation of nitric oxide signaling pathway in *Drosophila melanogaster* Malpighian tubules. Am. J. Physiol. Regul. Integr. Comp. Physiol. 273, R823–R827.

De Melo, E.B., da Silveia Gomes, A., Carvalho, I., 2006. α- and β-Glucosidase inhibitors: chemical structure and biological activity. Tetrahedron 62, 10277–10302.

Detmers, F.J.M., de Groot, B.L., Müller, E.M., Hinton, A., Konings, I.B.M., Sze, M., Flitsch, S.L., Grubmüller, H., Deen, P.M.T., 2006. Quaternary ammonium compounds as water channel blockers. Specificity, potency, and site of action. J. Biol. Chem. 281, 14207–14214.

Digan, M.E., Roberts, D.N., Enderlin, F.E., Woodworth, A.R., Kramer, S.J., 1992. Characterization of the precursor for *Manduca sexta* diuretic hormone Mas-DH. Proc. Natl. Acad. Sci. U.S.A. 89, 11074–11078.

Dircksen, H., 2009. Insect ion transport peptides are derived from alternatively spliced genes and differently expressed in the central and peripheral nervous system. J. Exp. Biol. 212, 401–412.

Dircksen, H., Tesfai, L.K., Albus, C., Nässel, D.R., 2008. Ion transport peptide splice forms in central and peripheral neurons throughout postembryogenesis of *Drosophila melanogaster*. J. Comp. Neurol. 509, 23–41.

Donini, A., O'Donnell, M.J., Orchard, I., 2008. Differential actions of diuretic factors on the Malpighian tubules of *Rhodnius prolixus*. J. Exp. Biol. 211, 42–48.

Dores, R.M., Dallmann, S.H., Herman, W.S., 1979. The regulation of post-eclosion and post-feeding diuresis in the monarch butterfly, *Danaus plexippus*. J. Insect Physiol. 25, 895–901.

Döring, F., Wischmeyer, E., Kühnlein, R.P., Jäckle, H., Karschin, A., 2002. Inwardly rectifying K^+ (Kir) channels in *Drosophila*. J. Biol. Chem. 277, 25554–25561.

Douglas, A.E., 2003. The nutritional physiology of aphids. Adv. Insect Physiol. 31, 73–140.

Douglas, A.E., 2006. Phloem-sap feeding by animals: problems and solutions. J. Exp. Bot. 57, 747–754.

Dow, J.A.T., 2009. Insights into the Malpighian tubule from functional genomics. J. Exp. Biol. 212, 435–445.

Dow, J.A.T., Romero, M.F., 2010. *Drosophila* provides rapid modeling of renal development, function, and disease. Am. J. Physiol. Renal Physiol. 299, F1237–F1244.

Dow, J.A.T., Maddrell, S.H.P., Davies, S.-A., Skaer, N.J.V., Kaiser, K., 1994. A novel role for the nitric-oxide-cGMP signaling pathway: the control of epithelial function in *Drosophila*. Am. J. Physiol. Regul. Integr. Comp. Physiol. 266, R1716–R1719.

Dow, J.A.T., Kell, D.C., Davies, S.A., Maddrell, S.H.P., Brown, D., 1995. A novel member of the major intrinsic protein family in *Drosophila*: are aquaporins involved in insect Malpighian (renal) tubule fluid secretion? J. Physiol. 489, 110p–111p.

Drake, L.L., Boudko, D.Y., Marinotti, O., Carpenter, V.K., Dawe, A.L., Hansen, I.A., 2010. The aquaporin gene family of the yellow fever mosquito, *Aedes aegypti*. PLoS One 5, e15578.

Duchesne, L., Hubert, J.-F., Verbavatz, J.-M., Thomas, D., Pietrantonio, V., 2003. Mosquito (*Aedes aegypti*) aquaporin, present in tracheolar cells, transports water, not glycerol, and forms orthogonal arrays in *Xenopus* oocyte membranes. Eur. J. Biochem. 270, 422–429.

Duman, J.G., Wu, D.W., Xu, L., Tursman, D., Olsen, T.M., 1991. Adaptation of insects to subzero temperatures. Q. Rev. Biol. 66, 387–410.

Echevarría, M., Ramírez-Lorca, R., Hernández, C.S., Gutiérrez, A., Méndes-Ferrer, S., Gonzáles, E., Toledo-Aral, J.J., Ilundáin, A.A., Whittembury, G., 2001. Identification of a new water channel (Rp-MIP) in the Malpighian tubules of the insect *Rhodnius prolixus*. Pflügers Arch. 442, 27–34.

Eigenheer, R.A., Nicolson, S.W., Schegg, K.M., Hull, J.J., Schooley, D.A., 2002. Identification of a potent antidiuretic factor acting on beetle Malpighian tubules. Proc. Natl. Acad. Sci. U.S.A. 99, 84–89.

Eigenheer, R.A., Wiehart, U.M., Nicolson, S.W., Schoof, L., Schegg, K.M., Hull, J.J., Schooley, D.A., 2003. Isolation, identification and localization of a second beetle antidiuretic peptide. Peptides 24, 27–34.

Elnitsky, M.A., Benoit, J.B., Denlinger, D.L., Lee Jr., R.E., 2008. Desiccation tolerance and drought acclimation in the Antarctic collembolan *Cryptopygus antarcticus*. J. Insect Physiol. 54, 1432–1439.

Endo, H., Nagasawa, H., Watanabe, T., 2000. Isolation of a cDNA encoding CHH-family peptide from the silkworm *Bombyx mori*. Insect Biochem. Mol. Biol. 30, 355–361.

Evans, J.M., Allan, A.K., Davies, S.A., Dow, J.A.T., 2005. Sulphonylurea sensitivity and enriched expression implicate inward rectifier K^+ channels in *Drosophila melanogaster* renal function. J. Exp. Biol. 208, 3771–3783.

Fields, P.G., Fleurat-Lessard, F., Lavenseau, L., Febvay, G., Peypelut, L., Bonnot, G., 1998. The effect of cold acclimation and deacclimation on cold tolerance, trehalose and free amino acid levels in *Sitophilus granarius* and *Cryptolestes ferrugineus* (Coleoptera). J. Insect Physiol. 44, 955–965.

Fu, D., Libson, A., Miercke, L.J.W., Weitzman, C., Nollert, P., Krucinski, J., Stroud, R.M., 2000. Structure of a glycerol-conducting channel and the basis of its selectivity. Science 290, 481–486.

Furuya, K., Schegg, K.M., Wang, H., King, D.S., Schooley, D.A., 1995. Isolation and identification of a diuretic hormone from the mealworm *Tenebrio molitor*. Proc. Natl. Acad. Sci. U.S.A. 92, 12323–12327.

Furuya, K., Schegg, K.M., Schooley, D.A., 1998. Isolation and identification of a second diuretic hormone from *Tenebrio molitor*. Peptides 19, 619–626.

Furuya, K., Milchak, R.J., Schegg, K.M., Zhang, J., Tobe, S.S., Coast, G.M., Schooley, D.A., 2000. Cockroach diuretic hormones: characterization of a calcitonin-like peptide in insects. Proc. Natl. Acad. Sci. U.S.A. 97, 6469–6474.

Gaede, K., Knülle, W., 1997. On the mechanism of water vapour sorption from unsaturated atmospheres by ticks. J. Exp. Biol. 200, 1491–1498.

Gámez, A.D., Gutiérres, A.M., García, R., Whittembury, G., 2012. Recent experiments towards a model for fluid secretion in *Rhodnius* Upper Malpighian Tubules (UMP). J. Insect Physiol. 58, 543–550.

Gautam, N.-K., Tapadia, M.G., 2010. Ecdysone signaling is required for proper organization and fluid secretion of stellate cells in the Malpighian tubules of *Drosophila melanogaster*. Int. J. Dev. Biol. 54, 635–642.

Gillen, C.M., Blair, C.R., Heilman, N.R., Somple, M., Stulberg, M., Thombre, R., Watson, N., Gillen, K.M., Itagaki, H., 2006. The cation-chloride cotransporter, masBSC, is widely expressed in *Manduca sexta* tissues. J. Insect Physiol. 52, 661–668.

Goh, S., Phillips, J.E., 1978. Dependence of prolonged water absorption by *in vitro* locust rectum on ion transport. J. Exp. Biol. 72, 25–41.

Goldbard, G.A., Sauer, J.R., Mills, R.R., 1970. Hormonal control of excretion in the American cockroach. II. Preliminary purification of diuretic and antidiuretic hormone. Comp. Gen. Pharm. 1, 82–86.

Gomes, D., Agasse, A., Thiébaud, P., Delrot, S., Gerós, H., Chaumont, F., 2009. Aquaporins are multifunctional water and solute transporters highly divergent in living organisms. Biochim. Biophys. Acta 1788, 1213–1228.

Gonen, T., Walz, T., 2006. The structure of aquaporins. Q. Rev. Biophys. 39, 361–396.

Gonen, T., Sliz, P., Kistler, J., Cheng, Y., Walz, T., 2004. Aquaporin-0 membrane junctions reveal the structure of a closed water pore. Nature 429, 193–197.

Goto, S.G., Philip, B.N., Teets, N.M., Kawarasaki, Y., Lee Jr., R.E., Denlinger, D.L., 2011. Functional characterization of an aquaporin in the Antarctic midge *Belgica antarctica*. J. Insect Physiol. 57, 1106–1114.

Guillam, M.-T., Beuron, F., Grandin, N., Hubert, J.-F., Boisseau, C., Cavalier, A., Couturier, A., Gouranton, J., Thomas, D., 1992. Expression of RNA isolated from the water-shunting complex of a sap-sucking insect increases the membrane permeability for water in *Xenopus* oocytes. Exp. Cell Res. 200, 301–305.

Haddoub, R., Rützler, M., Robin, A., Flitsch, S.L., 2009a. Design, synthesis and assaying of potential inhibitors. In: Beitz, E. (Ed.), Handbook of Experimental Pharmacology190, Springer-Verlag, Berlin, pp. 385–402.

Hadley, N.F., 1994. Water Relations of Terrestrial Arthropods. Academic Press, San Diego p. 356.

Hand, S.C., Menze, M.A., Toner, M., Boswell, L., Moore, D., 2011. LEA proteins during water stress: not just for plants anymore. Annu. Rev. Physiol. 73, 115–134.

Harries, W.E.C., Akhavan, D., Miercke, L.J.W., Khademi, S., Stroud, R.M., 2004. The Channel architecture of aquaporin 0 at a 2.2-Å resolution. Proc. Natl. Acad. Sci. U.S.A. 101, 14045–14050.

Hauser, F., Neupert, S., Williamson, M., Predel, R., Tanaka, Y., Grimmelikhuijzen, C.J.P., 2010. Genomics and peptidomics of neuropeptides and protein hormones present in the parasitic wasp *Nasonia vitripennis*. J. Proteome Res. 9, 5296–5310.

Hayes, T.K., Pannabecker, T.L., Hinckley, D.J., Holman, G.M., Nachman, R.J., Petzel, D.H., Beyenbach, K.W., 1989. Leucokinins, a new family of ion transport stimulators and inhibitors in insect Malpighian tubules. Life Sci. 44, 1259–1266.

Hayes, T.K., Holman, G.M., Pannabecker, T.L., Wright, M.S., Strey, A.A., Nachman, R.J., Hoel, D.F., Olson, J.K., Beyenbach, K.W., 1994. Culekinin depolarizing peptide: a mosquito leucokinin-like peptide that influences insect Malpighian tubule ion transport. Regul. Pept. 52, 235–248.

Hayward, S.A.L., Rinehart, J.P., Denlinger, D.L., 2004. Desiccation and rehydration elicit distinct heat shock protein transcript responses in flesh fly pupae. J. Exp. Biol. 207, 963–971.

Hazelton, S.R., Parker, S.W., Spring, J.H., 1988. Excretion in the house cricket (*Acheta domesticus*): fine structure of the Malpighian tubules. Tissue Cell 20, 443–460.

Hector, C.E., Bretz, C.A., Zhao, Y., Johnson, E.C., 2009. Functional differences between two CRF-related diuretic hormone receptors in *Drosophila*. J. Exp. Biol. 212, 3142–3247.

Hendrix, D.L., Wei, Y.-A., 1994. Bemisiose: an unusual trisaccharide in *Bemisia* honeydew. Carbohydr. Res. 253, 329–334.

Hendrix, D.L., Wei, Y.-A., Leggett, J.E., 1992. Homopteran honeydew sugar composition is determined by both the insect and the plant species. Comp. Biochem. Physiol. 101B, 23–27.

Hirama, C., Konno, K., Wasano, N., Nakamura, M., 2007. Differential effects of sugar-mimic alkaloids in mulberry latex on sugar metabolism and disaccharidase of Eri and do-mesticated silkworms: enzymatic adaptation of *Bombyx mori* to mulberry defense. Insect Biochem. Mol. Biol. 37, 1348–1358.

Hirata, T., Czapar, A., Brin, L., Haritonova, A., Bondeson, D.P., Linser, P., Cabrero, P., Thompson, J., Dow, J.A.T., Romero, M.F., 2012. Ion and solute transport by Prestin in *Drosophila* and *Anopheles*. J. Insect Physiol. 58, 563–569.

Ho, J.D., Yeh, R., Sandstromm, A., Chorny, I., Harries, W.E.C., Robbins, R.A., Miercke, L.J.W., Stroud, R.M., 2009. Crystal structure of human aquaporin 4 at 1.8 Å and its mechanism of conductance. Proc. Natl. Acad. Sci. U.S.A. 106, 7437–7442.

Holman, G.M., Cook, B.J., Nachman, R.J., 1986. Isolation, primary structure and synthesis of two neuropeptides from *Leucophaea maderae*: members of a new family of cephalomyotropins. Comp. Biochem. Physiol. 84C, 204–211.

Holman, G.M., Nachman, R.J., Schoofs, L., Hayes, T.K., Wright, M.S., DeLoof, A., 1991. The *Leucophaea maderae* hindgut preparation: a rapid and sensitive bioassay tool for the isolation of insect myotropins of other insect species. Insect Biochem. 21, 107–112.

Holmes, S.P., Barhoumi, R., Nachman, R.J., Pietrantonio, P.V., 2003. Functional analysis of a G protein-coupled receptor from the Southern cattle tick *Boophilus microplus* (Acari, Ixodidae) identifies it as the first arthropod myokinin receptor. Insect Mol. Biol. 12, 27–38.

Holtzhausen, W.D., Nicolson, S.W., 2007. Beetle diuretic peptides: the response of meal-worm (*Tenebrio molitor*) Malpighian tubules to synthetic peptides, and cross-reactivity studies with the dung beetle (*Orthophagus gazelle*). J. Insect Physiol. 53, 361–369.

Horsefield, R., Norde'n, K., Fellert, M., Backmark, A., Törnroth-Horsefield, S., van Scheltinga, A.C.T., Kvassman, J., Kjellbom, P., Johanson, U., Neutze, T., 2008. High-resolution X-ray structure of human aquaporin 5. Proc. Natl. Acad. Sci. U.S.A. 105, 13327–13332.

Huber, V.J., Tsujita, M., Yamazaki, M., Sakimura, K., Nakada, T., 2007. Identification of arylsulfonamides as aquaporin-4 inhibitors. Bioorg. Med. Chem. Lett. 17, 1270–1273.

Huber, V.J., Tsujita, M., Kwee, I.L., Nakada, T., 2009. Inhibition of aquaporin 4 by antiepileptic drugs. Bioorg. Med. Chem. 17, 418–424.

Hubert, J.-F., Thomas, D., Cavalier, A., Gouranton, J., 1989. Structural and biochemical observations on specialized membranes of the "filter chamber", a water-shunting com-plex in sap-sucking homopteran insects. Biol. Cell 66, 155–163.

Huesmann, G.R., Cheung, C.C., Loi, P.K., Lee, T.D., Swiderek, K.M., Tublitz, N.J., 1995. Amino acid sequence of CAP_{2b}, an insect cardioacceleratory peptide from the tobacco hawkworm *Manduca sexta*. FEBS Lett. 371, 311–314.

Ianowski, J.P., O'Donnell, M.J., 2001. Transepithelial potential in Malpighian tubules of *Rhodnius prolixus*: lumen-negative voltages and triphasic response to serotonin. J. Insect Physiol. 47, 411–421.

Ianowski, J.P., Christensen, R.J., O'Donnell, M.J., 2004. Na^+ competes with K^+ in bumetanide-sensitive transport by Malpighian tubules of *Rhodnius prolixus*. J. Exp. Biol. 207, 3707–3716.

Ianowsky, J.P., Paluzzi, J.-H., Te Brugge, V.A., Orchard, I., 2010. The antidiuretic neurohormone RhoprCAPA-2 downregulates fluid transport across the anterior midgut in the blood-feeding insect *Rhodnius prolixus*. Am. J. Physiol. Regul. Integr. Comp. Physiol. 298, R548–R557.

Ishiguru, S., Li, Y., Nakano, K., Tsumuki, H., Goto, M., 2007. Seasonal changes in glycerol content and cold hardiness in two ecotypes of the rice stem borer, *Chilo suppressalis*, exposed to the environment in the Shonai district, Japan. J. Insect Physiol. 53, 392–397.

Iversen, A., Cazzamali, G., Williamson, M., Hauser, F., Grimmelikhuijen, C.J.P., 2002. Molecular cloning and functional expression of a *Drosophila* receptor for the neuropeptides capa-1 and -2. Biochem. Biophys. Res. Commun. 299, 628–633.

Jeffs, L.B., Phillips, J.E., 1996. Pharmacological study of the second messengers that control rectal ion and fluid transport in the desert locust (*Schistocerca gregaria*). Arch. Insect Biochem. Physiol. 31, 169–184.

Joanisse, D.R., Storey, K.B., 1995. Temperature acclimation and seasonal responses by enzymes in cold-hardy gall insects. Arch. Insect Biochem. Physiol. 28, 339–349.

Johnson, E.C., Bohn, L.M., Taghert, P.H., 2004. *Drosophila CG8422* encodes a functional diuretic hormone receptor. J. Exp. Biol. 207, 743–748.

Johnson, E.C., Shafer, O.T., Trigg, J.S., Park, J., Schooley, D.A., Dow, J.A., Taghert, P.H., 2005. A novel diuretic hormone receptor in *Drosophila*: evidence for conservation of CGRP signaling. J. Exp. Biol. 208, 1239–1246.

Jüttner, F., Wessel, H.P., 2003. Isolation of di(hydroxymethyl) dihydroxypyrrolodine from cyanobacterial genus *Cylindrospermum* that effectively inhibits digestive glucosidases of aquatic insects and crustacean grazers. J. Phycol. 39, 26–32.

Kambara, K., Takematsu, Y., Azuma, M., Kobayashi, J., 2009. cDNA cloning gene expressed in the digestive track of the Formosan subterranean termite, *Coptotermes formosanus* Shiraki (Isoptera; Rhinotermitidae). Appl. Entomol. Zool. 44, 315–321.

Kanamori, Y., Saito, A., Hagiwara-Komoda, Y., Tanaka, D., Mitsumasu, K., Kikuta, S., Watanabe, M., Cornette, R., Kikawada, T., Okuda, T., 2010. The trehalose transporter 1 gene sequence conserved in insects and encodes proteins with different kinetic properties involved in trehalose import into peripheral tissues. Insect Biochem. Mol. Biol. 40, 30–37.

Karley, A.J., Ashford, D.A., Minto, L.M., Pritchard, J., Douglas, A.E., 2005. The significance of gut sucrase activity for osmoregulation in the pea aphid, *Acyrthosiphon pisum*. J. Insect Physiol. 51, 1313–1319.

Kataoka, H., Troetschler, R.G., Li, J.P., Kramer, S.J., Carney, R.L., Schooley, D.A., 1989. Isolation and identification of a diuretic hormone from tobacco hornworm, *Manduca sexta*. Proc. Natl. Acad. Sci. U.S.A. 86, 2976–2980.

Kataoka, N., Miyake, S., Azuma, M., 2009a. Aquaporin and aquaglyceroporin in silkworms, differentially expressed in the hindgut and midgut of *Bombyx mori*. Insect Mol. Biol. 18, 303–314.

Kataoka, N., Miyake, S., Azuma, M., 2009b. Molecular characterization of aquaporin and aquaglyceroporin in the alimentary canal of *Grapholita molesta* (the oriental fruit moth)—comparison with *Bombyx mori* aquaporins. J. Insect Biotechnol. Sericol. 72, 81–90.

Kaufman, W.R., Phillips, J.E., 1973. Ion and water balance in the ixodid tick *Dermacentor andersoni*. I Routes of ion and water excretion. J. Exp. Biol. 58, 523–536.

Kaufmann, N., Mathai, J.C., Hill, W.G., Dow, J.A.T., Zeidel, M.L., Brodsky, J.L., 2005. Developmental expression and biophysical characterization of a *Drosophila melanogaster* aquaporin. Am. J. Physiol. Cell Physiol. 289, C397–C407.

Kay, I., Patel, M., Coast, G.M., Totty, N.F., Mallet, A.I., Goldworthy, G.J., 1992. Isolation and characterization of a CRF-related diuretic peptide from *Periplaneta americana* L. Regul. Pept. 42, 111–122.

Kean, L., Cazenave, W., Costes, L., Broderick, K.E., Graham, S., Pollock, V.P., Davies, S.A., Veenstra, J.A., Dow, J.A.T., 2002. Two nitridergic peptides are encoded by the gene *capability* in *Drosophila melanogaster*. Am. J. Physiol. Regul. Integr. Comp. Physiol. 282, R1297–R1307.

Keeley, L.L., Hayes, T.K., 1987. Speculations on biotechnology applications for insect neuroendocrine research. Insect Biochem. 17, 639–651.

Keeley, L.L., Chung, J.S., Hayes, T.K., 1992. Diuretic and antifeedant actions by *Manduca sexta* diuretic hormone in lepidopteran larvae. Experientia 48, 1145–1148.

Kersch, C.N., Pietrantonio, P.V., 2011. Mosquito *Aedes aegypti* (L.) leucokinin receptor is critical for in vivo fluid excretion post blood feeding. FEBS Lett. 585, 3507–3512.

Khani, A., Moharrampour, S., Barzegar, M., 2007. Cold tolerance and trehalose accumulation in overwintering larvae of the codling moth *Cydia pomonella* (Lepidoptera: Tortricidae). J. Insect Physiol. 44, 955–965.

Kikawada, T., Minakawa, N., Watanabe, M., Okuda, T., 2005. Factors inducing successful anhydrobiosis in the African chironomid *Polypedilum vanderplanki*: significance of the larval tubular nest. Integr. Comp. Biol. 45, 710–714.

Kikawada, T., Nakahara, Y., Kanamori, Y., Iwata, K.-I., Watanabe, M., McGee, B., Tunnacliffe, A., Okuda, T., 2006. Dehydration-induced expression of LEA proteins in an anhydrobiotic chironomid. Biochem. Biophys. Res. Commun. 348, 56–61.

Kikawada, T., Saito, A., Kanamori, Y., Fujita, M., Snigórska, K., Watanabe, M., Okuda, T., 2008. Dehydration-inducible changes in expression of two aquaporins in the sleeping chironomid *Polypedilum vanderplanki*. Biochim. Biophys. Acta 1778, 514–520.

King, D., Meredith, J., Wang, Y.J., Phillips, J.E., 1999. Biological actions of synthetic locust ion transport peptide. Insect Biochem. Mol. Biol. 29, 11–18.

Konno, K., Ono, H., Nakamura, M., Tateishi, K., Hirayama, C., Tamura, Y., Hattori, M., Koyama, A., Kohno, K., 2006. Mulberry latex rich in antidiabetic sugar-mimic alkaloids forces dieting on caterpillars. Proc. Natl. Acad. Sci. U.S.A. 103, 1337–1341.

Koštál, V., Tollarova-Borovanská, M., 2009. The 70 kDa heat shock protein assists during the repair of chilling injury in the insect, *Pyrrhocoris apterus*. PLoS One 4, e4546.

Koštál, V., Doležal, P., Rozsypal, J., Moravcová, M., Zahradníčková, H., Šimek, P., 2011. Physiological and biochemical analysis of overwintering and cold tolerance in two central European populations of the spruce bark beetle, *Ips typographus*. J. Insect Physiol. 57, 1136–1146.

Küppers, E., Gleiser, C., Brito, V., Wachter, B., Pauly, T., Hirt, B., Grissmer, S., 2008. AQP4 expression in striatal primary cultures is regulated by dopamine—implications for proliferation of astrocytes. Eur. J. Neurosci. 28, 2173–2182.

Laenen, B., De Decker, N., Steels, P., Van Kerkhove, E., Nicolson, S., 2001. An antidiuretic factor in the forest ant: purification and physiological effects on Malpighian tubules. J. Insect Physiol. 47, 185–193.

Lange, A.B., 2009. Tyramine: from octopamine precursor to neuroactive chemical in insects. Gen. Comp. Endocrinol. 162, 18–26.

Lange, A.B., Orchard, I., Barrett, F.M., 1989. Changes in haemolymph serotonin levels associated with feeding in the blood-sucking bug, *Rhodnius prolixus*. J. Insect Physiol. 35, 393–397.

Lavigne, C., Embleton, J., Audy, P., King, R.R., Pelletier, Y., 2001. Partial purification of a novel insect antidiuretic factor from the Colorado potato beetle, *Leptinotarsa decemlineata* (Say) (Coleoptera: Chrysomelidae), which acts on Malpighian tubules. Insect Biochem. Mol. Biol. 31, 339–347.

Le Cahérec, F., Bron, P., Verbavatz, J.M., Garret, A., Morel, G., Cavalier, A., Bonnec, G., Thomas, D., Gouranton, J., Hubert, J.F., 1996. Incorporation of proteins into (*Xenopus*) oocytes by proteoliposome microinjection: functional characterization of a novel aquaporin. J. Cell Sci. 109, 1285–1295.

Le Cahérec, F., Guillam, M.-T., Beuron, F., Cavalier, A., Thomas, D., Gouranton, J., Hubert, J.-F., 1997. Aquaporin-related proteins in the filter chamber of homopteran insects. Cell Tissue Res. 290, 143–151.

Lee Jr., R.E., 2010. A primer on insect cold-tolerance. In: Denlinger, D.L., Lee Jr., R.E. (Eds.), Low Temperature Biology of Insects. Cambridge University Press, Cambridge, pp. 3–34.

Lee Jr., R.E., Costanzo, J.P., 1998. Biological ice nucleation and ice distribution in cold-hardy ectothermic animals. Annu. Rev. Physiol. 60, 55–72.

Lee, K.S., Kim, S.R., Lee, S.M., Lee, K.R., Sohn, H.D., Jin, B.R., 2001. Molecular cloning and expression of a cDNA encoding the aquaporin homologue from the firefly, *Pyrocoelia rufa*. Korean J. Entomol. 31, 269–279.

Li, H., Wang, H., Schegg, K.M., Schooley, D.A., 1997. Metabolism of an insect diuretic hormone by Malpighian tubules studied by liquid chromatography coupled with electrospray ionization mass spectrometry. Proc. Natl. Acad. Sci. U.S.A. 94, 13463–13468.

Li, B., Predel, R., Neupert, S., Hauser, F., Tanaka, Y., Cazzameli, G., Williamson, M., Arakane, Y., Verleyen, P., Schoofs, L., Schachtner, J., Grimmelikhuijzen, C.J.P., Park, Y., 2008. Genomics, transcriptomics, and peptidomics of neuropeptides and protein hormones in the red flour beetle *Tribolium castaneum*. Genome Res. 18, 113–122.

Liu, K., Tsujimoto, H., Cha, S.-J., Agre, P., Rasgon, J.L., 2011. Aquaporin water channel AgAQP1 in the malaria vector mosquito *Anopheles gambiae* during feeding and humidity adaptation. Proc. Natl. Acad. Sci. U.S.A. 108, 6062–6066.

Lopez-Martinez, G., Benoit, J.B., Rinehart, J.P., Elnitsky, M.A., Lee Jr., R.E., Denlinger, D.L., 2009. Dehydration, and overhydration alter patterns of gene expression in the Antarctic midge, *Belgica antarctica*. J. Comp. Physiol. B 179, 481–491.

Lovejoy, D.A., Balment, R.J., 1999. Evolution and physiology of the corticotropin-releasing factor (CRF) family of neuropeptides in vertebrates. Gen. Comp. Endocrinol. 115, 1–22.

Lu, H.-L., Kersch, C., Pietrantonio, P.V., 2011a. A kinin receptor is expressed in the Malpighian tubule stellate cells in the mosquito *Aedes aegypti* (L.): a new model needed to explain ion transport? Insect Biochem. Mol. Biol. 41, 135–140.

Lu, H.-L., Kersch, C.N., Taneja- Bageshwar, S., Pietrantonio, P.V., 2011b. A calcium bioluminescence assay for functional analysis of mosquito (*Aedes aegypti*) and tick (*Rhipicephalus microplus*) G protein-coupled receptors. J. Vis. Exp. 50, 2732.

Machin, J., 1983. Water vapor absorption in insects. Am. J. Physiol. Regul. Integr. Comp. Physiol. 244, R187–R192.

Macins, A., Meredith, J., Zhao, Y., Brock, H.W., Phillips, J.E., 1999. Occurrence of ion transport peptide (ITP) and ion transport-like peptide (ITP-L) in orthopteroids. Arch. Insect Biochem. Physiol. 40, 107–117.

MacKinnon, R., Yellen, G., 1990. Mutations affecting TEA blockade of ion permeation in voltage-activated K^+ channels. Science 250, 276–279.

Maddrell, S.H.P., 1963. Excretion in the blood-sucking bug, *Rhodnius prolixus* Stål. I. The control of diuresis. J. Exp. Biol. 40, 147–150.

Maddrell, S.H.P., 1966. The site of release of the diuretic hormone in *Rhodnius*—a new neurohaemal system in insects. J. Exp. Biol. 45, 499–508.

Maddrell, S.H.P., Pilcher, D.E.M., Gardiner, B.O.C., 1969. Stimulatory effect of 5-hydroxytryptamine (serotonin) on secretory by Malpighian tubules of insects. Nature 222, 784–785.

Maddrell, S.H.P., Herman, W.S., Mooney, R.L., Overton, J.A., 1991. 5-Hydroxyptyamine: a second diuretic hormone in *Rhodnius prolixus*. J. Exp. Biol. 156, 557–566.

Martini, S.V., Goldenberg, R.C., Fortes, F.S.A., Campos-de Carvalho, A.C., Falkenstein, D., Morales, M.M., 2004. *Rhodnius prolixus* Malpighian tubule's aquaporin expression is modulated by 5-hydroxytryptamine. Arch. Insect Biochem. Physiol. 57, 133–141.

Martini, S.V., Nascimento, S.B., Morales, M.M., 2007. *Rhodnius prolixus* Malpighian tubules and control of diuresis by neurohormones. An. Acad. Bras. Cienc. 79, 87–95.

Masler, R.P., Kelly, T.J., Menn, J.J., 1993. Insect neuropeptides: discovery and application in insect management. Arch. Insect Biochem. Physiol. 22, 87–111.

Massaro, R.C., Lee, L.W., Patel, A.B., Wu, D.S., Yu, M.-J., Scott, B.N., Schooley, D.A., Schegg, K.M., Beyenbach, K.W., 2004. The mechanism of action of the antidiuretic peptide Temo ADFa in Malpighian tubules of *Aedes aegypti*. J. Exp. Biol. 207, 2877–2888.

Mathew, L.G., Campbell, R.M., Yool, A.J., Fabrick, J.A., 2011. Identification and characterization of functional aquaporin water channel protein from alimentary tract of whitefly, *Bemisia tabaci*. Insect Biochem. Mol. Biol. 41, 178–190.

Mellor, H.R., Nolan, J., Pickering, L., Wormald, M.R., Platt, F.M., Dwek, R.A., Fleet, G.W.J., Butters, T.D., 2002. Preparation, biochemical characterization and biological properties of radiolabelled N-alkylated deoxynojirimycins. Biochem. J. 366, 225–233.

Meredith, J., Ring, M., Macins, A., Marschall, J., Cheng, N.N., Theilmann, D., Brock, H.W., Phillips, J.E., 1996. Locust ion transport peptide (ITP): primary structure, cDNA and expression in a baculovirus system. J. Exp. Biol. 199, 1053–1061.

Michaud, M.R., Benoit, J.B., Lopez-Martinez, G., Elnitsky, M.A., Lee Jr., R.E., Denlinger, D.L., 2008. Metabolomics reveals unique and shared metabolic changes in response to heat shock, freezing and desiccation in the Antarctic midge, *Belgica antarctica*. J. Insect Physiol. 54, 645–655.

Migliat, E., Meurice, N., DuBois, P., Fang, J.S., Somasekharan, S., Beckett, E., Flynn, G., Yool, A.J., 2009. Inhibition of aquaporin-1 and aquaporin-4 water permeability by a derivative of the loop diuretic bumetanide acting at an internal pore-occluding binding site. Mol. Pharmacol. 76, 105–112.

Miyake, S., Azuma, M., 2008. Developmental expression and the physiological role of aquaporin in the silk gland of *Bombyx mori*. J. Insect Biotechnol. Sericol. 77, 87–93.

Montagné, N., Desdevises, Y., Solyez, D., Toullec, J.-V., 2010. Molecular evolution of the crustacean hyperglycemic hormone family in ecdysozoans. BMC Evol. Biol. 10, 62.

Morgan, P.J., Mordue, W., 1984. 5-Hydroxytryptamine stimulates fluid secretion in locust Malpighian tubules independent of cAMP. Comp. Biochem. Physiol. C 79, 305–310.

Morrissey, R.E., Baust, J.G., 1976. The ontogeny of cold tolerance in the gall fly, *Eurosta solidaginis*. J. Insect Physiol. 22, 431–437.

Müller, E.M., Hub, J.S., Grubmüller, H., de Groot, B.L., 2008. Is TEA an inhibitor for human aquaporin-1? Pfluger Arch. 456, 663–669.

Nachman, R.J., Holman, G.M., Haddon, W.F., 1993. Leads for insect neuropeptide mimetic development. Arch. Insect Biochem. Physiol. 22, 181–197.

Nachman, R.J., Isaac, R.E., Coast, G.M., Holman, G.M., 1997. Aib-containing analogues of the insect kinin neuropeptide family demonstrate resistance to an insect angiotensin-converting enzyme and potent diuretic activity. Peptides 18, 53–57.

Nachman, R.J., Strey, A., Isaac, E., Pryor, N., Lopez, J.D., Deng, J.-G., Coast, G.M., 2002. Enhanced in vivo activity of peptidase-resistant analogs of the insect kinin neuropeptide family. Peptides 23, 735–745.

Nachman, R.J., Russell, W.K., Coast, G.M., Russell, D.H., Predel, R., 2005. Mass spectrometric assignment of Leu/Ile in neuropeptides from single neurohemal organ preparations in insects. Peptides 26, 2151–2156.

Nachman, R.J., Russell, W.K., Coast, G.M., Russell, D.H., Miller, J.A., Predel, R., 2006. Identification of PVK/CAP2b neuropeptides from single neurohemal organs of the stable fly and the horn fly via MALDI-TOF/TOF tandem mass spectrometry. Peptides 27, 521–526.

Nachman, R.J., Kaczmarek, K., Zabrocki, J., Coast, G.M., 2012. Active diuretic peptidomimetic insect kinin analogs that contain β-turn mimetic motif 4-aminopyroglutamate and lack of native peptide bonds. Peptides 34, 262–265.

Nakahara, Y., Watanabe, M., Fujita, A., Kanamori, Y., Tanaka, D., Iwata, K.-I., Furuku, T., Sakurai, M., Kikawada, T., Okuda, T., 2008. Effects of dehydration rate on physiological responses and survival after rehydration in larvae of the anhydrobiotic chironomid. J. Insect Physiol. 54, 1220–1225.

Nicchia, G.P., Mastrototaro, M., Rossi, A., Pisani, F., Tortorella, C., Ruggieri, M., Lia, A., Trojano, M., Frigeri, A., Svelto, M., 2009. Aquaporin-4 orthogonal arrays of particles are the target for neuromyelitis optica autoantibodies. Glia 57, 1363–1373.

Nicolson, S.W., 1980. Diuresis and its hormonal control in butterflies. J. Insect Physiol. 26, 841–846.

Nicolson, S.W., Millar, R.P., 1983. Effects of biogenic amines and hormones on butterfly Malpighian tubules: dopamine stimulates fluid secretion. J. Insect Physiol. 29, 611–615.

Nieminen, P., Paakkonen, T., Eerilä, H., Puukka, K., Riikonen, J., Lehto, V.-P., Mustonen, A.-M., 2012. Freezing tolerance and low molecular weight cryoprotectants in an invasive parasitic fly, the deer ked (Lipoptena cervi). J. Exp. Zool. 317A, 1–8.

Nollert, P., Harries, W.E.C., Fu, D., Meircke, L.J.W., Stroud, R.M., 2001. Atomic structure of glycerol channel and implications for substrate permeation in aqua(glycero)porins. FEBS Lett. 504, 112–117.

O'Connor, K.R., Beyenbach, K.W., 2001. Chloride channels in apical membrane patches of stellate cells of Malpighian tubules of Aedes aegypti. J. Exp. Biol. 204, 367–378.

O'Donnell, M.J., Dow, J.A.T., Huesmann, G.R., Tublitz, N.J., Maddrell, S.H.P., 1996. Separate control of anion and cation transport in Malpighian tubules of Drosophila melanogaster. J. Exp. Biol. 199, 1163–1175.

O'Donnell, M.J., Rheault, M.R., Davies, S.A., Rosay, P., Harvey, B.J., Maddrell, S.H.P., Kaiser, K., Dow, J.A., 1998. Hormonally controlled chloride movement across Drosophila tubules is via ion channels in stellate cells. Am. J. Physiol. Regul. Integr. Comp. Physiol. 274, R1039–R1049.

Okech, B.A., Boudko, D.Y., Linser, P.J., Harvey, W.R., 2008. Cationic pathway of pH in larvae of Anopheles gambiae. J. Exp. Biol. 211, 957–968.

Olsen, S.S., Cazzamali, G., Williamson, M., Grimmelikhuijen, C.J.P., Hauser, F., 2007. Identification of one capa and two pyrokinin receptors from the malaria mosquito Anopheles gambiae. Biochem. Biophys. Res. Commun. 362, 245–251.

Orchard, I., 1989. Serotonergic neurohaemal tissue in Rhodnius prolixus: synthesis, release and uptake of serotonin. J. Insect Physiol. 35, 943–947.

Orchard, I., 2006. Serotonin: a coordinator of feeding-related physiological events in the blood-gorging bug, Rhodnius prolixus. Comp. Biochem. Physiol. A Mol. Integr. Physiol. 144, 316–324.

Paluzzi, J.-P., Orchard, I., 2010. A second gene encodes the anti-diuretic hormone in the insect, *Rhodnius prolixus*. Mol. Cell. Endocrinol. 317, 53–63.

Paluzzi, J.-P., Russell, W.K., Nachman, R.J., Orchard, I., 2008. Isolation, cloning, and expression mapping of a gene encoding an antidiuretic hormone and other CAPA-related peptides in the disease vector, *Rhodnius prolixus*. Endocrinology 149, 4638–4646.

Paluzzi, J.-P., Park, Y., Nachman, R.J., Orchard, I., 2010. Isolation, expression analysis, and functional characterization of the first antidiuretic hormone receptor in insects. Proc. Natl. Acad. Sci. U.S.A. 107, 10290–10295.

Patel, M., Hayes, T.K., Coast, G.M., 1995. Evidence for the hormonal function of a CRF-related diuretic peptide (*Locusta*-DP) in *Locusta migratoria*. J. Exp. Biol. 198, 793–804.

Petzel, D.H., Berg, M.M., Beyenbach, K.W., 1987. Hormone-controlled cAMP-mediated fluid secretion in the yellow-fever mosquito. Am. J. Physiol. Regul. Integr. Comp. Physiol. 253, R701–R7011.

Philip, B.N., Kiss, A.J., Lee Jr., R.E., 2011. The protective role of aquaporins in the freeze-tolerant insect *Eurosta solidaginis*: functional characterization and tissue abundance of EsAQP1. J. Exp. Biol. 214, 848–857.

Phillips, J.E., Audsley, N., 1995. Neuropeptide control of ion and fluid transport across locust hindgut. Am. Zool. 35, 503–514.

Piermarini, P.M., Grogan, L.F., Lau, K., Wang, L., Beyenbach, K.W., 2010. A SLC4-like anion exchanger from renal tubules of the mosquito (*Aedes aegypti*): evidence for a novel role of stellate cells in diuretic fluid secretion. Am. J. Physiol. Regul. Integr. Comp. Physiol. 298, R642–R660.

Pietrantonio, P.V., Jagge, C., Keeley, L.L., Ross, L.S., 2000. Cloning of an aquaporin-like cDNA and *in situ* hybridization in adults of a mosquito *Aedes Aegypti* (Diptera: Culicidae). Insect Mol. Biol. 9, 407–418.

Pietrantonio, P.V., Jagge, C., Taneja-Bageshwar, S., Nachman, R.J., Barhoumi, R., 2005. The mosquito *Aedes aegypti* (L.) leucokinin receptor is a multiligand receptor for the three *Aedes* kinins. Insect Mol. Biol. 14, 55–67.

Pollock, V.P., Radford, J.C., Pyne, S., Hasan, G., Dow, J.A.T., Davies, S.-A., 2003. *norpA* and *itpr* mutants reveal roles for phospholipase C and inositol (1,4,5)-triphosphate receptor in *Drosophila melanogaster* renal function. J. Exp. Biol. 206, 901–911.

Pollock, V.P., McGettigan, J., Cabrero, P., Maudlin, I.M., Dow, J.A.T., Davies, S.-A., 2004. Conservation of capa peptide-induced nitric oxide signaling in Diptera. J. Exp. Biol. 207, 4135–4245.

Pompon, J., Quiring, D., Goyer, C., Giordanengro, P., Pelletier, Y., 2011. A phloem-sap feeder mixes phloem and xylem sap to regulate osmotic potential. J. Insect Physiol. 57, 1317–1322.

Predel, R., Wegener, C., 2006. Biology of the CAPA peptides in insects. Cell. Mol. Life Sci. 63, 2477–2490.

Predel, R., Linde, D., Rapus, J., Vettermann, S., Penzlin, H., 1995. Periviscerokinin (Pea-PVK): a novel myotropic neuropeptide from the peripympathetic organs of the American cockroach. Peptides 16, 61–66.

Predel, R., Kellner, R., Baggerman, G., Steinmetzer, T., Schoofs, L., 2000. Identification of novel periviscerokinins from single neurohaemal releas sites in insects. Eur. J. Biochem. 267, 3869–3873.

Predel, R., Russell, W.K., Neupert, S., Russell, D.H., Esquivel, J.F., Nachman, R.J., 2006. Identification of the first neuropeptides from the CNS of Hemiptera: CAPA peptides of the Southern green stinkbug *Nezara viridula* (L.). Peptides 27, 2670–2677.

Predel, R., Russell, W.K., Russell, D.H., Lopez, J., Esquivel, J., Nachman, R.J., 2008. Comparative peptidomics of four hemipteran species: pyrokinins, myosuppressin, corazonin, adipokinetic hormone, sNPF, and periviscerokinins. Peptides 29, 162–167.

Preston, G.M., Jung, J.S., Guggino, W.B., Agre, P., 1993. The mercury-sensitive residue at cysteine 189 in the CHIP28 water channel. J. Biol. Chem. 268, 17–20.

Price, D.R.G., Karley, A.J., Ashford, D.A., Isaacs, H.V., Pownall, M.E., Wilkinson, H.S., Gatehouse, J.A., Douglas, A.E., 2007. Molecular characterization of a candidate gut sucrase in the pea aphid *Acyrthosiphon pisum*. Insect Biochem. Mol. Biol. 37, 307–317.

Proux, J., Rougon, G., Cupo, A., 1982. Enhancement of excretion across locust tubules by a diuretic vasopressin-like hormone. Gen. Comp. Endocrinol. 47, 449–457.

Proux, B., Proux, J., Phillips, J.E., 1984. Antidiuretic action of a corpus cardiacum factor (CTSH) on long-term fluid absorption across locust recta *in vitro*. J. Exp. Biol. 113, 409–421.

Quinlan, M.C., O'Donnell, M.J., 1998. Anti-diuresis in the blood-feeding insect *Rhodnius prolixus* Stål: antagonistic actions of cAMP and cGMP and the role of organic transport. J. Insect Physiol. 44, 561–568.

Quinlan, M.C., Tublitz, N.J., O'Donnell, M.J., 1997. Anti-diuresis in the blood-feeding insect *Rhodnius prolixus* Stål: the peptide CAP_{2b} and cyclic GMP inhibit Malpighian tubule fluid secretion. J. Exp. Biol. 200, 2363–2367.

Radford, J.C., Davies, S.A., Dow, J.A., 2002. Systematic G-protein-coupled receptor analysis in *Drosophila melanogaster* identifies a leucokinin receptor with novel roles. J. Biol. Chem. 277, 38810–38817.

Radford, J.C., Terhzaz, S., Cabrero, P., Davies, S.-A., Dow, J.A.T., 2004. Functional characterization of the *Anopheles* leucokinins and their cognate G-protein coupled receptor. J. Exp. Biol. 207, 4573–4586.

Rafaeli, A., Moshitzky, P., Applebaum, S.W., 1987. Diuretic action and immunological cross-reactivity of corticotropin and locust diuretic hormone. Gen. Comp. Endocrinol. 67, 1–6.

Ramsay, J.A., 1954. Active transport of water by the Malpighian tubules of the stick insect, *Dixippus morosus* (Orthoptera, Phasmidae). J. Exp. Biol. 31, 104–113.

Rash, J.E., Yasumura, T., Hudson, C.S., Agre, P., Nielsen, S., 1998. Direct immunogold labeling of aquaporin-4 in square arrays of astrocyte and ependymocyte plasma membranes in rat brain and spinal cord. Proc. Natl. Acad. Sci. U.S.A. 95, 11981–11986.

Regulski, M., Tully, T., 1995. Molecular and biochemical characterization of *dNOS*: a *Drosophila* Ca^{2+}/calmodulin-dependent nitric oxide synthase. Proc. Natl. Acad. Sci. U.S.A. 92, 9072–9076.

Rheault, M.R., Okech, B.A., Keen, S.B.W., Miller, M.M., Meleshkevich, E.A., Linser, P.J., Boudko, D.Y., Harvey, W.R., 2007. Molecular cloning, phylogeny and localization of AgNHA1: the first Na^+/H^+ antiporter (NHA) from a metazoan, *Anopheles gambiae*. J. Exp. Biol. 210, 3848–3861.

Rhodes, J.D., Croghan, P.C., Dixon, A.F.G., 1997. Dietary sucrose and oligosaccharide synthesis in relation to osmoregulation in the pea aphid, *Acyrthosiphon pisum*. Physiol. Entomol. 22, 373–379.

Rinehart, J.P., Li, A., Yocum, G.D., Robich, R.M., Hayward, S.A.L., Denlinger, D.L., 2007. Up-regulation of heat shock proteins is essential for cold survival during insect diapause. Proc. Natl. Acad. Sci. U.S.A. 104, 11130–11137.

Roberts, V.A., Nachman, R.J., Coast, G.M., Hariharan, M., Chung, J.S., Holman, G.M., Williams, H., Tainer, J.A., 1997. Consensus chemistry and β-turn conformation of the active core of the insect kinin neuropeptide family. Chem. Biol. 4, 105–117.

Roller, L., Yamanaka, N., Watanabe, K., Daubnerová, I., Žitňaň, D., Kataoká, H., Tanaka, Y., 2008. The unique evolution of neuropeptide genes in the silkworm *Bombyx mori*. Insect Biochem. Mol. Biol. 38, 1147–1157.

Sakurai, M., Furuki, T., Akao, K.-I., Tanaka, D., Nakahara, Y., Kikawada, T., Watanabe, M., Okuda, T., 2008. Vitrification is essential for anhydrobiosis in an African chironomid, *Polypedilum vanderplanki*. Proc. Natl. Acad. Sci. U.S.A. 105, 5093–5098.

Salvucci, M.E., 2000. Effect of the α-glucosidase inhibitor, bromoconduritol, on carbohydrate metabolism in the silverleaf whitefly, *Bemisia argentifolii*. Arch. Insect Biochem. Physiol. 45, 117–128.

Sampredo, J.G., Uribe, S., 2004. Trehalose-enzyme interactions result in structure stabilization and activity inhibition. The role of viscosity. Mol. Cell. Biochem. 256/257, 319–327.

Savage, D.F., Stroud, R.M., 2007. Structural basis of aqauporin inhibition by mercury. J. Mol. Biol. 368, 607–617.

Savage, D.F., Egea, P., Robles-Colmenares, Y., O'Connell III, J.D., Stroud, R.M., 2003. Architecture and selectivity in aquaporins: 2.5 Å X-ray structure of aquaporin Z. PLoS Biol. 1, e72.

Savage, D.F., O'Connell, J.D.I.I.I., Miercke, L.J.W., Finer-Moore, J., Stroud, R.M., 2010. Structural context shapes aquaporin selectivity filter. Proc. Natl. Acad. Sci. U.S.A. 107, 17164–17169.

Scherkenbeck, J., Zdobinsky, T., 2009. Insect neuropeptides: structure, chemical modification and potential for insect control. Bioorg. Med. Chem. 17, 4071–4084.

Scherkenbeck, J., Antonicek, H.-P., Voselsang, K., Zdobinsky, T., Brücher, K., Rehländer, D., Chen, H., 2009. Receptor assay guided structure-activity studies of helicokinin neuropeptides and peptidomimetic analogues. J. Pept. Sci. 15, 783–789.

Schoofs, L., Holman, G.M., Proost, P., Van Damme, J., Hayes, T.K., De Loof, A., 1992. Locustakinin, a novel myotropic peptide from *Locusta migratoria*, isolation, primary structure and synthesis. Regul. Pept. 37, 49–57.

Schooley, D.A., Horodyski, F.M., Coast, G.M., 2005. Hormones controlling homeostasis in insects. In: Gilbert, L.I., Latrou, K., Gill, S.S. (Eds.), Comprehensive Molecular Insect Science. Elsevier, Amsterdam, pp. 493–550.

Schowalter, T.D., 2006. Insect Ecology: An ecosystem Approach. Academic Press, New York p. 572.

Scofield, A.M., Witham, P., Nash, R.J., Kite, G.D., Fellows, L.E., 1995. Castanospermine and other polyhydroxy alkaloids as inhibitors of insect glycosidases. Comp. Biochem. Physiol. A Physiol. 112, 187–196.

Seinsche, A., Dyker, H., Losel, P., Backhaus, D., Scherkenbeck, J., 2000. Effect of helicokinins and ACS inhibitors on water balance and development of *Heliothis virescens* larvae. J. Insect Physiol. 46, 1423–1431.

Sformo, T., Walters, K., Jeannet, K., Wowk, B., Fahy, G.M., Barnes, B.M., Duman, J.G., 2010. Deep supercoolong, vitrification and limited survival to −100 °C in the Alaskan beetle *Cucujus clavipes puniceus* (Coleoptera: Cucujidae) larvae. J. Exp. Biol. 231, 502–509.

Shakesby, A.J., Wallace, L.S., Isaacs, H.V., Pritchard, J., Roberts, D.M., Douglas, A.E., 2009. A water-specific aquaporin involved in aphid osmoregulation. Insect Biochem. Mol. Biol. 39, 1–10.

Shao, H.-B., Chu, L.-Y., Shao, M.-A., Zhao, C.-X., 2008. Advances in functiobal regulation mechanisms of plant aquaporins: their diversity, gene expression, localization, structure and roles in plant soil-water relations (Review). Mol. Membr. Biol. 25, 179–191.

Singer, M.A., Lindquist, S., 1998. Multiple effects of trehalose on protein folding in vitro and in vivo. Mol. Cell 1, 639–648.

Skaer, N.J.V., Nässel, D.R., Maddrell, S.H.P., Tublitz, N.J., 2002. Neurochemical fine tuning of a peripheral tissue: peptidergic and aminergic regulation of fluid secretion by Maplighian tubules in the tobacco hawkmoth *M. sexta*. J. Exp. Biol. 205, 1869–1880.

Smagghe, G., Mahdian, K., Zubrzak, P., Nachman, R.J., 2010. Antifeedant activity and high mortality in the pea aphid *Acyrthosiphon pisum* (Homoptera: Aphidae) induced by biostable insect kinin analogs. Peptides 31, 498–505.

Søgaard, R., Zeuthen, T., 2008. Test of blockers of AQP1 water permeability by a high-resolution method: no effects of tetraethylammonium ions or acetazolamide. Pflugers Arch. 456, 285–292.

Spring, J.H., Robichaux, S.R., Kaufmann, N., Brodsky, J.L., 2007. Localization of a *Drosophila* DRIP-like aquaporin in the Malpighian tubules of the house cricket, *Acheta domesticus*. Comp. Biochem. Physiol. A Mol. Integr. Physiol. 148, 92–100.

Spring, J.H., Robichaux, S.R., Hamlin, J.A., 2009. The role of aquaporins in excretion in insects. J. Exp. Biol. 212, 358–362.

Sterkel, M., Oliveira, P.L., Urlaub, H., Hernandez-Martinez, S., Rivera-Pomar, R., Ons, S., 2012. OKB, a novel family of brain-gut neuropeptides from insects. Insect Biochem. Mol. Biol. 42, 466–473.

Storey, K.B., 1997. Organic solutes in freezing tolerance. Comp. Biochem. Physiol. A Physiol. 117, 319–326.

Storey, K.B., Storey, J.M., 1988. Freeze tolerance in animals. Physiol. Rev. 68, 27–84.

Storey, K.B., Storey, J.M., 2012. Insect dold hardiness: metabolic, gene, and protein adaptation. Can. J. Zool. 90, 456–475.

Tammariello, S.P., Rinehart, J.P., Denlinger, D.L., 1999. Desiccation elicits heat shock protein transcription in the flesh fly, *Sarcophaga crassipalpis*, but enhance tolerance in high or low temperatures. J. Insect Physiol. 45, 933–938.

Taneja-Bageshwar, S., Strey, A., Zubrzak, P., Pietrantonio, P.V., Nachman, R.J., 2006. Comparative structure-activity analysis of insect kinin core analogs on recombinant kinin receptors from the Southern cattle tick *Boophilus microplus* (Acari: Ixodidae) and the mosquito *Aedes aegypti* (Diptera: Culicidae). Arch. Insect Biochem. Physiol. 62, 128–140.

Taneja-Bageshwar, S., Strey, A., Isaac, R.E., Coast, G.M., Zubrzak, P., Pietrantonio, P.V., Nachman, R.J., 2009. Biostable agonists that match or exceed activity of native kinins on recombinant arthropod GPCRs. Gen. Comp. Endocrinol. 162, 122–128.

Te Brugge, V.A., Orchard, I., 2008. Distribution and activity of a Dippu DH_{31}-like peptide in the large milkweed bug *Oncopeltus fasciatus*. Peptides 29, 206–213.

Te Brugge, V.A., Schooley, D.A., Orchard, I., 2002. The biological activity of diuretic factors in *Rhodnius prolixus*. Peptides 23, 671–681.

Te Brugge, V.A., Lombardi, V.C., Schooley, D.A., Orchard, I., 2005. Presence and activity of a Dippu-DH_{31}-like peptide in the blood-feeding bug, *Rhodnius prolixus*. Peptides 26, 29–42.

Te Brugge, V., Paluzzi, J.-P., Schooley, D.A., Orchard, I., 2011. Identification of the elusive peptidergic diuretic hormone in the blood-feeding bug *Rhodnius prolixus*: a CRF-related peptide. J. Exp. Biol. 214, 371–381.

Terhzaz, S., O'Donnell, F.C., Pollock, V.P., Kean, L., Davies, S.A., Veenstra, J.A., Dow, J.A., 1999. Isolation and characterization of a leucokinin-like peptide of *Drosophila melanogaster*. J. Exp. Biol. 202, 3667–3676.

Terhzaz, S., Cabrero, P., Robben, J.H., Radford, J.C., Hudson, B.D., Milligan, G., Dow, J.A.T., Davies, S.-A., 2012. Mechanism and function of *Drosophila* capa GPCR: a desiccation stress-response receptor with functional homology to human neuromedinU receptor. PLoS One 7, e29897.

Tobe, S.S., Zhang, J.R., Schooley, D.A., Coast, G.M., 2005. A study of signal transduction for the two diuretic peptides of *Diploptera punctata*. Peptides 26, 89–98.

Törnroth-Horsefield, S., Hedfalk, K., Fischer, G., Lindkvist-Petersson, K., Neutze, R., 2010. Structural insights into eukaryotic aquaporin regulation. FEBS Lett. 584, 2580–2588.

Veenstra, J.A., 1988. Effects of 5-hydroxytryptamine on the Malpighian tubules of *Aedes aegypti*. J. Insect Physiol. 34, 299–304.

Veenstra, J.A., 1994. Isolation and identification of three leucokinins from the mosquito *Aedes aegypti*. Biochem. Biophys. Res. Commun. 202, 715–719.

Veenstra, J.A., Pattillo, J.M., Petzel, D.H., 1997. A single cDNA encodes all three *Aedes* leucokinins, which stimulate both fluid secretion by the Malpighian tubules and hindgut contractions. J. Biol. Chem. 272, 10402–10407.

Verbavatz, J.-M., Ma, T., Gobin, R., Verkman, A.S., 1997. Absence of orthogonal arrays in kidney, brain and muscle from transgenic knockout mice lacking water channel aquaporin-4. J. Cell Sci. 110, 2855–2860.

Verkman, A.S., 2011. Aquaporins at a glance. J. Cell Sci. 124, 2107–2112.

Wallace, I.S., Shakesby, A.J., Hwang, J.H., Choi, W.G., Martínková, N., Douglas, A.E., Roberts, D.M., 2012. *Acyrthosiphon pisum* AQP2: a multifunctional insect aquaglyceroporin. Biochim. Biophys. Acta 1818, 627–635.

Wang, J., Kean, L., Yang, J., Allan, A.K., Davies, S.A., Herzyk, P., Dow, J.A.T., 2004. Function-informed transcriptome analysis of *Drosophila* renal tubule. Genome Biol. 5, R69.

Watanabe, M., Kikawada, T., Minagawa, N., Yukuhiro, F., Okuda, T., 2002. Mechanism allowing an insect to survive complete dehydration and extreme temperatures. J. Exp. Biol. 205, 2799–2802.

Wei, Y.-A., Hendrix, D.L., Nieman, R., 1997. Diglucomelezitose, a novel pentasaccharide in silverleaf whitefly honeydew. J. Agric. Food Chem. 45, 3481–3486.

Wieczorek, H., Beyenbach, K.W., Huss, M., Vitavska, O., 2009. Vacuolar-type proton pump in insect epithelia. J. Exp. Biol. 212, 1611–1619.

Wiehart, U.I.M., Nicolson, S.W., Eigenheer, R.A., Schooley, D.A., 2002. Antagonistic control of fluid secretion by the Malpighian tubules of *Tenebrio molitor*: effects of diuretic and antidiuretic peptides and their second messengers. J. Exp. Biol. 205, 493–501.

Wiehart, U.I.M., Klein, G., Steels, P., Nicolson, S.W., Van Kerkhove, E., 2003a. K^+ transport in Malpighian tubules of *Tenebrio molitor* L.: is a K_{ATP} channel involved? J. Exp. Biol. 206, 959–965.

Wiehart, U.I.M., Nicolson, S.W., Van Kerkhove, E., 2003b. The effects of endogenous diuretic and antidiuretic peptides and their second messengers in the Malpighian tubules of *Tenebrio molitor*: an electrophysiological study. J. Insect Physiol. 49, 955–965.

Wilkinson, T.L., Ashford, D.A., Pritchard, J., Douglas, D.E., 1997. Honeydew sugars and osmoregulation in the pea aphid *Acyrthosiphon pisum*. J. Exp. Biol. 200, 2137–3143.

Williams, L.B., Shorthouse, J.D., Lee Jr., R.E., 2002. Extreme resistance to desiccation and microclimate-related differences in cold-hardiness of gall wasps (Hymenoptera: Cynipidae) overwintering on roses in southern Canada. J. Exp. Biol. 205, 2115–2124.

Woodring, J., Wiedemann, R., Volkl, W., Hoffmann, K.H., 2006. Oligosaccharide synthesis regulates gut osmolality in the ant-attended aphid *Metpeurum fuscoviride* but not in the unattended aphid *Macrosiphoniella tanacetaria*. J. Appl. Entomol. 131, 1–7.

Worland, M.R., Grubor-Lajsic, G., Montiel, P.O., 1998. Partial desiccation induced by sub-zero temperatures as a component of the survival strategy of the Arctic collembolan *Onychiurus arcticus* (Tullberg). J. Insect Physiol. 44, 211–219.

Xiang, M.A., Lisner, P.J., Price, D.A., Harvey, W.R., 2012. Localization ot two Na^+- or K^+-H^+ antiporters, AgNHA1 and AgNHA2, in *Anopheles gambiae* larval Malpighian tubules and functional expression of AgNHA2 in yeast. J. Insect Physiol. 58, 570–579.

Xu, J.-C., Lytle, C., Zhu, T.T., Payne, J.A., Benz Jr., C., Forbush III, B., 1994. Molecular cloning and functional expression of the bumetanide-sensitive Na-K-Cl cotransporter. Proc. Natl. Acad. Sci. U.S.A. 91, 2201–2205.

Yamanaka, N., Roller, L., Žitňan, D., Satake, H., Mizoguchi, A., Kataoka, H., Tanaka, Y., 2011. *Bombyx* orcokinins are brain-gut peptides involved in the neuronal regulation of ecdysteroidogenesis. J. Comp. Neurol. 519, 238–246.

Yang, B., Zhang, H., Verkman, A.S., 2008. Lack of aquaporin-4 water transport inhibition by antiepileptics and arylsulfonamides. Bioorg. Med. Chem. 16, 7489–7493.

Yi, S.-X., Benoit, J.B., Elnitsky, M.A., Kaufmann, N., Brodsku, J.L., Zeidel, M.L., Denlinger, D.L., Lee Jr., R.E., 2011. Function and immono-localization of aquaporins in the Antarctic midge *Belgica antarctica*. J. Insect Physiol. 57, 1096–1105.

Yoder, J.A., Benoit, J.B., Denlinger, D.L., Rivers, D.B., 2006. Stress-induced accumulation of glycerol in the flesh fly, *Sarcophaga bullata*: evidence indicating anti-desiccant and cryo-protectant functions of this polyol and a role for the brain in coordinating the response. J. Insect Physiol. 52, 202–214.

Yool, A.J., Brokl, O.H., Pannabecker, T.L., Dantzler, W.H., Stamer, W.D., 2002. Tetra-ethylammonium block of water flux in aquaporin-1 channels expressed in kidney thin limbs of Henle's loop and a kidney-derived cell line. BMC Physiol. 2, 4.

Yu, M.-J., Beyenbach, K.W., 2004. Effects of leucokinin-VIII on *Aedes* Malpighian tubule segments lacking stellate cells. J. Exp. Biol. 207, 519–526.

Zandawala, M., 2012. Calcitonin-like diuretic hormones in insects. Insect Biohem. Mol. Biol. 42, 816–825.

Zandawala, M., Paluzzi, J.-P., Orchard, I., 2011. Isolation and characterization of the cDNA encoding DH_{31} in the kissing bug, *Rhodnius prolixus*. Mol. Cell. Endocrinol. 331, 79–88.

Zdobinsky, T., Scherkenbeck, J., Zerbe, O., Antonicek, H., Chen, H., 2009. Structures of micelle-bound selected insect neuropeptides and analogues: implication for receptor selection. ChemBioChem 10, 2644–2653.

Zhang, J.-F., Zheng, Y.-G., Shen, Y.-C., 2007. Inhibitory effect of valienamine on the en-zymatic activity of honeybee (*Apis cerana* Fabr.) α-glucosidase. Pest. Biochem. Physiol. 87, 73–77.

Zhang, G., Storey, J.M., Storey, K.B., 2011. Chaperone proteins and winter survival by a freeze tolerant insect. J. Insect Physiol. 57, 1115–1122.

> CHAPTER TWO

Chemical Biology of the Nicotinic Insecticide Receptor

Motohiro Tomizawa

Graduate School of Medical Sciences, Nagoya City University, Nagoya, Japan

Contents

Abstract

Binding site interactions of nicotinic insecticides have been resolved by comparative chemical and structural neurobiology approaches using mollusk acetylcholine-binding proteins which serve as structural surrogates of the extracellular ligand-binding domain of the insect and vertebrate nicotinic acetylcholine receptor (nAChR) subtypes. Neonicotinoids with an electronegative pharmacophore are embraced by a reversed position in the binding pocket compared with a nicotinoid cationic functionality. A single dominant binding orientation causes the high affinity for neonicotinoids at the

Advances in Insect Physiology, Volume 44
ISBN 978-0-12-394389-7
http://dx.doi.org/10.1016/B978-0-12-394389-7.00002-8

insect nAChR homologue. However, the inferior potency of neonicotinoids at the vertebrate nAChR model is attributable to multiple binding conformations conflicting in the agonist-binding pocket. These findings in molecular recognition conferring potency and selectivity facilitated illustrative studies on nAChR structure-guided insecticide design, yielding outstandingly potent and selective candidate compounds with novel pharmacophores.

ABBREVIATIONS

ACh acetylcholine
AChBP ACh-binding protein
DCTHIA descyano-THIA
DNIMI desnitro-IMI
EPI epibatidine
ESP electrostatic potential
IMI imidacloprid
MS mass spectrometry
nAChR nicotinic ACh receptor
NIC nicotine
SAR structure–activity relationship
THIA thiacloprid

1. INTRODUCTION

Until the mid-twentieth century, botanical and inorganic insecticides played principal roles in pest insect management in agriculture, although their effectiveness was inadequate. Subsequently, discoveries of extremely high insecticidal activities of nerve-active compounds (organochlorines, organophosphates, methylcarbamates, and pyrethroids) opened an era of synthetic insecticides (Casida, 2009; Casida and Quistad, 1998). However, the effectiveness of these insecticides regrettably has diminished over time due to the emergence of resistant insect strains with enhanced detoxification mechanisms and/or less sensitive targets in their nervous system, thereby creating a critical demand for a new chemical type of neuroactive insecticide with a different, yet highly, sensitive targets (Casida and Quistad, 1998; ffrench-Constant, et al., 2004).

Nicotine (NIC; Fig. 2.1) in the form of tobacco extracts was reported in 1690 as the first plant-derived insecticide, followed by the pyrethrins from pyrethrum flowers and rotenone from Derris roots in the early 1800s (Casida

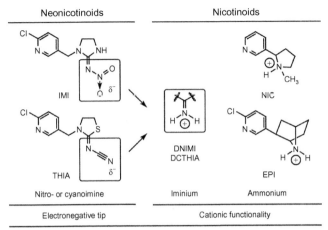

Nicotinic agonist pharmacophores

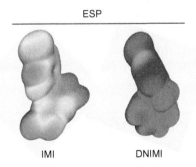

Figure 2.1 Neonicotinoid and nicotinoid chemotypes of nicotinic agonists with an electronegative pharmacophore and cationic functionality, respectively (*upper*). Electrostatic potential (ESP) mapping on the molecular surfaces of representative neonicotinoid IMI and nicotinoid DNIMI (protonated at physiological pH) was obtained by B3LYP/6-311G** (Tomizawa et al., 2003) (*lower*). ESP surfaces are exhibited in red for negative and are graded through orange, yellow, and green to blue for positive with an overall energy range of −60 to 160 kcal/mol. (See Color Plate 6 at the back of the book.)

and Quistad, 1998). The NIC was, for centuries, the best available agent to prevent piercing–sucking insects from damaging crops, although by current standards this botanical alkaloid was hazardous to people and actually not very effective. Attempts to discover novel nicotinic insecticides based on the NIC structure were not successful (Yamamoto, 1965). However, screening chemicals of novel structures in the 1970s, and then optimization of a lead compound, gave a fascinatingly potent 2-nitromethylene-thiazine compound designated as nithiazine (Kollmeyer et al., 1999). This nitromethylene heterocyclic insecticide was more effective and safer than

NIC, yet was unfortunately photolabile and therefore not appropriate for crop protection. Further optimization introduced a chloropyridinylmethyl substituent and also replaced the nitromethylene ($=CHNO_2$) group with a nitroimine ($=NNO_2$) or cyanoimine ($=NCN$) moiety conferring greatly enhanced photostability while conserving much of the insecticidal potency (Kagabu, 2011). This new class of insecticides was given the name neonicotinoids based on similarities to NIC and related compounds (nicotinoids) in structure and action as agonists of the nicotinic acetylcholine receptor (nAChR) (Tomizawa and Yamamoto, 1992, 1993). Selectivity is pivotal for insecticide use, combining high potency for pest insects with low risk for humans and wildlife. Fortunately, the fact that neonicotinoids are much more active on insects than mammals lead eventually to their very extensive use (currently accounting for more than one-fourth of the global insecticide market) (Jeschke et al., 2011; Matsuda et al., 2009; Tomizawa and Casida, 2005, 2009), as well as to an intense curiosity in their mechanism of selective action. The goal of this chapter is to introduce the progress that has been made in defining the chemical basis of nicotinic insecticide potency and selectivity at the target molecule, thereby encouraging receptor structure-guided insecticide design.

2. NICOTINIC INSECTICIDES

2.1. Chemotypes

Neonicotinoids and nicotinoids are structurally similar in some aspects but not others (Fig. 2.1). Neonicotinoid insecticides represented here by imidacloprid (IMI) and thiacloprid (THIA) and nicotinoids exemplified by NIC and epibatidine (EPI) (isolated from the skin of an Ecuadorian poisonous frog) (Spande et al., 1992) have in common a pyridin-3-yl moiety with or without chlorine at the sixth position. These two chemotypes are distinct in being nonprotonated and predominantly protonated, respectively, under physiological conditions. IMI guanidine moiety has pK_a values of 1.56 and 11.12 for protonation and deprotonation, respectively, indicating less than 0.0002% protonation at physiological pH (Chamberlain et al., 1996; Tomizawa et al., 2003). The neonicotinoid nitrogen atom bridging to the chloropyridinylmethyl substituent is of sp^2 nature in contrast to the sp^3 protonatable amine nitrogen of NIC or EPI. The neonicotinoid is a coplanar between the guanidine or amidine plane and nitro or cyano substituent, yielding electronic conjugation to facilitate partial negative charge (δ^-) flow towards the tip atom, thereby enabling H-bonding and π-stacking with the receptor subsites (Kagabu and Matsuno, 1997; Tomizawa et al., 2000, 2003). The equivalent region of positive charge (δ^+) to balance the negative charge on the tip is not localized in any specific atom but is rather dispersed in the whole guanidine or amidine moiety (Tomizawa et al., 2000, 2003; Wang et al., 2007). Therefore, this unique molecular system serves as the neonicotinoid electronegative pharmacophore. In contrast, the nicotinoid chemotype has a basic nitrogen atom which is protonated (ammonium ion) at physiological pH: that is, cationic functionality. It is noteworthy that the desnitro-IMI and descyano-THIA (DNIMI or DCTHIA, respectively) are protonated (iminium ion) and as such were categorized as nicotinoids (Tomizawa et al., 2000).

2.2. Target site selectivity

Neonicotinoids are selective for insect pests (aphids, leafhoppers, and other sensitive insect species), while nicotinoids are selective for vertebrates. This inverse selectivity profile is based largely on the differential sensitivity of the insect and vertebrate nAChR subtypes (Table 2.1), which is attributable to their respective unique chemical features. Neonicotinoids with a nitro- or

Table 2.1 Opposite toxicity profile of neonicotinoids and nicotinoids

| | Intrinsic toxicity, LD$_{50}$ (mg/kg) | | Binding affinity, IC$_{50}$ (nM) | | | |
| | | | nAChR[a] | | AChBP[b] | |
Compound	Insect[c]	Mouse[d]	Insect	α4β2	Aplysia[e]	Lymnaea[f]
Neonicotinoid						
IMI	0.02	45	4.3	2600	19	1000
THIA	0.03	28	2.7	900	3.9	220
Nicotinoid						
DNIMI	>5	8.0	1500	8.2	15	18
DCTHIA	>5	1.1	200	4.4	1.0	16
(±)-EPI	>25	0.08	430	0.04	1.0	0.3
(−)-NIC	>50	7.0	4000	7.0	30	100

[a]Native fruit fly and recombinant chick α4β2 nAChRs assayed with [^{3}H]IMI and [^{3}H]NIC, respectively.
[b]K_D value determined by an adaptation of a scintillation proximity assay.
[c]Intrathoracic injection into houseflies pretreated with a cytochrome P450 inhibitor.
[d]Intraperitoneal injection.
[e]Y55W mutant evaluated with [^{3}H]acetamiprid ((E)-N^1-(6-chloropyridin-3-ylmethyl)-N^2-cyano-N^1-methylacetamidine).
[f]Assayed with [^{3}H]EPI.

cyanoimine electronegative pharmacophore have high affinity for the insect nAChR but exhibit rather low potency at the vertebrate receptor. Similar to NIC and EPI providing a cationic functionality, DNIMI and DCTHIA have diminished affinity to the insect receptor while simultaneously acquire enhanced agonist potency at the vertebrate nAChR (Tomizawa and Casida, 1999; Tomizawa et al., 2000, 2001a).

2.3. Mode of insecticidal action

In insects, the nAChR is widely and predominantly distributed in the neuropil regions of the central nervous system, thus being an important target for insecticide action. Protonated NIC is poor in penetrating the "ion-impermeable" barrier surrounding the insect central nervous system. In contrast, the cationic nature of NIC is an essential requirement for interaction with the nAChR (Fig. 2.2). The inferior insecticidal activity of NIC is therefore attributed to these two contradictory features (Tomizawa and Yamamoto, 1992; Yamamoto, 1965). The penetration of neonicotinoids into the insect central nervous system is related to their hydrophobicity and is greater than that of the protonatable nicotinoids (Yamamoto et al., 1998). Accordingly, the neonicotinoids overcome the protonation obstacle of NIC in their translocation and target site interaction (Tomizawa and Casida, 2003; Yamamoto et al., 1995).

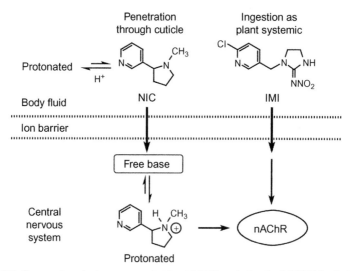

Figure 2.2 Comparison between neonicotinoid (IMI) and nicotinoid (NIC) in their mode of insecticidal action.

3. STRUCTURES OF nAChR AND ACETYLCHOLINE-BINDING PROTEIN

Neonicotinoids and nicotinoids are agonists of nAChRs, which are prototypical agonist-regulated ion channels responsible for rapid excitatory neurotransmission. They belong to the superfamily of Cys-loop receptors, which also includes γ-aminobutyric acid, glycine, and serotonin type-3 receptors. The vertebrate nAChR is a pentameric transmembrane structure consisting of diverse subtypes assembled from different sets of subunits expressed in skeletal muscle or electric organ of *Torpedo* (electric ray) (α1, β1, γ (ε), and δ), neurons (α2–α10 and β2–β4), and sensory epithelia (α9 and α10) (Changeux and Edelstein, 2005). The insect counterparts also have diverse nAChR subunits across many species (Millar and Denholm, 2007; Thany et al., 2006). However, the pentameric stoichiometries of the various insect nAChRs have not been resolved, and they can be examined functionally only as recombinant hybrids consisting of various insect α subunits and a vertebrate β2 subunit (Bertrand et al., 1994; Lansdell and Millar, 2000; Liu et al., 2005; Shimomura et al., 2006; Tomizawa et al., 2005; Yao et al., 2008). Native insect nAChRs are partially understood being based on chemical approaches using neonicotinoid structures including affinity chromatography and photoaffinity labelling. These approaches facilitated the purification and identification of the neonicotinoid binding subunit, but not the definition of the binding site interactions (Latli et al., 1997; Tomizawa and Casida, 1997, 2001; Tomizawa et al., 1996, 2001b).

The functional architecture of the *Torpedo* nAChR was visualized by cryo-electron microscopy (Fig. 2.3) (Unwin, 2005), although not with an adequate resolution to understand the recognition properties of the ligand-binding sites. The nicotinic agonist or competitive antagonist binding pocket is localized at interfacial regions between subunits. Specific subunit combinations, therefore, confer differences in sensitivity to ACh and in pharmacological profiles. Understanding drug–receptor interactions was greatly facilitated by the discovery and crystallization of soluble ACh-binding proteins (AChBPs) from the freshwater snail *Lymnaea stagnalis* and the saltwater mollusk *Aplysia californica* which serve as structural surrogates for the extracellular ligand-binding domain of the nAChR (Brejc et al., 2001; Celie et al., 2004; Hansen et al., 2005) (Fig. 2.3). The AChBP can be expressed in functional form when combined with transmembrane spans of a Cys-loop receptor (Bouzat et al., 2004). The nAChR is hetero- or

Heteropentameric nAChR

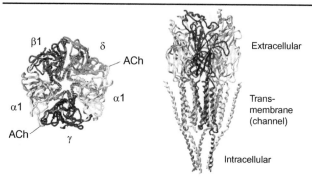

Homopentameric AChBP

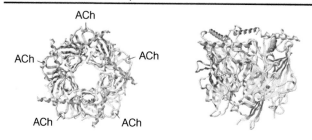

An interfacial ligand-binding pocket of AChBP

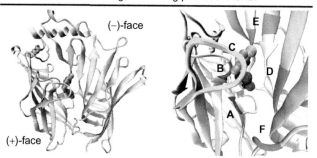

Figure 2.3 Protein structures of *Torpedo marmorata* heteropentameric nAChR (*top*) and *A. californica* homopentameric AChBP (*middle*) (views from top (*left*) and side (*right*) in each protein) based on refined electron microscopy and crystallography data (PDB ID codes 2BG9 and 2BYN, respectively) (Hansen et al., 2005; Unwin, 2005). AChBP lacks the transmembrane domain (ion channel pore) and cytoplasmic end of the nAChR. An interfacial ligand-binding pocket of AChBP (extracted from 3C79 pentamer (Talley et al., 2008)) (overview, *bottom left*) consisting of six regions (loops A–F as highlighted by colours) (zoomed-in, *bottom right*). Loops A–C domains are from the (+)-face principal subunit, while loops D–F areas are from the (−)-face partnering subunit. IMI molecule is also displayed as a landmark. (See Color Plate 7 at the back of the book.)

homopentameric with two or five ACh-binding sites, respectively, while AChBP, with a defined high-resolution crystal structure, is homopentameric encompassing five ligand-binding sites. Subunit interfacial ligand-binding pocket of nAChR and AChBP consists of six discontinuous regions (loops A–F); that is, amino acids on the six loops configure inner-wall of the pocket (Fig. 2.3). *Aplysia* AChBP is highly sensitive to neonicotinoids and nicotinoids (Table 2.1). Interestingly, the two chemotypes of agonists act at the same *Aplysia* site in an analogous way (Tomizawa et al., 2007a). In a marked contrast, the *Lymnaea* AChBP subtype has a lower affinity for neonicotinoids than nicotinoids (Tomizawa et al., 2008a). These observations indicate that the two AChBP subtypes from mollusks have a distinct pharmacology suggestive of the nAChRs from species as divergent as insects and vertebrates. Hence, the *Aplysia* AChBP serves as a plausible dual structural surrogate for interactions of neonicotinoids and nicotinoids with the insect and vertebrate nAChR, respectively. Furthermore, the *Lymnaea* AChBP may be a homologue of the vertebrate nAChR (Tomizawa and Casida, 2009).

4. DEFINING NICOTINIC INSECTICIDE BINDING SURFACES

Preliminary attempts to understand agonist–receptor interactions involve site-directed mutagenesis or chimeragenesis, which help to analyse the role of specific region(s) or amino acid(s) in the pharmacological response. However, in this approach, diminished or enhanced biological responses could be directly attributed to modified interacting determinant side chains in the binding site or indirectly arise from altered conformational states of the receptor. The structural biology approach of high-resolution X-ray crystallography reveals orientations of functional amino acids in ligand-bound state and conformational rearrangements of the protein upon ligand occupation. Crystallographic structures may not reveal the exact binding orientations observed in physiological medium because crystal packing and the position of the symmetry-related molecule may influence the selection of conformational states. Alternatively, the chemical biology strategy, including incorporation of unnatural amino acids or photoaffinity labelling, defines ligand–receptor recognition properties in a physiologically relevant, aqueous solution environment. Photoaffinity labelling may preferentially involve the more reactive residues, and stable derivatization leads to a possible interpretative bias because other proximal side chains may be far

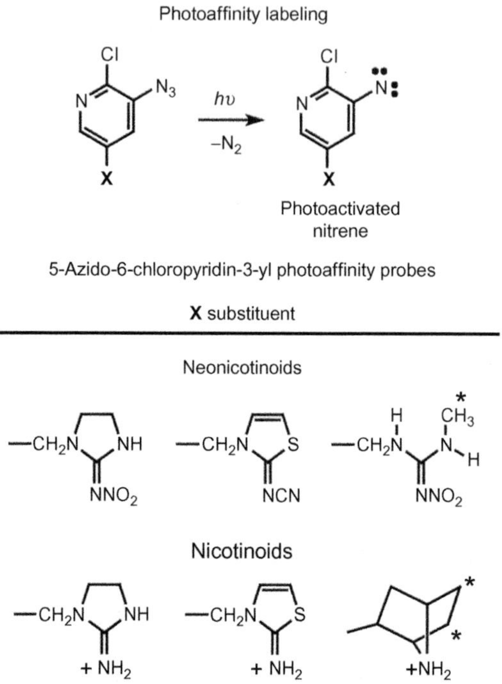

Photoaffinity labeling

Photoactivated
nitrene

5-Azido-6-chloropyridin-3-yl photoaffinity probes

X substituent

Neonicotinoids

Nicotinoids

Figure 2.4 Scheme for photoaffinity labelling (*upper*) and chemical structures of 5-azido-6-chloropyridin-3-yl neonicotinoid and nicotinoid photoprobes (*lower*). In theory, the agonist probe binds to the specific site and afterward the reactive nitrene intermediate, generated by photoirradiation at 300 nm, reacts covalently with the target molecule. Asterisks specify the positions of tritium introduced.

less reactive. Accordingly, comparisons of crystal- and solution-based structures offer a more comprehensive and precise picture of multiple binding site interactions.

4.1. Photoaffinity labelling

The photoaffinity labelling studies employed two chemotypes of nicotinic photoprobes that share the 5-azido-6-chloropyridin-3-yl structural moiety but have distinct pharmacophores: that is, nitro- or cyanoimine neonicotinoids versus desnitro or descyano iminium or ammonium nicotinoids (Fig. 2.4). The azido substituent of the photoprobes does not appreciably alter the affinity at the insect and vertebrate $\alpha4\beta2$ nAChRs and AChBPs (Kagabu et al., 2000; Tomizawa et al., 2001b, 2007a,b, 2008a, 2009; Zhang et al., 2002, 2003). Photoaffinity labelling with mass spectrometry (MS) technology of *Aplysia* AChBP enabled precise comparison of the binding site interactions of neonicotinoids versus nicotinoids. Both probe chemotypes adequately and specifically modify the AChBP with up to one agonist molecule for each subunit as based on the analysis of the intact derivatized protein (Tomizawa et al., 2007a,b, 2008a). Both neonicotinoid and nicotinoid photoprobes label *Aplysia* AChBP at only one position exactly at the interface between loop C, Tyr[195] on the principal or (+)-face subunit, and loop E, Met[116] on the partnering or (−)-face subunit (Table 2.2 and Fig. 2.5). Tryptic fragments of [³H]photoprobe-derivatized AChBP, which were subjected to HPLC separation and then to Edman sequencing, revealed two sites of labelling (peptides containing Tyr[195] and Met[116]) in equal frequency (Tomizawa et al., 2007b). Furthermore, a Met- or Tyr-scanning approach on the loops C and E regions involving 17 AChBP mutants was also designed with respect to azidochloropyridinyl nicotinoid and neonicotinoid labelling: that is, focusing on the influence of the neighbouring amino acids and the efficiency of photoderivatization. Ultimately, the results defined the specific site undergoing photoderivatization and the precise position and conformation of the bound agonist ligand. The direction of the 5-azido/nitrene-6-chloropyridin-3-yl moiety of the photoaffinity probe is independent of that for the potentially reacting amino acid side chains (Tomizawa et al., 2009).

4.2. Crystallography

Aplysia AChBP sensitive to neonicotinoids was crystallized. The high-resolution crystal structures of AChBP–neonicotinoid complexes with IMI and THIA (Talley et al., 2008) and other findings for agonist-occupied AChBP crystal structures (Celie et al., 2004; Hansen et al., 2005) are consistent with the results of photoaffinity labelling in a physiological condition

Table 2.2 Photoaffinity probes and photoderivatized site(s) in AChBP subtypes

	Aplysia		*Lymnaea*	
Probe chemotype[a]	Loop	Site	Loop	Site
Neonicotinoids	C	Tyr[195]	C	Tyr[192]
	E	Met[116]	F	Tyr[164]
Nicotinoids	C	Tyr[195]	C	Try[192]
	E	Met[116]		

[a]The same results were obtained with each of the neonicotinoid probes and each of the nicotinoid probes (see chemical structures in Fig. 2.4). *Aplysia* AChBP Tyr[195] and Met[116] are spatial neighbours to the azido substituent of the probes, whereas *Lymnaea* AChBP Tyr[192] and Tyr[164] are distinct and distant sites. *Lymnaea* AChBP Tyr[192] corresponds to *Aplysia* AChBP Tyr[195].

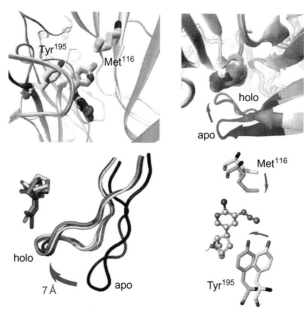

Figure 2.5 Agonist-binding site at the *Aplysia* AChBP as a suitable insect nAChR surrogate. Representative insecticide IMI is nestled in the interfacial agonist-binding pocket between the primary or (+)-face (silver) and complementary or (−)-face (pink) subunits (*upper left*). Sites for photoderivatization (Tyr[195] and Met[116] on the principal and partnering subunits, respectively) are featured. Conformational rearrangement of the loop C domain induced by agonist occupation based on crystal structures apo- and agonist-bound (holo) *Aplysia* AChBPs. Typical image for the interfacial binding pocket embracing the IMI molecule (PDB ID code 3C79) which is superimposed onto the apo form (2BYN) (*upper right*). The loop C region of the ligand vacant apo form (blue, 2BYN) is overlaid with those of the IMI- (red, 3C79), THIA- (green, 3C84), and EPI-bound (yellow, 2BYQ) forms (Hansen et al., 2005; Talley et al., 2008) (*lower left*). Comparisons of geometries for Tyr[195] and Met[116] between the azidoepibatidine–AChBP complex (docked onto 2BYQ) and the apo form (2BYN) (*lower right*). Photoderivatization sites Tyr[195] and Met[116] in the photoprobe-bound conformation (aquamarine) are superimposed onto those in the apo structure (yellow) (Tomizawa et al., 2007b). (See Color Plate 8 at the back of the book.)

with 5-azido-6-chloropyridin-3-yl neonicotinoid and nicotinoid photoprobes (Tomizawa et al., 2007a,b, 2008a). Analysis of data obtained from the crystal structures characterized the atomic interactions critical to the neonicotinoids and provided a crystallographic template comparable to those of nicotinoid agonists. Common water molecule positions are captured in the crystal structures liganded with neonicotinoids and nicotinoids. A water or solvent molecule is observed near the pyridine nitrogen of NIC or EPI (Celie et al., 2004; Hansen et al., 2005) and IMI or THIA (Talley et al.,

2008) bridging to the loop E (and/or loop B) amino acids. An additional water bridge around the tip nitrogen of the THIA cyano substituent conceivably enhances the interaction of this electronegative tip with loop C (Talley et al., 2008). Comparing AChBP–ligand complexes with an apo structure provides direct evidence of differential conformational rearrangement, particularly of the loop C domain that is induced by agonist and antagonist occupation (Hansen et al., 2005). The loop C movement, which tightly caps the binding pocket (characteristic of agonist occupation), is also proved for the high-affinity *Aplysia* AChBP–neonicotinoid complexes (about 7 Å difference between apo- and agonist-bound states) (Talley et al., 2008). Thus, the neonicotinoid electronegative pharmacophore interacts with the closed loop C region (Fig. 2.5). This can be rationalized as an initial event for the ligand-induced channel opening mechanism of the nAChR (Hansen et al., 2005), but it remains unclear how agonist occupation is coupled to the distant transmembrane-localized ion gating site. *Lymnaea* AChBP, poorly sensitive to neonicotinoids, was also co-crystallized with IMI and clothianidin insecticides, although there is little if any loop C movement (Ihara et al., 2008). The elicited conformational changes by the agonist occupation are also supported by fluorescence spectroscopy, deuterium–hydrogen exchange MS, and NMR spectroscopy analyses of AChBP (Gao et al., 2005, 2006; Shi et al., 2006). The photoaffinity labelling with neonicotinoid and nicotinoid agonist probes clearly indicated that the loop C Tyr[195] is oriented towards the loop E Met[116], and upon probe binding, loop C draws inward to enclose the photoprobe molecule (Tomizawa et al., 2007a,b) (Fig. 2.5). Similar conclusions based on photolabelling studies were drawn for tightening the interfacial binding domain (involving primary and complementary subunits) after ligand interaction in *Torpedo* nAChR (Grutter et al., 2002; Mourot et al., 2006; Nirthanan et al., 2005).

4.3. Agonist-binding site interactions

Neonicotinoid and nicotinoid binding site interactions in chemical or atomic resolution have been defined by both photoaffinity labelling and X-ray crystallography investigations using AChBP subtypes (Talley et al., 2008; Tomizawa et al., 2007a,b, 2008a, 2009). These findings ultimately established the two structural models for interfacial agonist-binding domains of nAChR subtypes of the green peach aphid (*Myzus persicae*) α2β1 (for neonicotinoids) and of the chick α4β2 (for nicotinoids) (Fig. 2.6). The IMI chloropyridinyl chlorine atom can have favourable van der Waals

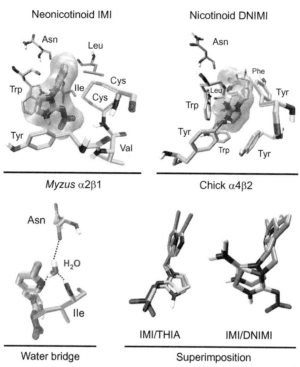

Neonicotinoid IMI   Nicotinoid DNIMI

Myzus α2β1   Chick α4β2

IMI/THIA   IMI/DNIMI

Water bridge   Superimposition

Figure 2.6 Structural models for binding site interactions of IMI and DNIMI with the α–β subunit interfacial agonist-binding pocket of nicotinic receptors based on chemical and structural biology investigations. Representative neonicotinoid IMI and nicotinoid DNIMI are embedded in the aphid (*Myzus persicae*) α2β1 and chick α4β2 interfaces, respectively (*upper*). Amino acids in lime-green or pink are from aphid α2 or chick α4 subunit, and those in orange or cyan are from aphid β1 or chick β2 subunit, respectively. A water molecule near the pyridine nitrogen atom, captured in AChBP-IMI or AChBP-THIA crystal structure (PDB ID code 3C79 or 3C84, respectively) (Talley et al., 2008), is superimposed onto this IMI-bound structure (*lower left*). Consistently, a water or solvent bridge is also observed in the AChBP crystals liganded with nicotinoids NIC and EPI (Celie et al., 2004; Hansen et al., 2005). Binding conformations of IMI, THIA, and DNIMI as observed in the agonist-binding pocket are compared (*lower right*). IMI and THIA are nicely superimposable, whereas IMI and DNIMI (or THIA and DCTHIA (not shown)) pharmacophores are reversed relative to each other. DNIMI- and EPI-bound conformations are suitably overlaid (not shown) (Tomizawa et al., 2007a, 2008a). (See Color Plate 9 at the back of the book.)

interactions and/or halogen bonding with the backbone of loop E amino acids such as Asn and Leu. The pyridine nitrogen atom forms a water bridge to the backbone NH of Ile and the carbonyl oxygen of Asn (loop E) (Fig. 2.6). In an identical manner, chlorothiazolyl nitrogen and tetra-hydrofuryl oxygen atoms of neonicotinoids serve as hydrogen acceptors. A structure–activity relationship (SAR) study, focusing on neonicotinoid probes with different H-bonding points in location and capability, also demonstrates that the position of the hydrogen-accepting point of the nicotinic agonist and orientation of the intervening water molecules are firmly restricted in the conserved directions (Ohno et al., 2009a). Interestingly, the electronically conjugated guanidine plane primarily π-stacks with the loop C Tyr aromatic side chain and also interacts via stacking or hydrophobic interactions with the loop B Trp indole moiety. The nitro oxygen or cyano nitrogen tip undergoes H-bonding with the loop C Cys and/or Val backbone. On the other hand, the *Aplysia* AChBP loop D Gln side chain (spatially corresponding to Arg, Lys, or Asn on the insect loop D) also faces towards the tip nitro oxygen atom of IMI but not to the cyano nitrogen of THIA in the crystal structures (Talley et al., 2008). Similarly in *Lymnaea* AChBP, IMI nitro oxygen is close enough to the loop D Gln residue (but nitro oxygen of clothianidin is not) (Ihara et al., 2008). These results suggest that the loop D amino acid may reinforce the interaction of IMI nitro oxygen tip with the insect receptor. Alternatively, the insect nAChR loop D amino acid is predicted to undergo hydrogen networks with the loop C Cys on the α subunit in the neonicotinoid-bound complex, thereby stabilizing the neonicotinoid interaction with loop C in the closed-conformation (Tomizawa et al., 2007a). Relative to the nicotinoid DNIMI, the chloropyridinyl moiety interacts with the loop E amino acids (Asn, Phe, and Leu) of the chick α4β2 receptor (Fig. 2.6). Importantly, the nicotinoid cationic functionality (iminium or ammonium head) critically contacts the carbonyl oxygen of the loop B Trp via H-bonding, and this interaction is stabilized by cation-π contacts with the loop B Trp indole and other aromatic residues of loops A, C, and D. Accordingly, the nicotinoid cationic functionality is nestled in a reverse direction compared with the neonicotinoid electronegative pharmacophore (Tomizawa et al., 2007a,b) (Fig. 2.6).

5. MULTIPLE BINDING CONFORMATIONS CONFERRING SELECTIVE RECEPTOR INTERACTION

The structural determinants of nAChR subtype selectivity have been studied for a family of peptide antagonists with a binding region extending over a large interfacial surface to embrace a unique moiety of the antagonist (Dutertre et al., 2007; Hansen et al., 2005; Talley et al., 2006; Ulens et al., 2006). However, the molecular mechanism of selectivity for small agonist molecules is less well resolved because most of the key amino acids in the nAChR-binding pocket are conserved in all of the receptor subtypes and species. The amino acids forming the binding pockets are structurally or functionally consistent not only in the diverse nAChR subtypes but also in the AChBPs, yet there is considerable neonicotinoid selectivity. Chimaera hybrid receptor consisting of insect α subunit and vertebrate β2 subunit, wherein insect loop D, E, or F sequence is inserted, enhances IMI-elicited agonist responses. However, the influence of the chimeragenesis is modest and of similar magnitude among the three chimaera hybrid receptors (Shimomura et al., 2006; Yao et al., 2008). This result points out a distinct limitation for the mutagenesis or chimeragenesis approach to pinpoint the specific region or amino acid(s) for neonicotinoid selectivity.

Photoaffinity labelling in physiological medium with *Lymnaea* AChBP, a suitable homologue for neonicotinoid-insensitive vertebrate receptor, successfully reveals the determinant for the inferior neonicotinoid affinity, thereby ultimately deciphering the mechanism of neonicotinoid selectivity (Tomizawa et al., 2008a). In sharp contrast to *Aplysia* AChBP in both solution- and crystal-based investigations, neonicotinoid photoaffinity labelling of *Lymnaea* AChBP specifically yields two distinct and distant modification sites at loop F Tyr[164] and loop C Tyr[192]. The stoichiometry of photoincorporation is one ligand molecule per binding site as based on the MS analysis of intact-modified subunit protein. However, only site Tyr[192] is pinpointed in nicotinoid photolabelling (Table 2.2), leading to the proposed unique neonicotinoid binding conformations (Tomizawa et al., 2008a) (Fig. 2.7). *Lymnaea* AChBP accommodates the neonicotinoids in two distinct bound conformations. One binding orientation is completely inverted compared with the common conformation (which is the one observed in the *Aplysia* subtype). Based on the results described above with *Lymnaea* AChBP, a model for interfacial agonist-binding domain of α4β2 nAChR subtype was developed. As with *Lymnaea* AChBP, the chick α4β2 receptor model yields two bound conformations for IMI or THIA that are once again inverted relative to each other (Tomizawa et al., 2008a). Therefore, a blend of two very disparate binding conformations competing at the *Lymnaea* AChBP and vertebrate nAChR coincides with the inferior affinity of neonicotinoids at these sites, possibly contributing to the poor binding constant, which reflects a weighted average of a multiplicity of binding orientations. Only a single tight binding conformation at the *Aplysia* AChBP as in the insect nAChR model confers high neonicotinoid sensitivity. In nicotinoids, a single binding orientation is conserved for all AChBP and nAChR subtypes (Tomizawa et al., 2008a). The final binding constant represents a

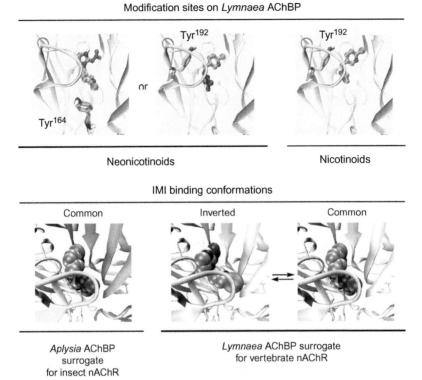

Figure 2.7 Molecular recognition conferring selective receptor interaction. Identified photoderivatization sites on the *Lymnaea* AChBP as a structural homologue of the vertebrate nAChR (*upper*). Neonicotinoid photoprobe specifically modifies either Tyr[192] on the primary face or Tyr[164] on the complementary face, whereas nicotinoid photoprobe exclusively derivatizes Tyr[192] (Tomizawa et al., 2008a). Comparative binding conformations of IMI embraced by *Aplysia* AChBP as an insect nAChR homologue and by *Lymnaea* AChBP as a vertebrate nAChR surrogate (*lower*). IMI binds in two conformations designated "common" and "inverted". (See Color Plate 10 at the back of the book.)

combination of multiple individual constants unique to different conformations. The same agonist molecule can also take different binding directions at other Cys-loop receptors depending upon the nature and position of the aromatic amino acids (Dougherty, 2008; Mu et al., 2003; Padgett et al., 2007).

6. RESISTANCE IN PESTS

Recently, the effectiveness of neonicotinoids has diminished due to the emergence of resistant strains including the silverleaf whitefly, *Bemisia tabaci*; the aphid, *M. persicae*; the brown planthopper, *Nilaparvata lugens*, or

the Colorado potato beetle, *Leptinotarsa decemlineata*. The neonicotinoid resistance is principally attributable to the enhanced detoxification by cytochrome P450 monooxygenases (Karunker et al., 2009; Puinean et al.,
2010). *B. tabaci* CYP6CM1vQ is responsible for 5-hydroxylation (with
some olefin formation) on the IMI imidazolidine ring (Karunker et al.,
2009). In addition, nAChR mutation may partly be associated with reduced
IMI mortality and receptor sensitivity, involving Tyr to Ser (adjacent to loop
B Trp) of α subunit in *Nilaparvata* or Arg to Thr (neighbouring loop D Trp)
of β subunit in *Myzus* (Bass et al., 2011; Liu et al., 2005).

7. RECEPTOR STRUCTURE-GUIDED INSECTICIDE DESIGN

The resolved binding site structure in the neonicotinoid-bound state
has ultimately facilitated the molecular designs of novel insecticidal nicotinic
compounds with unique pharmacophore(s), undergoing distinct binding
mechanism(s). This approach should expand the insecticidal spectrum of the
present neonicotinoids and may circumvent possible development of resistance
caused by activated detoxification systems and modified target sites. Studies in
designing nicotinic agonists with novel chemotype pharmacophores based on
combining the chemorational approach with the receptor binding site structure
yielded several candidate compounds with excellent insecticidal effectiveness
rivalling those of the present neonicotinoids and other classes of commercial
insecticides (Tomizawa and Casida, 2009; Tomizawa et al., 2011a). Additionally, extensive efforts in chemorational tactics successfully led to discover novel
neonicotinoid insecticides with *cis*-nitromethylene and cyanosulfoximine
moieties (Shao et al., 2011; Zhu et al., 2011) (Fig. 2.8).

7.1. Strategy for pharmacophore modification

The neonicotinoid nitroimine pharmacophore has two functional points responsible for interaction with the receptor subsite(s): that is, hydrogen-
acceptable tip nitro oxygen atom and imine π-electron system. Interestingly,

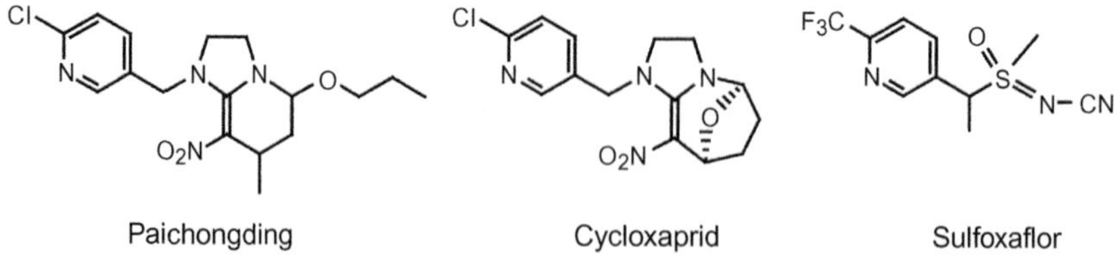

| Paichongding | Cycloxaprid | Sulfoxaflor |

Figure 2.8 Chemical structures of newly developed neonicotinoid insecticides.

the nitroso analogue (=NNO) retains the potency of the nitroimine (=NNO$_2$) compounds, thus defining the functional oxygen tip (Fig. 2.9) (Tomizawa et al., 2003). However, the formylimine (=NC(O)H) congeners show a great reduction in potency. The obvious potency difference between the two functional groups (=NNO vs. =NC(O)H) can be attributed to their pharmacophore orientations as based on their crystal structures and/or quantum mechanics calculations. The =NNO tip oxygen substantially faces the descending direction (active form) for H-bonding formation with the subsite. In contrast, the =NC(O)H oxygen possibly assumes two flexible directions under biological conditions: that is, the alternative upward oxygen orientation (inactive form) and the active one as with the =NNO tip (Ohno et al., 2009b; Tomizawa et al., 2003) (Fig. 2.9). Accordingly, the flexibility of the tip oxygen orientations in the binding pocket may determine the final binding constant of the =NC(O)H analogue as an average. Intriguingly, the =NC(O)H moiety can be replaced by extended and/or hydrophobic substituents, encouraging to explore novel chemotype pharmacophores.

7.2. Unique niche revealed by receptor structure

The IMI-binding site is located at an interfacial region between α and β subunits of the insect nAChR (Fig. 2.10). The neonicotinoid nitro oxygen or cyano nitrogen tip H-bonds to the loop C Cys (and/or adjacent one) backbone on the α subunit face (Fig. 2.6) (Talley et al., 2008; Tomizawa et al., 2007a,b, 2008a). The IMI–receptor structure, intriguingly, reveals a unique niche extending from the bound IMI nitro tip oxygen towards the loop D Arg on the β subunit (about 4–6 Å depth). This space provides room for one aromatic ring structure. Moreover, the loop D Arg on the insect β subunit is spatially equivalent to Thr with a short side chain on the β2 subunit of the vertebrate neuronal nAChR, suggesting a difference between insect and vertebrate loop D cavities in their depths and functional residues (Fig. 2.10). Therefore, this observation leads to designing prototype analogues with extended acylimine and aryloxycarbonylimine functional moieties proposed to specifically interact with the loop D region of the insect nAChRs (Fig. 2.9) (Ohno et al., 2009b; Tomizawa and Casida, 2009; Tomizawa et al., 2008b, 2011a). Most of the nicotinic agonists are embraced by the binding pocket consisting of five domains (loops A–E) (Fig. 2.3), except for benzylidene derivatives of anabaseine that also anchor to the loop F domain (Hibbs et al., 2009). This implies a possibility of designing insecticidal agonists targeting the loop F region.

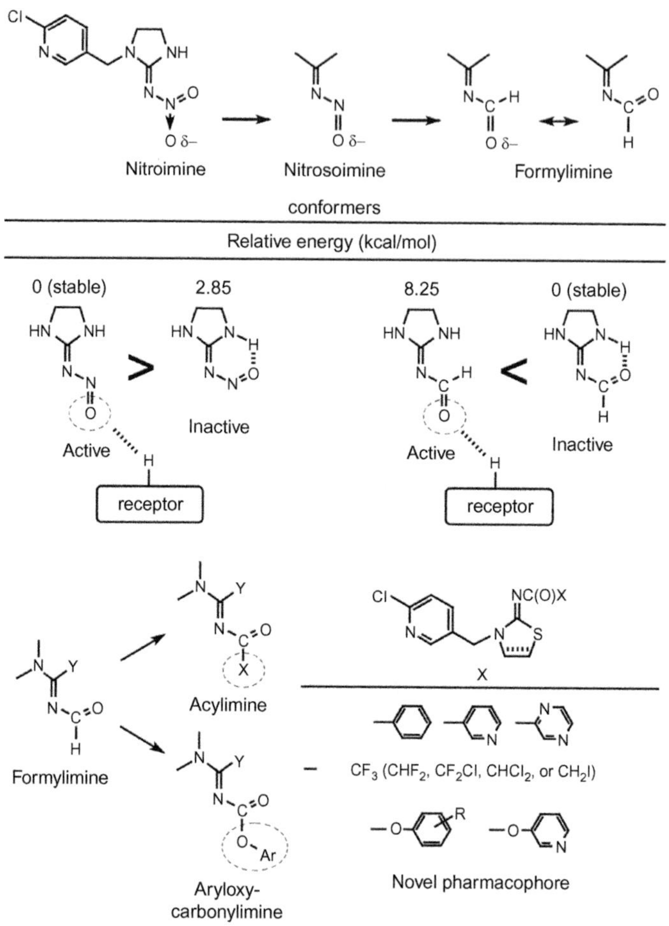

Figure 2.9 Chemical structures of nitroimine IMI and systematic variants of the electronegative pharmacophores (*top*). Relative electronic energies of possible conformers of 2-nitrosoimine and 2-formylimine model structures (calculated at B3LYP/6-311+G (2d,2p)) (Ohno et al., 2009b) (*middle*). The relative energies can be compared only within each conformer set. Active conformation (tip oxygen atom in the descending direction) enables to form hydrogen bonding with the receptor subsite, whereas the alternative one may not (inactive). Nitrosoimino- and formylimino-imidazolidine neonicotinoid compounds have lower receptor potency than those of the corresponding thiazolidine analogues. The existing ratio to each conformer may be different between imidazolidine and thiazolidine analogues (Ohno et al., 2009b). The formylimine moiety can chemorationally be modified with acylimine and aryloxycarbonylimine pharmacophores (*bottom*). Preferred geometries of the =NC(O)X and =NC(O)OAr model structures were calculated at B3LYP/6-311+G(2d,2p) and the substituent faces toward the descending direction (not shown) (Ohno et al., 2009b).

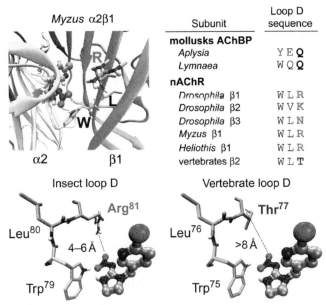

Subunit	Loop D sequence
mollusks AChBP	
Aplysia	Y E Q
Lymnaea	W Q Q
nAChR	
Drosophila β1	W L R
Drosophila β2	W V K
Drosophila β3	W L N
Myzus β1	W L R
Heliothis β1	W L R
vertebrates β2	W L T

Figure 2.10 Insect nAChR structural model in IMI-bound state reveals a unique cavity extending from the bound IMI nitro oxygen tip towards the loop D region. IMI is docked into the aphid (*Myzus persicae*) α2β1 interfacial agonist-binding pocket (Tomizawa et al., 2008b) (*upper left*). Three amino acid residues (W (Trp), L (Leu), and R (Arg)) on the β1 subunit make up the loop D cavity. Alignment of the loop D amino acid sequences from two mollusk AChBP subtypes, five insect β subunits (represented by fruit fly *Drosophila melanogaster*, green peach aphid *Myzus persicae*, and tobacco budworm *Heliothis virescens*), and the vertebrate (human, rodent, and chick) β2 subunit of the nAChR (*upper right*). The loop D Arg on the *Myzus* β1 subunit is spatially equivalent to Thr (T) with a shorter side chain on the β2 subunit of the vertebrate neuronal nAChR, suggesting a difference between insect and vertebrate loop D cavities in their depths and functional residues (*lower*). (See Color Plate 11 at the back of the book.)

7.3. *N*-Acylimine analogues

Replacement of the nitro- or cyanoimine pharmacophore by extended substituents may provide point(s) for hydrogen acceptance and/or van der Waals contacts at the targeted binding cavity (loop D). According to this hypothesis, prototype benzoylimine, pyridinoylimine, and pyrazinoylimine derivatives were prepared to evaluate the binding potency to the insect nAChR (Fig. 2.9) (Ohno et al., 2009b; Tomizawa et al., 2008b). The benzoyl compound shows moderate potency, thus indicating that an aromatic ring can be accommodated into the target niche. The 3-pyridinoyl analogue with an H-acceptor nitrogen atom has a much higher affinity (2- and 4-pyridinoyl isomers are twice less active than the 3-isomer) than that of the benzoyl compound. This enhancement is more pronounced with the 2,5-pyrazinoyl analogue with two H-acceptor nitrogen atoms (Tomizawa et al., 2008b). Therefore, these findings clearly underscore that H-accepting nitrogen atoms play a crucial role in recognition by the amino acid(s) at the target subsite. Surprisingly, the N-trifluoroacetylimine ($=NC(O)CF_3$) analogue, providing both van der Waals contacting and H-bonding abilities, exhibits high affinity comparable to those of IMI and THIA. However, the N-acetylimine ($=NC(O)CH_3$) analogue has greatly diminished receptor potency, straightforwardly indicating that the fluorine atoms play a decisive role on the binding mechanism (Ohno et al., 2009b; Tomizawa et al., 2003, 2008b). These illustrative acylimine compounds with pyrazine and CF_3 substituents show high intrinsic insecticidal activity rivalling those of commercial neonicotinoids (Tomizawa et al., 2008b). The common brown house mosquito (*Culex quinquefasciatus*) provides an example of a major pest with increasing difficulty to control because of the selection of strains resistant to the major pyrethroid, organophosphate, and methylcarbamate insecticides. Neonicotinoids are possible replacements if they have suitable potency and little or no cross-resistance (Corbel et al., 2004; Liu et al., 2004). The CF_3 neonicotinoid was compared with permethrin, the standard mosquito control agent for insecticidal potency on fourth instar larvae. Although the CF_3 compound is twofold less active than permethrin as a larvicide on the susceptible colony, it is seven times more potent on the permethrin-resistant colony (Tomizawa et al., 2008b). The target site cross-resistance of pyrethroids and DDT or of organophosphates and methylcarbamates does not carry over to the neonicotinoids (Corbel et al., 2004; Liu et al., 2004), making them candidates for further development as mosquito control agents. Moreover, the trifluoroacetyl and haloacetyl derivatives have unique biological properties attributable to their enhanced hydrophobic nature (discussed later).

Selectivity of the pyrazinoyl and trifluoroacetyl analogues was determined by comparing receptor potency at the insect and vertebrate nAChRs and by their toxicity to flies and mice (Table 2.3) (Tomizawa et al., 2008b). As with IMI and THIA, the two compounds have low potency at the vertebrate $\alpha4\beta2$ receptor and low toxicity to mice. In addition, the binding affinities of the pyrazinoyl and trifluoroacetyl analogues to a hybrid receptor, consisting of the aphid (*Myzus*) $\alpha2$ subunit and the rat $\beta2$ subunit, are much lower than those of IMI and THIA (Tomizawa et al., 2005, 2008b), strongly indicating that the insect α subunit plays a critical role in the recognition of IMI and THIA with nitro- and cyanoimine pharmacophores. In sharp contrast, the insect β subunit with loop D Arg (or functionally analogous amino acid) is clearly important for embracing the extended N-pyrazinoylimine or

Table 2.3 Selectivity of neonicotinoids with nitro- or cyanoimine and acylimine pharmacophores between insects and vertebrates

Insecticide	Binding to nAChRs, IC_{50} (nM)			Intrinsic toxicity, LD_{50} (mg/kg)	
	Fruit fly native[a]	Hybrid aphid α2/rat β2[a]	Chick α4β2[b]	Housefly[c]	Mouse[d]
IMI	4.3	3.6	2600	0.021	45
THIA	2.7	10	900	0.032	28
Pyrazine	1.5	720	900	0.035	>24[e]
CF_3	3.1	300	500	0.027	>36[e]

[a]Assayed with [^{3}H]IMI.
[b]Determined with [^{3}H]NIC.
[c]Intrathoracic injection into houseflies pretreated with a cytochrome P450 inhibitor.
[d]Intraperitoneal injection.
[e]No lethality (0%) was observed at the indicated dose (maximal dose administered due to solubility limitations in the vehicle).

N-trifluoroacetylimine moiety. Accordingly, the difference between insect and vertebrate loop D niches in their functional amino acid residues serves as a determinant for the target site selectivity of the N-acylimino neonicotinoids, although the selectivity mechanism for IMI and THIA depends on the multiple binding conformations (Tomizawa and Casida, 2009, 2011; Tomizawa et al., 2008a).

In a molecular dynamics simulation of the insect nAChR structural model liganded with the N-pyrazinoylimine analogue, two pyrazine nitrogen atoms and the =NC(O) oxygen undergo H-bonding with the loop D Arg guanidine NH_2 and Trp indole NH (Fig. 2.11). Similarly, the N-trifluoroacetylimine compound interacts with loop C and D regions: that is, the fluorine atoms H-bond to loop D Arg and Trp (directly or possibly via water bridges) and to loop C Cys and also make van der Waals contact with the Trp side chain; the NC(O) oxygen H-bonds with the Trp indole NH (Tomizawa et al., 2008b). Binding mechanisms of the chloropyridine and amidine moieties are identical to those of IMI and THIA (Tomizawa and Casida, 2009; Tomizawa et al., 2007b, 2008a). Accordingly, the *in silico* binding site interactions are unambiguously consistent with the observed SAR.

7.4. *N*-Haloacetylimine analogues

An exhaustive SAR study of neonicotinoid analogues with a =NC(O)CF$_3$ or =CHC(O)CF$_3$ substituent was then conducted to examine the hypothesis that the trifluoroacetyl pharmacophore also enhances the

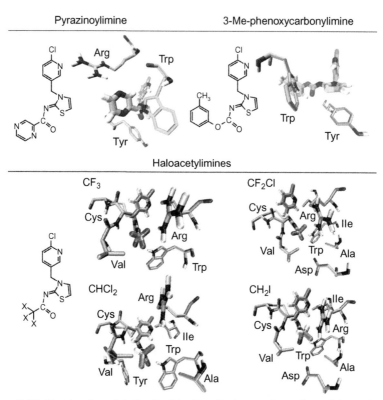

Figure 2.11 Structural models for the binding site interactions of candidate nicotinic insecticides with the α–β interfacial agonist-binding pocket of the insect nAChR (*Myzus* α2β1), featuring the functions of novel pharmacophores in the regional binding niche (particularly, loops C–D and their adjacent areas). The relevant amino acids in pink are from α2 subunit and in lime-green are from β1 subunit. The binding interactions of the 2-imino-3-(6-chloropyridin-3-ylmethyl)-thiazoline moiety are identical to those of 2-imino-3-(6-chloropyridin-3-ylmethyl)-imidazolidine moiety of IMI (see Fig. 2.6) (Ohno et al., 2009b, 2010b; Tomizawa et al., 2008b, 2011b). (See Color Plate 12 at the back of the back.)

hydrophobicity and thereby confers the improved insecticidal effectiveness relative to those of the standard nitro or cyano neonicotinoids (Ohno et al., 2010a). The $=NC(O)CF_3$ analogues are highly active both in receptor binding and in intrinsic insecticidal potencies comparable to those of the $=NNO_2$ and $=NCN$ compounds (Table 2.4) (Ohno et al., 2010a; Tomizawa et al., 2003, 2008b). The $=CHC(O)CF_3$ analogues show outstanding insecticidal activity attributable to superior receptor potency. Strikingly, compounds with $=NC(O)CF_3$ and $=CHC(O)CF_3$ pharmacophores are excellent insecticides by topical application without synergist treatment, rivalling the potency of

Table 2.4 Summarized SAR of trifluoroacetyl and haloacetyl neonicotinoid insecticides

Pharmacophore	Binding to housefly nAChR IC_{50} (nM)[a]	Contact toxicities to houseflies, LD_{50} (μg/g female)[b] Synergist	Alone[d]	Hydrophobicity log P_{ow}[c]
NNO_2 or NCN	4–50	0.2–0.6	>100	0.5–1.2
$NC(O)CF_3$	5–31	0.04–0.5	1.1–17	1.9–3.1
$NC(O)X^e$	4–28	0.025–0.1	3.3–7.1	2.8–3.3
$NC(O)X^f$	6–130	0.12–2.2 or >10	>100	1.8–3.6
$CHC(O)CF_3$	1–3	0.004–0.12	0.8–3.1	1.7–2.1

[a]Housefly brain nAChR assayed with [^{3}H]IMI.
[b]Topical application in the absence or the presence of a cytochrome P450 inhibitor (synergist).
[c]Log P coefficients between 1-octanol and water.
[d]Chlorpyrifos as a standard insecticide had a LD_{50} value of 1.9 μg/g female, as compared with those of other important insecticides propoxur, dieldrin, and DDT of 23, 0.7, and 14 μg/g female, respectively.
[e]X = CF_2Cl and CHF_2.
[f]X = CCl_3, $CHCl_2$, CH_2Cl, CH_2Br, and CH_2I.

other standard insecticides. In contrast, the corresponding =NNO$_2$ or = NCN neonicotinoids are entirely inactive under the same condition. Neonicotinoids generally have low log P values (from −0.7 to 1.3), which explain their exceptional plant-systemic properties (Jeschke et al., 2011; Tomizawa and Casida, 2005). The =NC(O)CF$_3$ and =CHC(O)CF$_3$ substituents greatly enhance the hydrophobicity of nitro- or cyanoimine neonicotinoids (Ohno et al., 2010a). Therefore, the increased hydrophobicity of trifluoroacetyl neonicotinoids largely improves the penetrability of the compound through the insect integument and consequently their insecticidal effectiveness. In addition, the =NC(O)CF$_3$ neoncotinoid appears to retain adequate photostability, yet the =CHC(O)CF$_3$ analogue is conceivably photolabile (Ohno et al., 2010a).

Furthermore, extensive SARs of N-haloacetylimine neonicotinoid derivatives (with CHF$_2$, CF$_2$Cl, CCl$_3$, CHCl$_2$, CH$_2$Cl, CH$_2$Br, and CH$_2$I haloacetyl heads and thiazoline/thiazolidine heterocycles) were examined to clarify the function of halogen atoms in the binding niche, and consequently the insecticidal activity (Table 2.4) (Tomizawa et al., 2011b). The analogues with the CHF$_2$ substituent retain the receptor potency of CF$_3$ compounds. Interestingly, compounds having the CF$_2$Cl yielded the highest binding affinity among the test compounds. The analogues with the CCl$_3$ substituent have diminished nAChR potency, whereas compounds bearing the $CHCl_2$ group showed higher activity than those of the CCl_3 analogues. Unexpectedly, the mono-halogenated compounds with substituents such as CH_2Cl, CH_2Br, and CH_2I also display relatively high receptor potency. These observations suggest that the fluorine atoms of N-acetylimino group make a crucial contribution to the binding interactions with the subsite via H-bonding and/or hydrophobic contact. Moreover, the bulky tri-chloromethyl head may be too crowded in the subsite, while one or two chlorine atom(s) (or one bromine or iodine atom) on the acetyl moiety may be favoured in the binding niche by presumably undergoing van der Waals and/or halogen-bond interactions.

In silico binding site interactions of N-haloacetylmine neonicotinoids (representative compounds with $=NC(O)X$ wherein $X = CF_3$, CF_2Cl, $CHCl_2$, and CH_2I moieties) were then established with an insect nAChR structural model based on the SAR findings (Tomizawa et al., 2011b) and chemical and structural biology investigations (Fig. 2.11) (Talley et al., 2008; Tomizawa et al., 2007a,b, 2008a,b). The CF_3 fluorine atoms H-bond with loop D Arg NH_2, Trp indole NH, and backbone NH of loop C Cys and Val (see also Section 7.3). Subsequently, in the CF_2Cl species, the simulations show these same interactions although additionally chlorine atom has favourable halogen-bond or van der Waals interactions with a variety of side chains including Trp, Ile, and Cys. In some rotamers, the chlorine atom rotates slightly and, in addition, interacts with Asp ($\beta1$ subunit), Ala ($\beta1$ subunit), and loop C Val. For the $CHCl_2$ system, van der Waals interactions between a chlorine atom and Arg as well as with the loop C backbone (Tyr, Val, and Cys) and possibly a halogen-bond interaction with Cys sulphur were observed. The other chlorine atom is also close enough to Trp, Arg, Ile, and Ala side chains for favourable van der Waals interactions. In the CH_2I compound, van der Waals interactions between the iodine and Trp, Arg, and Ala side chains, as well as a possible halogen-bond interaction with Asp side chain oxygen, are suggested. The loop C Cys and Val side chains and backbones are also within the range of van der Waals interactions. A slight rotation leads to a possible halogen-bond interaction with the Cys sulphur and an additional van der Waals interaction with loop E Ile side chain.

The CF_3, CHF_2, and CF_2Cl compounds with high log P values show excellent insecticidal activity via the dermal route, which is greatly higher than those of the control nitro- and cyanoimine neonicotinoids, and comparable to those of other chemotype insecticides (Table 2.4) (Ohno et al., 2010a; Tomizawa et al., 2008b). Highly lipophilic CCl_3 compounds, showing moderate nAChR affinities, have similar toxicity profile to that of IMI.

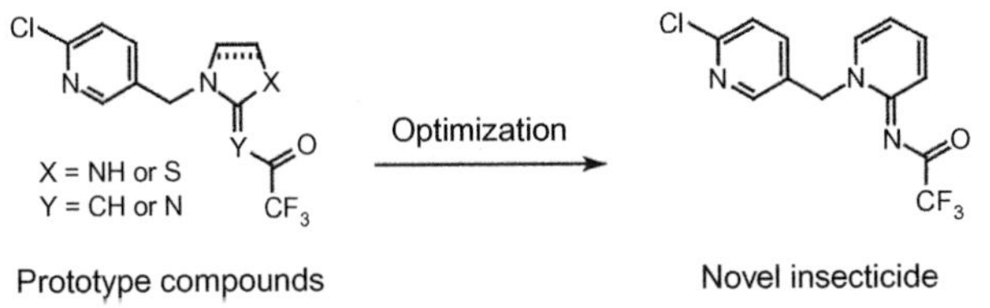

Figure 2.12 Prototype compounds (Ohno et al., 2010a; Tomizawa et al., 2003, 2008b, 2011b) and novel neonicotinoid insecticide (Kagabu et al., 2012) with a trifluoroacetyl pharmacophore. Binding mechanism of the promising new insecticide is undoubtedly consistent with that of prototype CF$_3$ compound as illustrated in Fig. 2.11.

Analogues with a CHCl$_2$ substituent having relatively high log P values and binding potencies, however, generate largely inferior insecticidal activity. Furthermore, the CH$_2$Cl, CH$_2$Br, and CH$_2$I compounds, with comparable binding affinity to that of IMI, are not toxic to flies, presumably due to substantial detoxification (dehalogenation) mechanism(s) in addition to oxidative pathway(s). Kagabu et al. (2012) ultimately discovered a promising novel neonicotinoid insecticide with a trifluoroacetylimine pharmacophore (Fig. 2.12), and the optimized compound should be identical in its binding mechanism to that of the prototype analogues (Ohno et al., 2010a; Tomizawa et al., 2003, 2008b, 2011b).

7.5. *N*-Phenoxycarbonylimine variants

Neonicotinoid variants with the aryloxycarbonylimine moiety may assume a dissimilar binding conformation to that of acylimine analogues due to the flexibility of the *O*-hinged (rotatable) aryl ring (Ohno et al., 2009b). The pyridin-3-oxycarbonylimine analogue and phenoxycarbonylimine analogues (Fig. 2.9) were synthesized to define the SAR and then examine the binding site interactions at the insect nAChR. The plain phenoxycarbonylimine compound is highly potent, while the hydrogen-acceptable pyridin-3-oxycarbonylimine analogue is less active, conceivably due to the decreased π-electron density of the pyridine rather than providing a H-bonding point. Thus, the aromatic ring π-electron density may play an important role in the interaction with the regional subsite. This hypothesis was examined by the SAR approach with phenoxylcarbonylimine compounds containing electron-withdrawing and electron-donating substituents that should clarify the function of the aromatic π-electron system. For instance, receptor potency of the unsubstituted phenoxy compound is clearly reinforced by introducing an electron-donating 3-methyl or 3-methoxy substituent and is greatly reduced by substitution with an electron-withdrawing 4-CF_3 or 4-NO_2. Therefore, the π-electron density on the benzene ring plays a key role in the interaction with a regional binding domain (Ohno et al., 2009b, 2010b). The receptor potency of the 3-methylphenoxy compound is equal to that of IMI, although it is much less toxic to insects, presumably due to metabolic detoxification (hydrolysis) (Ohno et al., 2010b). Finally, the SAR results predict binding site interactions featuring the phenoxy ring of the neonicotinoid and the receptor loop D Trp indole plane forming a T-shaped aromatic interaction (Fig. 2.11) (Ohno et al., 2009b, 2010b). The face-to-edge aromatic interaction can provide as much stabilization as the more standard π-stacking (Bissantz et al., 2010; McGaughey et al., 1998). In contrast, simulated molecular recognition of the methoxycarbonylimine ($=NC(O)OCH_3$) derivatives is almost identical to that of nitromine ($=NNO_2$) compounds, and expectedly, these derivatives show high receptor potency (Ohno et al., 2009b).

7.6. Bis-neonicotinoid derivatives

The bis-pharmacophore drug design approach has been applied to neonicotinoid insecticide, yielding $N^3,N^{3'}$-alkylene-tethered bis-imidacloprid(bis-IMI)derivatives alkylene-tethered bis-imidacloprid (bis-IMI) derivatives (Fig. 2.13) (Kagabu et al., 2002, 2004, 2006). The dimeric IMI analogues with hexamethylene, heptamethylene, and octamethylene linkers have plant-systemic insecticidal properties (Mori et al., 2010). The number of bis-IMI methylene units determines the binding affinity to the insect nAChR (Kagabu et al., 2002, 2004, 2006; Ohno et al., 2009c), and the heptamethylene analogue shows optimal potency (Ohno et al., 2009c). The bis compounds acting directly at the insect nAChR elicit initial excitation and subsequent blockade (Kagabu et al., 2002), although they serve as antagonists when co-applied with the agonist ACh (Ihara et al., 2007) to the American cockroach abdominal ganglion. The molecular recognition of bis-quaternary ammonium compound (suberyldicholine) and curariform ligands (*d*-tubocurarine and metocurine) has been predicted (Carter et al., 2007; Gao et al., 2003) using mollusk AChBP as a structural surrogate of the nAChR (Brejc et al., 2001; Celie et al., 2004; Hansen et al., 2005). Interestingly, the bis-quaternary ammonium ligand conformations are recognized as having an extended geometry rather than being tightly folded, and the two ammonium heads contact two isolated subsites consisting of aromatic amino acid side chains

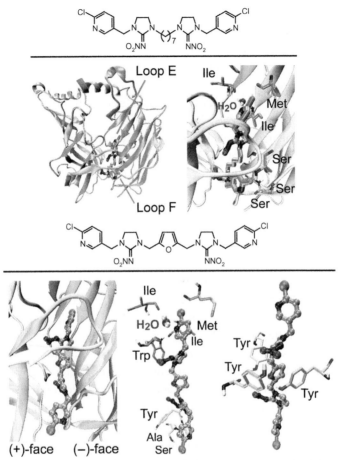

Figure 2.13 Predicted binding site interactions of bis-IMI insecticides bearing a heptamethylene linker (*upper*) and a furan-2,5-dimethylene fulcrum (*lower*) with the *Aplysia* AChBP (PDB ID code 3C79 (Talley et al., 2008)) as an insect nAChR homologue (Kagabu et al., 2010; Ohno et al., 2009c). The bis-IMIs are docked in the subunit interfacial binding pocket between the (+)-face (lime-green) and (−)-face (orange) subunits, wherein the chloropyridine moieties contact loops E and F domains. One of five interfacial binding pockets is extracted and shown from the homopentameric AChBP structure. The water molecule (Ohno et al., 2009a; Talley et al., 2008), bridging between pyridine nitrogen and related amino acids, is displayed. The heptamethylene spacer optimally bridges these two subsites, yet the linker itself binds in a relatively non-specific manner. In contrast, the furan ring is specifically embraced by a hydrophobic pocket, consisting of three aromatic amino acids, and is stabilized via H-bonding. (See Color Plate 13 at the back of the book.)

in AChBP (Carter et al., 2007) and also in acetylcholinesterase (Harel et al., 1993). It is, therefore, fascinating to consider the structural uniqueness of divalent neonicotinoid molecular recognition.

The IMI dimer with a heptamethylene linker, showing optimal binding potency among the C2–C10 alkylene-tethered analogues, interacts in a unique manner with the insect nAChR structural homologue. The chloropyridine moieties are embraced by two distinct and distant domains (loops E and F), and the heptamethylene spacer suitably bridges these two subsites in an optimal configuration of alkylene units, which is determined by regional binding domains involving various stabilization interactions (Fig. 2.13). Consistent with this simulation, the bis-phenyl analogue replacing the bis-chloropyridine substituents has no biological activity. Therefore, definitely one and possibly two chloropyridine substituents play an important role in the molecular recognition. The binding conformations seen in the molecular dynamics simulations show motions involving H-bonding interactions (forming and breaking) between the downward IMI moiety and the loop F niche. Other interactions in this area may be mediated via water(s) as these residues are near the protein surface (Ohno et al., 2009c).

The bis-IMI heptamethylene spacer itself binds in a relatively nonspecific manner, so inducing a further structural modification for the linker moiety. A bis-IMI analogue with a functional ether or heteroaromatic spacer moiety, which undergoes specific interaction(s) as a fulcrum in the unique receptor cavity, may accordingly enhance the potency relative to the alkylene-tethered derivatives. The compound with a diethylene ether spacer showed higher binding potency to the insect nAChR than the compound with a dipropylene ether, suggesting that an oxygen atom in an optimal position may play an important role in molecular recognition (Kagabu et al., 2010). Intriguingly, bis-IMI with a furan-2,5-dimethylene fulcrum is outstandingly potent, and the thiophene analogue also shows relatively high affinity (Kagabu et al., 2010). Therefore, unique and specific interactions are expected for a rational fulcrum, in addition to the two chloropyridine moieties, conceivably consolidating the binding interactions via the three major receptor cavities. *In silico* binding site interactions for the compound with a fran-2,5-dimethylene fulcrum were explored with the insect nAChR structural surrogate *Aplysia* AChBP (Fig. 2.13). Uniquely, the furan moiety is embraced by a hydrophobic pocket consisting of three aromatic side chains: that is, loop A Tyr, loop C Tyr, and loop D Tyr. The loop D Tyr of *Aplysia* AChBP is replaced by Trp in the nAChRs (the Tyr of AChBP is functionally consistent with the Trp of receptor in making hydrophobic interactions with furan ring). More importantly, the furan oxygen favourably H-bonds to the loop C Tyr OH, and in this manner stabilizes the ligand in the hydrophobic cavity. Therefore, these interactions are responsible for the high receptor potency of furan analogue.

8. CONCLUSIONS

The selective toxicity of neonicotinoids and nicotinoids to insects and mammals, respectively, is attributed to a large extent to differences in their interactions with the target site. Potency and selectivity of nicotinic insecticides have ultimately been defined by chemical biology approaches. The two AChBP subtypes proved to be suitable structural surrogates because they have differential sensitivity to the two chemotypes of nicotinic agonists, resulting in pharmacological profiles reminiscent of insect and mammalian nAChRs. For the neonicotinoids, the nitroguanidine or the cyanoamidine pharmacophore is nestled in a reversed position in the binding hole relative to the cationic functionality of nicotinoids. A single dominant binding orientation is attributable to the high affinity of neonicotinoids to the insect nAChR and to different pharmacophore positionings for nicotinoids at the mammalian nAChR. However, the inferior potency of neonicotinoids at the mammalian nAChR homologue is associated with rivalling multiple binding conformations in the agonist-binding pocket. This contributes to the poor binding constant, being reflected by a weighted average of a multiplicity of binding orientations. Outstandingly, potent and selective nicotinic insecticides were synthesized by pharmacophore modifications. Neonicotinoids with extended and/or hydrophobic substituents specifically fit the unique niche of the insect nAChR which is not accessible to the present nitroimino or cyanoimino pharmacophore. The SAR study of bis-IMI analogues with diverse spacer moieties reveals an intriguing molecular recognition mode involving the three major cavities (anchoring to the loops E and F domains and fitting a unique hydrophobic trap). Consequently, these findings in molecular recognition regarding potency and selectivity of nicotinic agonists facilitate further discovery of novel nicotinic insecticides with unique biological properties, high effectiveness, and maximal safety.

ACKNOWLEDGEMENTS

I greatly appreciate Dr. John E. Casida of the University of California at Berkeley for his enlightenment and everlasting supports. I also extend my sincere gratitude to collaborators from other research institutions who contributed to concepts and findings. I am deeply indebted to Dr. Shinzo Kagabu, the father of neonicotinoid insecticide, for his support and encouragement.

REFERENCES

Bass, C., Puinean, A.M., Andrews, M., Cutler, P., Daniels, M., Elias, J., Paul, V.L., Crossthwaite, A.J., Denholm, I., Field, L.M., Foster, S.P., Lind, R., Williamson, M.S., Slater, R., 2011. Mutation of a nicotinic acetylcholine receptor β subunit is associated with resistance to neonicotinoid insecticides in the aphid *Myzus persicae*. BMC Neurosci. 12, 51.

Bertrand, D., Ballivet, M., Gomez, M., Bertrand, S., Phannavong, B., Gundelfinger, E.D., 1994. Physiological properties of neuronal nicotinic receptors reconstituted from the vertebrate β2 subunit and *Drosophila* α subunits. Eur. J. Neurosci. 6, 869–875.

Bissantz, C., Kuhn, B., Stahl, M., 2010. A medicinal chemist's guide to molecular interactions. J. Med. Chem. 53, 5061–5084.

Bouzat, C., Gumilar, F., Spitzmaul, G., Wang, H.-L., Rayes, D., Hansen, S.B., Taylor, P., Sine, S.M., 2004. Coupling of agonist binding to channel gating in an ACh binding protein linked to an ion channel. Nature 430, 896–900.

Brejc, K., van Dijk, W.J., Klaassen, R.V., Schuurmans, M., van der Oost, J., Smit, A.B., Sixma, T.K., 2001. Crystal structure of an ACh-binding protein reveals the ligand binding domain of nicotinic receptors. Nature 411, 269–276.

Carter, C.R.J., Cao, L., Kawai, H., Smith, P.A., Dryden, W.F., Raftery, M.A., Dunn, S.M.J., 2007. Chain length dependence of the interactions of bisquaternary ligands with the *Torpedo* nicotinic acetylcholine receptor. Biochem. Pharmacol. 73, 417–426.

Casida, J.E., 2009. Pest toxicology: the primary mechanisms of pesticide action. Chem. Res. Toxicol. 22, 609–619.

Casida, J.E., Quistad, G.B., 1998. Golden age of insecticide research: past, present, or future. Annu. Rev. Entomol. 43, 1–16.

Celie, P.H.N., van Rossum-Fikkert, S.E., van Dijk, W.J., Brejc, K., Smit, A.B., Sixma, T.K., 2004. Nicotine and carbamylcholine binding to nicotinic acetylcholine receptors as studied in AChBP crystal structures. Neuron 41, 907–914.

Chamberlain, K., Evans, A.A., Bromilow, R.H., 1996. 1-Octanol/water partition coefficient (K_{ow}) and pK_a for ionisable pesticides measured by a pH-metric method. Pestic. Sci. 47, 265–271.

Changeux, J.-P., Edelstein, S.J., 2005. Nicotinic Acetylcholine Receptors: From Molecular Biology to Cognition. Odile Jacob, New York.

Corbel, V., Duchon, S., Zaim, M., Hougard, J.-M., 2004. Dinotefuran: a potential neonicotinoid insecticide against resistant mosquitoes. J. Med. Entomol. 41, 712–717.

Dougherty, D.A., 2008. Cys-loop neuroreceptors: structure to the rescue. Chem. Rev. 108, 1642–1653.

Dutertre, S., Ulens, C., Büttner, R., Fish, A., van Elk, R., Kendel, Y., Hopping, G., Alewood, P.F., Schroeder, C., Nicke, A., Smit, A.B., Sixma, T.K., Lewis, R.J., 2007. AChBP-targeted α-conotoxin correlates distinct binding orientations with nAChR subtype selectivity. EMBO J. 26, 3858–3867.

ffrench-Constant, R.H., Daborn, P.J., Goff, G.L., 2004. The genetics and genomics of insecticide resistance. Trends Genet. 20, 163–170.

Gao, F., Bren, N., Little, A., Wang, H.-L., Hansen, S.B., Talley, T.T., Taylor, P., Sine, S.M., 2003. Curariform antagonists bind in different orientations to acetylcholine-binding protein. J. Biol. Chem. 278, 23020–23026.

Gao, F., Bren, N., Burghardt, T.P., Hansen, S., Henchman, R.H., Taylor, P., McCammon, J.A., Sine, S.M., 2005. Agonist mediated conformational changes in acetylcholine-binding protein revealed by simulation and intrinsic tryptophan fluorescence. J. Biol. Chem. 280, 8443–8451.

Gao, F., Mer, G., Tonelli, M., Hansen, S.B., Burghardt, T.P., Taylor, P., Sine, S.M., 2006. Solution NMR of acetylcholine binding protein reveals agonist-mediated conformational change of the C-loop. Mol. Pharmacol. 70, 1230–1235.

Grutter, T., Bertrand, S., Kotzyba-Hibert, F., Bertrand, D., Goeldner, M., 2002. Structural reorganization of the acetylcholine binding site of the *Torpedo* nicotinic receptor as revealed by dynamic photoaffinity labeling. Chembiochem 3, 652–658.

Hansen, S.B., Sulzenbacher, G., Huxford, T., Marchot, P., Taylor, P., Bourne, Y., 2005. Structures of *Aplysia* AChBP complexes with nicotinic agonists and antagonists reveal distinctive binding interfaces and conformations. EMBO J. 24, 3635–3646.

Harel, M., Schalk, I., Ehret-Sabatier, L., Bouet, F., Goeldner, M., Hirth, C., Axelsen, P.H., Silman, I., Sussman, J.L., 1993. Quaternary ligand binding to aromatic residues in the active-site gorge of acetylcholinesterase. Proc. Natl. Acad. Sci. U.S.A. 90, 9031–9035.

Hibbs, R.E., Sulzenbacher, G., Shi, J., Talley, T.T., Conrod, S., Kem, W.R., Taylor, P., Marchot, P., Bourne, Y., 2009. Structural determinants for interaction of partial agonists with acetylcholine binding protein and neuronal α7 nicotinic acetylcholine receptor. EMBO J. 28, 3040–3051.

Ihara, M., Hirata, K., Ishida, C., Kagabu, S., Matsuda, K., 2007. Blocking actions of alkylene-tethered bis-neonicotinoids on nicotinic acetylcholine receptors expressed by terminal abdominal ganglion neurons of *Periplaneta americana*. Neurosci. Lett. 425, 137–140.

Ihara, M., Okajima, T., Yamashita, A., Oda, T., Hirata, K., Nishiwaki, H., Morimoto, T., Akamatsu, M., Ashikawa, Y., Kuroda, S., Mega, R., Kuramitsu, S., Sattelle, D.B., Matsuda, K., 2008. Crystal structures of *Lymnaea stagnalis* AChBP in complex with neonicotinoid insecticides imidacloprid and clothianidin. Invert. Neurosci. 8, 71–81.

Jeschke, P., Nauen, R., Schindler, M., Elbert, A., 2011. Overview of the status and global strategy for neonicotinoids. J. Agric. Food Chem. 59, 2897–2908.

Kagabu, S., 2011. Discovery of imidacloprid and further developments from strategic molecular designs. J. Agric. Food Chem. 59, 2887–2896.

Kagabu, S., Matsuno, H., 1997. Chloronicotinyl insecticides. 8. Crystal and molecular structures of imidacloprid and analogous compounds. J. Agric. Food Chem. 45, 276–281.

Kagabu, S., Maienfisch, P., Zhang, A., Granda-Minones, J., Haettenschwiler, J., Kayser, H., Maetzke, T., Casida, J.E., 2000. 5-Azidoimidacloprid and an acyclic analogue as candidate photoaffinity probes for mammalian and insect nicotinic acetylcholine receptors. J. Med. Chem. 43, 5003–5009.

Kagabu, S., Iwaya, K., Konishi, H., Sakai, A., Itazu, Y., Kiriyama, K., Nishimura, K., 2002. Synthesis of alkylene-tethered bis-imidacloprid derivatives as highly insecticidal and nerve-exciting agents with potent affinity to [^{3}H]imidacloprid-binding sites on nicotinic acetylcholine receptor. J. Pestic. Sci. 27, 249–256.

Kagabu, S., Itazu, Y., Nishimura, K., 2004. Preparation of alkylenetethered acyclic divalent neonicotinoids and their insecticidal and neuroblocking activities for American cockroach (*Periplaneta americana* L.). J. Pestic. Sci. 29, 40–42.

Kagabu, S., Itazu, Y., Nishimura, K., 2006. 1,6-Bis[1-(2-chloro- 5-thiazolylmethyl)-2-nitroiminoimidazolidin-3-yl]hexane and 1,3,5-tris[1- (6-chloronicotinyl)-2-nitroiminoimidazolidin-3-ylmethyl]benzene-synthesis and insecticidal and neuroblocking activities in American cockroaches *Periplaneta americana*. J. Pestic. Sci. 31, 146–149.

Kagabu, S., Ohno, I., Tomizawa, M., Durkin, K.A., Matsuura, R., Uchiyama, D., Nagae, N., Kumazawa, S., 2010. Furan-2,5-dimethylene-tethered bis-imidacloprid insecticide conferring high potency. J. Agric. Food Chem. 58, 11832–11836.

Kagabu, S., Mitomi, M., Kitsuda, S., Horikoshi, R., Nomura, M., Onozaki, Y. 2012. Preparation of amine compounds as noxious organism control agents. PCT Int. Appl. WO 2012029672.

Karunker, I., Morou, E., Nikou, D., Nauen, R., Sertchook, R., Stevenson, B.J., Paine, M.J.I., Morin, S., Vontas, J., 2009. Structural model and functional characterization of the *Bemisia tabaci* CYP6CM1vQ, a cytochrome P450 associated with high levels of imidacloprid resistance. Insect Biochem. Mol. Biol. 39, 697–706.

Kollmeyer, W.D., Flattum, R.F., Foster, J.P., Powell, J.E., Schroeder, M.E., Soloway, S.B., 1999. Discovery of the nitromethylene heterocycle insecticides. In: Yamamoto, I., Casida, J.E. (Eds.), Nicotinoid Insecticides and the Nicotinic Acetylcholine Receptor. Springer, Tokyo, pp. 71–89.

Lansdell, S.J., Millar, N.S., 2000. The influence of nicotinic receptor subunit composition upon agonist, α-bungarotoxin and insecticide (imidacloprid) binding affinity. Neuropharmacology 39, 671–679.

Latli, B., Tomizawa, M., Casida, J.E., 1997. Synthesis of a novel [^{125}I]neonicotinoid photoaffinity probe for the *Drosophila* nicotinic acetylcholine receptor. Bioconjug. Chem. 8, 7–14.

Liu, H., Cupp, E.W., Guo, A., Liu, N., 2004. Insecticide resistance in Alabama and Florida mosquito strains of *Aedes albopictus*. J. Med. Entomol. 41, 946–952.

Liu, Z., Williamson, M.S., Lansdell, S.J., Denholm, I., Han, Z., Millar, N.S., 2005. A nicotinic acetylcholine receptor mutation conferring target-site resistance to imidacloprid in *Nilaparvata lugens* (brown planthopper). Proc. Natl. Acad. Sci. U.S.A. 102, 8420–8425.

Matsuda, K., Kananoka, S., Akamatsu, M., Sattelle, D.B., 2009. Diverse actions and target-site selectivity of neonicotinoids: structural insights. Mol. Pharmacol. 76, 1–10.

McGaughey, G.B., Gagné, M., Rappé, A.K., 1998. π-Staking interactions: alive and well in proteins. J. Biol. Chem. 273, 15458–15463.

Millar, N.S., Denholm, I., 2007. Nicotinic acetylcholine receptors: targets for commercially important insecticides. Invert. Neurosci. 7, 53–66.

Mori, M., Kikuchi, M., Ohno, I., Kagabu, S., 2010. Unexpected plant-systemic insecticidal properties of alkylene-tethered bis-imidacloprid derivatives. J. Pestic. Sci. 35, 124–130.

Mourot, A., Rodrigo, J., Kotzyba-Hibert, F., Bertrand, S., Bertrand, D., Goeldner, M., 2006. Probing the reorganization of the nicotinic acetylcholine receptor during desensitization by time-resolved covalent labeling using [^{3}H]AC5, a photoactivatable agonist. Mol. Pharmacol. 69, 452–461.

Mu, T.-W., Lester, H.A., Dougherty, D.A., 2003. Different binding orientations for the same agonist at homologous receptors: a lock and key or a simple wedge. J. Am. Chem. Soc. 125, 6850–6851.

Nirthanan, S., Ziebell, M.R., Chiara, D.C., Hong, F., Cohen, J.B., 2005. Photolabeling the *Torpedo* nicotinic acetylcholine receptor with 4-azido-2,3,5,6-tetrafluorobenzoylcholine, a partial agonist. Biochemistry 44, 13447–13456.

Ohno, I., Tomizawa, M., Durkin, K.A., Casida, J.E., Kagabu, S., 2009a. Neonicotinoid substituents forming a water-bridge at the nicotinic acetylcholine receptor. J. Agric. Food Chem. 57, 2436–2440.

Ohno, I., Tomizawa, M., Durkin, K.A., Naruse, Y., Casida, J.E., Kagabu, S., 2009b. Molecular features of neonicotinoid pharmacophore variants interacting with the insect nicotinic receptor. Chem. Res. Toxicol. 22, 476–482.

Ohno, I., Tomizawa, M., Durkin, K.A., Casida, J.E., Kagabu, S., 2009c. Bis-neonicotinoid insecticides: observed and predicted binding interactions with the nicotinic receptor. Bioorg. Med. Chem. Lett. 19, 3449–3452.

Ohno, I., Tomizawa, M., Aoshima, A., Kumazawa, S., Kagabu, S., 2010a. Trifluoroacetyl neonicotinoid insecticides with enhanced hydrophobicity and effectiveness. J. Agric. Food Chem. 58, 4999–5003.

Ohno, I., Tomizawa, M., Miyazu, N., Kushibiki, G., Noda, K., Hasebe, Y., Durkin, K.A., Miyake, T., Kagabu, S., 2010b. Structural features of phenoxycarbonylimino neonicotinoids acting at the insect nicotinic receptor. Bioorg. Med. Chem. Lett. 20, 5933–5935.

Padgett, C.L., Hanek, A.P., Lester, H.A., Dougherty, D.A., Lummis, S.C.R., 2007. Unnatural amino acid mutagenesis of the GABA$_A$ receptor binding site residues reveals a novel cation-π interaction between GABA and β2Tyr97. J. Neurosci. 27, 886–892.

Puinean, A.M., Foster, S.P., Oliphant, L., Denholm, I., Field, L.M., Millar, N.S., Williamson, M.S., Bass, C., 2010. Amplification of a cytochrome P450 gene is associated with resistance to neonicotinoid insecticides in the aphid Myzus persicae. PLoS Genet. 6 (6), e1000999.

Shao, X., Lee, P.W., Liu, Z., Xu, X., Li, Z., Qian, X., 2011. cis-Configuration: a new tactic/rationale for neonicotinoid molecular design. J. Agric. Food Chem. 59, 2943–2949.

Shi, J., Koeppe, J.R., Komives, E.A., Taylor, P., 2006. Ligand-induced conformational changes in the acetylcholine binding protein analyzed by hydrogen–deuterium exchange mass spectrometry. J. Biol. Chem. 281, 12170–12177.

Shimomura, M., Yokota, M., Ihara, M., Akamatsu, M., Sattelle, D.B., Matsuda, K., 2006. Role in the selectivity of neonicotinoids of insect-specific basic residues in loop D of the nicotinic acetylcholine receptor agonist binding site. Mol. Pharmacol. 70, 1255–1263.

Spande, T.F., Garraffo, H.M., Edwards, M.W., Yeh, H.J.C., Pannell, L., Daly, J.W., 1992. Epibatidine: a novel (chloropyridyl)azabicycloheptane with potent analgesic activity from an Ecuadoran poison frog. J. Am. Chem. Soc. 114, 3475–3478.

Talley, T.T., Olivera, B.M., Han, K.-H., Christensen, S.B., Dowell, C., Tsigelny, I., Ho, K.-Y., Taylor, P., McIntosh, J.M., 2006. α-Conotoxin OmIA is a potent ligand for the acetylcholine-binding protein as well as α3β2 and α7 nicotinic acetylcholine receptors. J. Biol. Chem. 281, 24678–24686.

Talley, T.T., Harel, M., Hibbs, R.H., Radić, Z., Tomizawa, M., Casida, J.E., Taylor, P., 2008. Atomic interactions of neonicotinoid agonists with AChBP: molecular recognition of the distinctive electronegative pharmacophore. Proc. Natl. Acad. Sci. U.S.A. 105, 7606–7611.

Thany, S.H., Lenaers, G., Raymond-Delpech, V., Sattelle, D.B., Lapied, B., 2006. Exploring the pharmacological properties of insect nicotinic acetylcholine receptors. Trends Pharmacol. Sci. 28, 14–22.

Tomizawa, M., Casida, J.E., 1997. [^{125}I]Azidonicotinoid photoaffinity labeling of insecticide binding subunit of Drosophila nicotinic acetylcholine receptor. Neurosci. Lett. 237, 61–64.

Tomizawa, M., Casida, J.E., 1999. Minor structural changes in nicotinoid insecticides confer differential subtype selectivity for mammalian nicotinic acetylcholine receptors. Br. J. Pharmacol. 127, 115–122.

Tomizawa, M., Casida, J.E., 2001. Structure and diversity of insect nicotinic acetylcholine receptors. Pest Manag. Sci. 57, 914–922.

Tomizawa, M., Casida, J.E., 2003. Selective toxicity of neonicotinoids attributable to specificity of insect and mammalian nicotinic receptors. Annu. Rev. Entomol. 48, 339–364.

Tomizawa, M., Casida, J.E., 2005. Neonicotinoid insecticide toxicology: mechanisms of selective action. Annu. Rev. Pharmacol. Toxicol. 45, 247–268.

Tomizawa, M., Casida, J.E., 2009. Molecular recognition of neonicotinoid insecticides: the determinants of life or death. Acc. Chem. Res. 42, 260–269.

Tomizawa, M., Casida, J.E., 2011. Unique neonicotinoid binding conformations conferring selective receptor interactions. J. Agric. Food Chem. 59, 2825–2828.

Tomizawa, M., Yamamoto, I., 1992. Binding of nicotinoids and the related compounds to the insect nicotinic acetylcholine receptor. J. Pestic. Sci. 17, 231–236.

Tomizawa, M., Yamamoto, I., 1993. Structure–activity relationships of nicotinoids and imidacloprid analogs. J. Pestic. Sci. 18, 91–98.

Tomizawa, M., Latli, B., Casida, J.E., 1996. Novel neonicotinoid-agarose affinity column for Drosophila and Musca nicotinic acetylcholine receptors. J. Neurochem. 67, 1669–1676.

Tomizawa, M., Lee, D.L., Casida, J.E., 2000. Neonicotinoid insecticides: molecular features conferring selectivity for insect versus mammalian nicotinic receptors. J. Agric. Food Chem. 48, 6016–6024.

Tomizawa, M., Cowan, A., Casida, J.E., 2001a. Analgesic and toxic effects of neonicotinoid insecticides in mice. Toxicol. Appl. Pharmacol. 177, 77–83.

Tomizawa, M., Wen, Z., Chin, H.-L., Morimoto, H., Kayser, H., Casida, J.E., 2001b. Photoaffinity labeling of insect nicotinic acetylcholine receptors with a novel [³H]azidoneonicotinoid. J. Neurochem. 78, 1359–1366.

Tomizawa, M., Zhang, N., Durkin, K.A., Olmstead, M.M., Casida, J.E., 2003. The neonicotinoid electronegative pharmacophore plays the crucial role in the high affinity and selectivity for the *Drosophila* nicotinic receptor: an anomaly for the nicotinoid cation-π interaction model. Biochemistry 42, 7819–7827.

Tomizawa, M., Millar, N.S., Casida, J.E., 2005. Pharmacological profiles of recombinant and native insect nicotinic acetylcholine receptors. Insect Biochem. Mol. Biol. 35, 1347–1355.

Tomizawa, M., Talley, T.T., Maltby, D., Durkin, K.A., Medzihradszky, K.F., Burlingame, A.L., Taylor, P., Casida, J.E., 2007a. Mapping the elusive neonicotinoid binding site. Proc. Natl. Acad. Sci. U.S.A. 104, 9075–9080.

Tomizawa, M., Maltby, D., Medzihradszky, K.F., Zhang, N., Durkin, K.A., Presley, J., Talley, T.T., Taylor, P., Burlingame, A.L., Casida, J.E., 2007b. Defining nicotinic agonist binding surfaces through photoaffinity labeling. Biochemistry 46, 8798–8806.

Tomizawa, M., Maltby, D., Talley, T.T., Durkin, K.A., Medzihradszky, K.F., Burlingame, A.L., Taylor, P., Casida, J.E., 2008a. Atypical nicotinic agonist bound conformations conferring subtype selectivity. Proc. Natl. Acad. Sci. U.S.A. 105, 1728–1732.

Tomizawa, M., Kagabu, S., Ohno, I., Durkin, K.A., Casida, J.E., 2008b. Potency and selectivity of trifluoroacetylimino and pyrazinoylimino nicotinic insecticides and their fit at a unique binding site niche. J. Med. Chem. 51, 4213–4218.

Tomizawa, M., Talley, T.T., Park, J.F., Maltby, D., Medzihradszky, K.F., Durkin, K.A., Cornejo-Bravo, J.M., Burlingame, A.L., Casida, J.E., Taylor, P., 2009. Nicotinic agonist binding site mapped by methionine- and tyrosine-scanning coupled with azidochloropyridinyl photoaffinity labeling. J. Med. Chem. 52, 3735–3741.

Tomizawa, M., Kagabu, S., Casida, J.E., 2011a. Receptor structure-guided neonicotinoid design. J. Agric. Food Chem. 59, 2918–2922.

Tomizawa, M., Durkin, K.A., Ohno, I., Nagura, K., Manabe, M., Kumazawa, S., Kagabu, S., 2011b. N-Haloacetylimino neonicotinoids: potency and molecular recognition at the insect nicotinic receptor. Bioorg. Med. Chem. Lett. 21, 3583–3586.

Ulens, C., Hogg, R.C., Celie, P.H., Bertrand, D., Tsetlin, V., Smit, A.B., Sixma, T.K., 2006. Structural determinants of selective α-conotoxin binding to a nicotinic acetylcholine receptor homolog AChBP. Proc. Natl. Acad. Sci. U.S.A. 103, 3615–3620.

Unwin, N., 2005. Refined structure of the nicotinic acetylcholine receptor at 4 Å resolution. J. Mol. Biol. 346, 967–989.

Wang, Y., Cheng, J., Qian, X., Li, Z., 2007. Actions between neonicotinoids and key residues of insect nAChR based on an *ab initio* quantum chemistry study: hydrogen bonding and cooperative π-π interaction. Bioorg. Med. Chem. 15, 2624–2630.

Yamamoto, I., 1965. Nicotinoids as insecticides. In: Metcalf, R.L. (Ed.), Advances in Pest Control Research, vol. 6. Wiley, New York, pp. 231–260.

Yamamoto, I., Yabuta, G., Tomizawa, M., Saito, T., Miyamoto, T., Kagabu, S., 1995. Molecular mechanism for selective toxicity of nicotinoids and neonicotinoids. J. Pestic. Sci. 20, 33–40.

Yamamoto, I., Tomizawa, M., Saito, T., Miyamoto, T., Walcott, E.C., Sumikawa, K., 1998. Structural factors contributing to insecticidal and selective actions of neonicotinoids. Arch. Insect Biochem. Physiol. 37, 24–32.

Yao, X., Song, F., Chen, F., Zhang, Y., Gu, J., Liu, S., Liu, Z., 2008. Amino acids within loops D, E, and F of insect nicotinic acetylcholine receptor β subunits influence neonicotinoid selectivity. Insect Biochem. Mol. Biol. 38, 834–840.

Zhang, N., Tomizawa, M., Casida, J.E., 2002. Structural features of azidopyridinyl neonicotinoid probes conferring high affinity and selectivity for mammalian $\alpha4\beta2$ and *Drosophila* nicotinic receptors. J. Med. Chem. 45, 2832–2840.

Zhang, N., Tomizawa, M., Casida, J.E., 2003. 5-Azidoepibatidine: an exceptionally potent photoaffinity ligand for neuronal $\alpha4\beta2$ and $\alpha7$ nicotinic acetylcholine receptors. Bioorg. Med. Chem. Lett. 13, 525–527.

Zhu, Y., Loso, M.R., Watson, G.B., Sparks, T.C., Rogers, R.B., Huang, J.X., Gerwick, B.C., Babcock, J.M., Kelley, D., Hegde, V.B., Nugent, B.M., Renga, J.M., Denholm, I., Gorman, K., DeBoer, G.J., Hasler, J., Meade, T., Thomas, J.D., 2011. Discovery and characterization of sulfoxaflor, a novel insecticide targeting sap-feeding pests. J. Agric. Food Chem. 59, 2950–2957.

CHAPTER THREE

Nicotinic Acetylcholine Receptors as Spinosyn Targets for Insect Pest Management

Chaoxian Geng, Gerald B. Watson, Thomas C. Sparks
Dow AgroSciences, Discovery Research, Indianapolis, Indiana 46268, USA

Contents

Advances in Insect Physiology, Volume 44
ISBN 978-0-12-394389-7
http://dx.doi.org/10.1016/B978-0-12-394389-7.00003-X

101

Abstract

The spinosyns are insecticidal natural products originated from *Saccharopolyspora spinosa*. Spinosad, the first commercial product, is a mixture of two naturally occurring spinosyns (A & D). A second product, spinetoram, is a mixture composed of two synthetically modified spinosyns that, compared to spinosad, provides improved insecticidal potency and a broader pest insect spectrum. The spinosyns act on a subgroup of insect nicotinic acetylcholine receptors (nAChR), which are distinct from those targeted by the neonicotinoid insecticides. Selection of *Drosophila* for resistance to spinosad helped pinpoint the Dα6 as the nAChR subunit involved in the insecticidal action of the spinosyns. Cases of both target-site- and metabolism-based spinosyn resistance in insect pests in the field have been reported, and programmes have been developed to manage spinosyn resistance. This review highlights the discovery of the spinosyns, elucidation of the spinosyn target site, spinosyn–receptor interaction, and progress in spinosyn resistance management.

ABBREVIATIONS

1KITE 1000 Insect Transcriptome Evolution
ACh acetylcholine
AChBP acetylcholine binding protein
ClogP calculated logP
icd incompletely dominant
IPM Integrated Pest Management
IRAC Insecticide Resistance Action Committee
IRM Insecticide resistance management
JH juvenile hormone
MET mitochondrial electron transport
MLR multiple linear regression
MR molecular radius
MSU Michigan State University
nAChR nicotinic acetylcholine receptor
NP natural product
QSAR quantitative structure–activity relationships
rec recessive
RR resistance ratio
SAR structure–activity relationship
SPAM average molecular span

1. INTRODUCTION

In the past 20 years, insect nicotinic acetylcholine receptors (nAChRs) have risen in prominence as targets for insecticides. Today, a number of commercial classes of insecticides, including the neonicotinoids, nereistoxin analogues, spinosyns, and sulfoximines (Sparks et al., 2012; Zhu et al., 2011), all target insect nAChRs in some way (Millar and Denholm, 2007). As target sites for insecticides, the nAChRs currently account for the largest share of the total global insecticide market (Fig. 3.1), primarily due to the preeminence of the neonicotinoids for sap-feeding insect control (Jeschke and Nauen, 2008). The market share for the nAChR-targeting insecticides is far larger than either of the next two most prevalent target sites, the enzyme acetylcholinesterase which is the target site for the organophosphate and the carbamate insecticides, and the voltage-gated sodium channel, the target site for the pyrethroids and sodium channel blockers (e.g., indoxacarb and metaflumizone).

An expanding segment of the nAChR-acting insecticides is the spinosyn class of insecticides, which is composed of two products, spinosad and spinetoram (Figs. 3.2 and 3.3). More than 20 years after their discovery, the spinosyns remain a unique class of insect control agents, possessing very unusual chemical structures (Kirst et al., 1992; Sparks et al., 1999), coupled with a novel mode of action at the nAChR (Orr et al., 2009; Watson et al., 2010). Equally

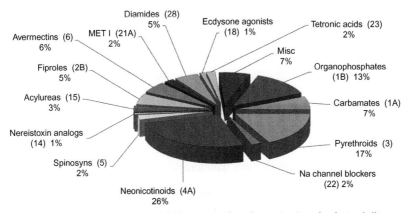

Figure 3.1 Percentage of total global insecticide sales 2010 [total = $12.6 billion end user dollars]. Numbers in ( ) are IRAC mode of action group classification. *Source: Agranova (2012).* (See Color Plate 14 at the back of the book.)

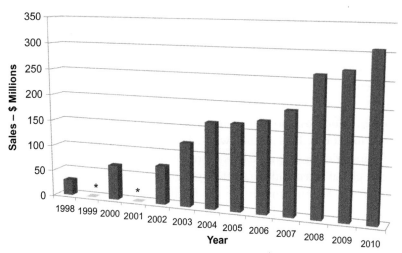

Figure 3.2 Combined sales of spinosad and spinetoram. *No data available for 1999 or 2001. *Source: Agranova (2012).*

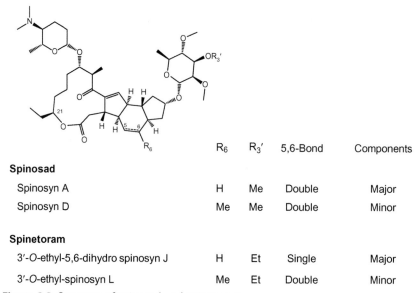

	R$_6$	R$_3'$	5,6-Bond	Components
Spinosad				
Spinosyn A	H	Me	Double	Major
Spinosyn D	Me	Me	Double	Minor
Spinetoram				
3'-O-ethyl-5,6-dihydro spinosyn J	H	Et	Single	Major
3'-O-ethyl-spinosyn L	Me	Et	Double	Minor

Figure 3.3 Structure of spinosad and spinetoram.

important, the spinosyns also exhibit attributes that are increasingly critical for insect control agents, including high levels of efficacy against a broad range of important insect pests, low impact on many beneficial insect species, and foremost, low impact on the environment (Dripps et al., 2011; Thompson et al., 2000). This review will explore the mode of action of the spinosyns, including the elucidation of their nAChR target site, the discovery of the current spinosyn commercial molecules, and facets of the structure–activity relationships (SARs) for this chemistry. Also covered are aspects of resistance to the spinosyns, cross-resistance, and resistance management.

2. DISCOVERY OF nAChRs

The discovery of vertebrate nAChRs has been recounted in an excellent book by Changeux and Edelstein (2005). One of the most significant accomplishments towards molecular identification of nAChRs was the isolation of the functional membrane fragments from the electric organs of *Electrophorus electricus* and *Torpedo marmorata*, which were recognized as rich sources of nAChRs. The isolated membrane fragments resealed *in vitro* to form closed membrane vesicles, or "microsacs" as revealed by electron microscopy. These microsacs retained the capability of increasing their permeability to Na^+ and K^+ in response to cholinergic agonists. Moreover, the permeability response could be blocked by known nAChR antagonists and cation channel blockers. These findings indicated that the microsacs retained virtually all the properties of the intact nAChRs and provided a foundation for receptor purification and further characterization. Extraction of nAChRs from the isolated membrane fragments under nondenaturing conditions helped preserve the biological activity of the nAChRs. It was demonstrated that the purified nAChRs, when reconstituted into membrane structures, retained the specific ligand-binding capability and displayed cation ion selectivity. These studies firmly established that the purified receptor oligomer contained not only the binding site for acetylcholine (Ach) but also the associated ion channel. In the presence of sodium dodecyl sulphate, the purified receptor complex dissociated into several subunits, as demonstrated by electrophoresis (Meunier et al., 1972; Reiter et al., 1972). It was subsequently discovered that there were several different nAChR subunits (Weill et al., 1974) and these could form heteropentamers (Karlin, 2002; Raftery et al., 1980). Protein micro sequencing of the purified subunits of *Torpedo californica* demonstrated that the pentamers contained four different peptides (α, β, γ, and δ) with high homology at their amino terminal portions where sequence data were available (Raftery et al., 1980). Molecular cloning and cDNA sequencing revealed the primary structures of the nAChR subunits of *Torpedo* (Numa et al., 1983), and soon nucleic acid sequencing helped reveal the primary sequences of calf and human nAChR subunits as well (Noda et al., 1983).

3. INSECT nAChRs

Advancement in genome sequencing technology and sequence analysis tools incited a swift expansion of the understanding of the nAChR families of vertebrates, as well as that of invertebrates (Jones and Sattelle, 2010; Sattelle, 2009). In comparison with nAChR families of vertebrates, or that of *Caenorhabditis elegans*, insect nAChR families are often compact with, for example, seven α subunits and three β subunits identified in the fruit fly *Drosophila melanogaster* genome (Sattelle et al., 2005). However, alternative splicing and RNA editing provided the possibility for large diversity of individual subunits (Grauso et al., 2002; Sattelle et al., 2005). Moreover, various combinations of the subunits could form different functional heteropentamers (Sattelle et al., 2005), which could serve as unique targets for structurally distinct compounds (see Sections 8.3 and 8.4). Comparison of the nAChRs in several insect species, where genome data were available, revealed that in addition to the shared subunit groups among these insect species there were obvious divergent subunits (Sattelle, 2009). Because some of the nAChRs play a critical role in the fast synaptic transmission, they have proven to be important targets for insect control agents (Jeschke and Nauen, 2012; Millar and Denholm, 2007). The spinosyns are one class of insecticides targeting nAChRs (Watson et al., 2010). Although neonicotinoids also target insect nAChRs, target-site cross-resistance has not been observed between spinosyn targets and neonicotinoids targets (see below; also see Chapter 2). Indeed, the neonicotinoids, spinosyns, and nereistoxin analogues have all been placed in separate mode of action groupings by the Insecticide Resistance Action Committee (IRAC) (Nauen et al., 2012) because of their distinct actions on nAChRs.

4. DISCOVERY OF NATURAL INSECTICIDAL SPINOSYNS

Natural products have been and continue to be good sources of insecticidal compounds and often serve as models or inspiration for new synthetic insecticides (Table 3.1). In addition, there are a number of classes of insecticides that were discovered through other approaches, which hypothetically could also have originated from a known natural product model (Table 3.1). The discovery of new insecticidal natural products is difficult and can be quite a lengthy process. The spinosyns exemplify the successful exploitation of natural products for the discovery of new and unique insect control agents. The discovery of the spinosyns was the result of the

Table 3.1 Natural products as sources and/or models for insecticides

IRAC group	Class	NP or NP model	NP based?	Example—synthetic or semisynthetic product
1A	Carbamates	Physostigmine	Yes	Propoxur
1B	Organophosphates	Hydantoin phosphonate	No	Chlorpyrifos
2A	Cyclodiene	Picrotoxinin	No	Chlordane
3A	Pyrethroids	**Pyrethrum**	Yes	Cypermethrin
4A	Neonicotinoids	**Nicotine**	No	Imidacloprid
5	***Spinosyns***	***Spinosad***	*Yes*	***Spinetoram***
6	Avermectins	**Abamectin**	Yes	**Emamectin benzoate**
6	Mibemycins	**Milbemctin**	Yes	**Lepimectin**
7	JH mimics (juvenoids)	Juvenile hormone	Yes	Methoprene
13	Uncouplers	Deoxypyrrolomycin	Yes	Chlorfenapyr
14	Nereistoxin analogues	Nereistoxin	Yes	Cartap
18	Ecdysone agonists	20-OH-ecdysone	No	Methoxyfenozide
19	Octopamine agonists	Octopamine	No	Amitraz
20C	MET III acaricides	β-Methoxyacrylates	Yes	Fluacrypyrim
21	MET I acaricides	**Rotenone**	No	Fenazaquin
28	Diamides	**Ryania**		No
UN	Azadirachtin	**Azadirachtin**	–	None
–	Sabadilla	**Sabadilla**	–	None

Bold designates natural product or natural product-based insecticides.

convergence of two critical components important in any successful natural products-based discovery programme. The first was a novel source of natural products, and the second was the application of a novel screening method. In the case of the spinosyns, the novel source was a soil sample from an abandoned sugar mill rum still in the Virgin Islands (Thompson et al., 2000). After fermentation of the soil sample and extraction of the fermentation broth, a sample of the broth was found to be insecticidally active when assayed in a, as then, new miniaturized mosquito (the yellow fever mosquito, *Aedes aegypti*) larvicide bioassay (Kirst, 2010; Kirst et al., 1992; Thompson and Sparks, 2002; Thompson et al., 2000). The microorganism producing the insecticidally active broth was determined to be a newly discovered species of actinomycete, *Saccharopolyspora spinosa* (Kirst, 2010; Kirst et al., 1992; Mertz and Yao, 1990; Thompson et al., 2000). Characterization of the insecticidal factors led to the identification of a new class of macrocyclic lactones possessing a novel tetracycle composed of a 12-member macrocycle fused to a 5-6-5 ring system with two sugars attached (Kirst et al., 1992; Sparks et al., 1999; Fig. 3.3). Multiple factors (later called spinosyns) were identified from the *S. spinosa* fermentation broth (Kirst et al., 1992). The two most abundant and active of these naturally occurring spinosyns were spinosyns A and D, which are the principal components of the commercial product spinosad (Kirst, 2010; Sparks et al., 1999; Thompson et al., 2000). The discovery of the spinosyns, and ultimate development of spinosad, encompassed 15 years from the collection of the soil sample in 1982 to the initial insecticide registration in 1997 (Thompson et al., 2000). At the time of its introduction, spinosad provided a level of efficacy equivalent to many synthetic insecticides, coupled with very favourable mammalian and environmental toxicological profiles. These favourable attributes were ultimately recognized by the Environmental Protection Agency, which awarded spinosad the Designing Greener Chemicals Award in 1999.

5. IDENTIFICATION OF SPINOSYN TARGET

5.1. Chemical mutagenesis using *Drosophila*

Aimed at discovering the spinosyn target site, scientists explored chemical mutagenesis as an approach to generate spinosyn-resistant mutants (Orr et al., 2006). The model organism, *D. melanogaster*, was chosen for mutagenesis primarily because of the availability of well-developed molecular and genetics tools. Moreover, spinosyns were highly potent against *D. melanogaster* causing mortality with a rather steep dose-dependent response. Therefore, under the chosen discriminating doses, few of the wild-type flies could survive, making the mutant selection process rather straightforward. Chemical mutagenesis using ethyl methane sulfonate was successfully conducted and selection with spinosad recovered individuals with significantly increased spinosad resistance (Orr et al., 2006; Watson et al., 2010). Five mutant lines of *D. melanogaster*, highly resistant to spinosyns, were established from the surviving mutant individuals.

5.2. Resistance mapping

Using standard recombination with selectable genetic markers, the resistance mutation was localized in a region of approximately 220 genes on the left arm of the second chromosome (Watson et al., 2010). Pairwise crosses among the five established lines were conducted, and the progeny showed resistance to spinosyns at levels similar to that of the parental individual lines. The lack of complementation among these five lines to restore the spinosyn susceptibility indicated that the mutations were probably in the same gene for all the five lines. The region containing the resistance mutations was further narrowed down, by mapping with deficiency and duplication strains, to a well-defined segment on the second chromosome which contained a nAChR subunit gene, $D\alpha6$, and 28 other genes (Watson et al., 2010).

5.3. Identification of the gene, when mutated, causing spinosyn resistance

Experimental data from several approaches, including electrophysiology, candidate gene sequencing, and functional expression of the candidate gene in a heterologous system, converged to reveal a target site for the spinosyns. It was previously shown that spinosyn A could activate nAChRs from cockroach neurons and the activation could be blocked by alpha-bungarotoxin (Salgado and Saar, 2004). The activation of nAChRs by spinosyns would be expected to be excitatory *in vivo*, could account for the observed excitatory insect symptoms after spinosyn treatment, and was correlated with the insecticidal activity of the spinosyns (e.g., Fig. 3.6). Additionally, wild-type *D. melanogaster* larvae ganglia produced increased nerve firing in response to spinosyn A. However, ganglia from the mutant *D. melanogaster* lines did not respond to spinosyn (Fig. 3.4) (Watson et al., 2010). These data, when considered in total, indicated that $D\alpha6$ was a reasonable candidate gene because it was the only nAChR-related gene in the resistance region defined by deficiency mapping in the mutant lines. In another study, it was also shown that spinosyn A had potent effects on the function of the GABA receptors of small-diameter cockroach neurons (Watson, 2001). However, the mapped resistance region in *D. melanogaster* did not contain any known GABA receptor genes.

Selected genes in this mapped region were sequenced, and loss-of-function mutations were identified in the $D\alpha6$ gene in all the five mutant lines, and either of two distinct mutations called DAS1 and DAS2 was present among these five lines (Watson et al., 2010). The DAS1 mutation caused an altered splicing donor site resulting in a truncated peptide

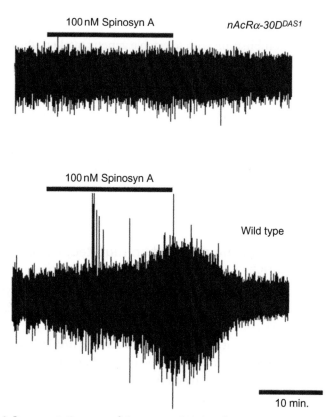

Figure 3.4 Representative nerve firing traces showing the response to spinosyn A from homozygous nAcRa-30DDAS1 (above) and wild-type (below) *D. melanogaster* larvae ganglia. Spinosyn A application to ganglia is indicated by horizontal line. The nerve firing of nAcRa-30DDAS1 ganglia was not affected by the application of 100 nM spinosyn A (similar results from 9 of 10 ganglia). Nerve firing from wild-type ganglia was increased substantially upon the application of 100 nM spinosyn A (similar results from nine of nine ganglia). *Reprinted with permission from Watson et al. (2010).*

containing only exon 1 and 22 extraneous amino acids encoded by the intronic sequence. The DAS2 mutation caused a premature stop codon at Trp458 and again resulting in a truncated peptide that is probably non-functional. Additional *D. melanogaster* spinosyn-resistant mutant lines were subsequently isolated and characterized. Sequence analyses identified mutations in the Dα6 gene from these mutant lines and supported the conclusion that spinosyn resistance could come from the lack of functional Dα6 (Table 3.2) (Watson et al., 2010).

Perry et al. (2007) sourced several *D. melanogaster* strains and made the crosses to obtain the progeny for analysis of resistance to spinosad.

Table 3.2 Dα6 mutations confering spinosyn resistance

nAcRα-30D allele	Mutation and predicted consequence
DAS1	Altered exon 1/intron 1 splice donor site; null mutation
DAS2	W458stop; null mutation
CB1	W203R; alteration in agonist binding domain
CB2	Altered intron 4/exon 5 splice acceptor; truncated peptide
CB3	P347L; altered second intracellular loop between TM3 and TM4
CB4	Q26stop; truncated peptide
CB5	Mutation 5' to intron 5/exon 5 splice acceptor leading to splicing defect; truncated peptide
CB6	Unknown
CB7	P146S; alteration in extracellular domain
CB8	P262L; altered intracellular loop between TM1 and TM2
CB9	R354stop; lack of TM4
CB10	Altered intron 11/exon 12 splice acceptor; lack of TM4
CB11	Mutation in RNA editing site I134F; edited version is I134V; alteration in extracellular domain
CB12	W86stop; truncated peptide
CB13	C168S; necessary for disulfide bridge loop; alteration in extracellular domain
CB14	L238stop; truncated peptide

TM, transmembrane.
Reprinted with permission from Watson et al. (2010)

Interestingly, a deficiency line over the CyO balancer [Df(2L)s1402/CyO] displayed a high level of resistance to spinosad (>1000-fold calculated resistance ratio (RR)) in comparison with a second chromosome marker line (*cn bw sp*). However, in the heterozygous combinations with the marker strain (*cn bw sp*), the spinosad RR was less than twofold with CyO or Df (2L)s1402. A second deficiency strain [Df(2L)Exel6025/CyO] with Dα6 and 11 other genes deleted also showed resistance to spinosad. These results indicated that the CyO strain carried the mutation that conferred the spinosad resistance, and the authors pursued the identification of the mutation in the CyO balancer (FlyBase). The CyO balancer contains several inversions, and the breakpoints of these inversions could have disrupted the function of spinosad target gene. The molecular evidence that one of the inversions on the CyO balancer caused a Dα6 truncation came from sequencing inverse-PCR products of the genomic DNA at the *Dα6* locus. The truncated Dα6 lacks the transmembrane domain 3 (TM3), the cytoplasmic loop, TM4, and the C-terminal tail domains and, therefore, is unlikely to be functional. These data independently confirmed the importance of functional Dα6 for spinosyn sensitivity in *D. melanogaster*.

Heterologous expression of wild-type Dα6 in the African clawed frog, *Xenopus laevis*, oocytes with Dα5 and ric3 coexpression resulted in nicotine-sensitive, acetylcholine-sensitive, as well as spinosyn-sensitive current (Fig. 3.5; Watson et al., 2010). However, the mutated Dα6 in combination with Dα5 and ric3 in the heterologous expression system did not respond to spinosyns (Gerald Watson et al., unpublished) demonstrating that, indeed, the mutated Dα6 was likely nonfunctional. Taken together, these results clearly established that Dα6 was a primary target of spinosyn.

5.4. nAChRs as spinosyn target in insect pest species

5.4.1 Pxα6 and Bdα6, spinosyn targets in the diamondback moth and the oriental fruit fly

A spinosyn-resistant diamondback moth (*Plutella xylostella*) population was collected from a field in Hawaii, and the resistant insects were further subjected to selection in the laboratory for spinosad resistance. The resulting strain ultimately became >18,000-fold resistant to spinosad (Rinkevich et al., 2010). The nAChR subunit Pxα6, which is a homolog of *Dα6*, mapped to the resistance linkage (Baxter et al., 2010). Genomic sequence analyses revealed a mutation that caused an altered splicing donor site (GT to *AT*) between exons 9 and 10 in *Pxα6*. This altered splicing removed the 140 aa cytoplasmic loop between TM3 and TM4, 19 aa of TM4 as well as the extracellular carboxy-terminus (Baxter et al., 2010). The truncated Pxα6 peptide resulting from this altered splicing is predicted to be non-functional (Baxter et al., 2010; Rinkevich et al., 2010). Thus, it is very likely that *P. xylostella* nAChR subunit Pxα6 is the target site for spinosyns.

Recently, Hsu et al. (2012) reported the association of spinosyn resistance with mutated nAChR subunit gene *Bdα6* in the oriental fruit fly, *Bactrocera dorsalis*. The lab-selected *B. dorsalis* strain showed a high level of resistance to spinosyn (>2000-fold) in comparison with the parental susceptible strain. However, these spinosyn-resistant insects did not show cross-resistance to imidacloprid or fipronil. Biochemical studies indicated that observed resistance

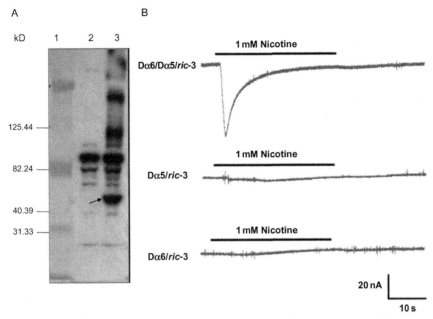

Figure 3.5 Coexpression of Dα6/Dα5/*C. elegans* ric-3 in *Xenopus* oocytes. (A)Western blot of Dα6 expression in oocytes; lane 1, molecular weight marker; lane 2, uninjected oocytes; lane 3, oocytes injected with Dα6, Dα5, and ric-3. Coinjection of cRNA encoding Dα6, Dα5, and ric-3 resulted in the expression of Dα6 which was detected by a polyclonal antibody as an approximately 53.2-kDa band not present in the uninjected controls. (B) Nicotine sensitivity of oocytes expressing Dα6/Dα5/ric-3. Top: Oocytes expressing Dα6/Dα5/ric-3 were capable of responding to nicotine with a fast-onset, desensitizing current. Oocytes from the same donor frog injected with Dα5/ric-3 (middle) or Dα6/ric-3 (bottom) failed to consistently express nicotine-sensitive receptors (nicotine applied to each oocyte as indicated by the horizontal bar). *Reprinted with permission from Watson et al. (2010).*

was unlikely because of altered metabolism. Molecular analyses revealed that the resistant strain contained only truncated *Bdα6* transcripts, resulting in a non-functional *B. dorsalis* Bdα6 nAChR subunit (Hsu et al., 2012).

5.4.2 Spinosyn target(s) to be identified in other insect species

There are other reported spinosyn-resistant insects where metabolism-based resistance mechanisms have been ruled out, and target-site-based resistance mechanisms are, therefore, suspected. Examples of this include the western flower thrips, *Frankliniella occidentalis* (Bielza et al., 2007a,b; Zhang et al., 2008), and the tobacco budworm, *Heliothis virescens* (Roe et al., 2010; Young et al., 2001). However, the actual mechanisms of spinosyn resistance in these cases have not been established. The development of sequencing technology and sequencing analysis tools has made it feasible to survey many insect genomes and transcriptomes. For example, the Insect and Other Arthropod Genome Sequencing Initiative (i5k) plans to finish sequencing 5000 insect and related species by 2016, and the 1000 Insect Transcriptome Evolution (1KITE) initiative anticipates completion of transcriptome sequencing for 1000 selected insect species by end of 2012 (www.1KITE. org). It is anticipated that these sequencing initiatives will significantly expedite the genome information-gathering process and help develop tools to simplify sequence analysis. Comparing the genomes and transcriptomes between the susceptible- and spinosyn-resistant strains within a species might shed light on spinosyn targets in these insect species.

5.5. Spinosyn target in the housefly

Selection for spinosad resistance in the housefly, *Musca domestica*, resulted in greater than 150-fold resistance to spinosad in 10 generations (Shono and Scott, 2003), and the resistance was inherited as a recessive trait on chromosome 1. Experimental data demonstrated that the resistance was not due to enhanced metabolism of spinosad, because pretreatment with the metabolism inhibitors piperonyl butoxide, diethyl meleate or S,S,S-tributyl-phosphortrithioate did not alter the spinosad RR (Shono and Scott, 2003). $Md\alpha6$, localized on chromosome 1 and a homolog of *Drosophila*, $D\alpha6$, was anticipated to be the gene of interest as the potential spinosyn target site. However, sequencing of the genomic DNA of the $Md\alpha6$ did not identify alterations that could account for the spinosyn resistance in the resistant line when compared with the susceptible line (Scott, 2008). Two other nAChR subunit genes, $Md\alpha5$ and $Md\beta3$, were also demonstrated as not being involved in the spinosyn resistance (Gao et al., 2007) and therefore were also ruled out as potential spinosyn targets in the housefly (Gao et al., 2007; Scott, 2008). To date, the molecular target site for this instance of spionosyn resistance in housefly remains unknown.

6. SPINOSYN–RECEPTOR INTERACTION

6.1. nAChR as a spinosyn target site

Initial studies on the mode of action of the spinosyns discovered that insecticidal spinosyns produce excitatory symptoms of likely neural origin in susceptible insects (Salgado, 1998). Additionally, spinosyns elicited neuronal excitation when applied to excised ganglia from various insect species (Salgado, 1997, 1998; Salgado et al., 1998). Taken as a whole, these findings suggested that the mechanism of toxicity for the spinosyns likely involved the overexcitation of neurons by, at that time, an unknown mechanism.

Since that time, a growing body of evidence supports the nAChR as a significant target site for the spinosyns in insects. Initial electrophysiological studies on isolated neurons from the American cockroach (*Periplaneta americana*) demonstrated that insecticidal spinosyns activated inward currents that were sensitive to nAChR antagonists, suggesting that the currents were induced by activation of nAChRs (Salgado and Saar, 2004; Salgado and Sparks, 2010). Studies ultimately revealed that the spinosyns selectively interact with a nondesensitizing subtype of nAChR response, which distinguishes the spinosyns from other nAChR-active insecticides (Salgado and Saar, 2004; Salgado and Sparks, 2010). The inward current induced by the spinosyns would be anticipated to depolarize neurons, making it consistent with an excitatory effect on the insect nervous system, and also consistent with many of the excitatory symptoms observed in insects poisoned by the spinosyns (Salgado, 1998). Further evidence correlating the spinosyn-induced current with toxicity was produced by comparing the SAR of the spinosyn and spinosoid-induced currents with their toxicity to *H. virescens* neonates (Fig. 3.6, adapted from Salgado et al., 1997) (Salgado and Sparks, 2010). The significant relationship between currents and toxicity strongly suggested that the activation of nAChRs could underlie spinosyn toxicity.

However, unlike many other insecticides acting at nAChRs, the spinosyns were not found to displace known nAChR-acting radioligands, suggesting that the spinosyns must be acting through a novel binding site at the insect nAChR to activate currents (Orr et al., 2009). Identification and confirmation of a molecular target site for the spinosyns required the identification and heterologous expression of nAChR subunits that were sensitive to the spinosyns.

6.2. Heterologous expression of spinosyn-sensitive nAChRs

Methods used for the study of nAChR subunit function and pharmacology include heterologous expression in cell lines and in oocytes obtained from frogs such as *X. laevis* (see review by Millar, 2009). For a number of years, various subunits from vertebrate nAChRs have been extensively studied and characterized using these techniques (Forsayeth et al., 1990; Mishina et al., 1984). Unfortunately, comparable heterologous expression, especially functional expression, of insect nAChRs using only insect nAChR subunits has

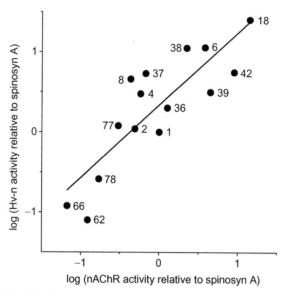

Figure 3.6 Relationship between insecticidal activity against *H. virescens* neonate larvae and relative potency against nAChR for a selected set of spinosyns and semisynthetic analogues. Numbers of the compounds correspond to those in Table 3.3. *Figure adapted from Salgado and Sparks (2010).*

proved exceedingly difficult. Nonetheless, it has been known for some time that some insect nAChR subunits will functionally express when coexpressed with vertebrate β nAChR subunits (Bertrand et al., 1994; Lansdell and Millar, 2000). Functional expression of these insect/vertebrate nAChR heteromers has proved useful for the characterization of a number of insecticidal compounds. For example, the nAChR that results from oocyte coexpression of the *D. melanogaster* Dα2 nAChR subunit with a vertebrate (rat or chick) β2 nAChR subunit has been used extensively to characterize the neonicotinoids imidacloprid and clothianidin (Ihara et al., 2004; Matsuda et al., 1998), nereistoxin derivatives (Raymond Delpech et al., 2003), and, more recently, the sulfoximine insecticide sulfoxaflor (Watson et al., 2011).

However, not all nAChR-acting insecticides interact with nAChRs composed of, for example, Dα2 type nAChR subunits. As previously stated, recent evidence indicates that, for *D. melanogaster*, the spinosyn insecticides are selective for nAChRs that contain the Dα6 subunit (Perry et al., 2007; Watson et al., 2010). For example, adult *D. melanogaster* with mutations in the Dα6 subunit has greatly reduced sensitivity to insecticidal spinosyns (Perry et al., 2007; Watson et al., 2010). Several independent mutations in the *Dα6* gene were subsequently identified that would be anticipated to yield incomplete or nonfunctional versions of the Dα6 subunit (Watson et al., 2010). These results strongly suggest that nAChRs that incorporate the Dα6 nAChR subunit are necessary for potent insecticidal activity of the spinosyns, and therefore, Dα6 represents a likely target site for the chemistry. This further suggests that the spinosyns must interact with Dα6 subunits directly, or indirectly, to elicit their insecticidal actions.

Dα6 is one of three *D. melanogaster* nAChR subunits with high sequence homology to the vertebrate α7 nAChR subunit (Grauso et al., 2002). The other two *D. melanogaster* nAChR subunits, Dα5 and Dα7, were not altered in spinosyn-resistant *D. melanogaster* (Watson et al., 2010). A major difference between Dα5 and the other two subunits is the presence of a much longer N-terminus of the gene sequence for Dα5 (Grauso et al., 2002). The role of the longer N-terminus is unknown but could be important for any number of receptor processes such as nAChR assembly or membrane translocation.

As vertebrate α7 nAChR subunits are known to express in homomeric form, it was initially assumed that, due to their sequence homology with α7, Dα6 might also express in homomeric form. However, repeated attempts to express Dα6 in oocytes failed to yield functional nAChRs (Lansdell et al., 2012; Watson et al., 2010). Further, coexpression of Dα6 with the *C. elegans* nAChR chaperone protein ric3 (Halevi et al., 2002) also failed to improve functional nAChR expression (Watson et al., 2010). A breakthrough for the expression of a spinosyn-sensitive nAChR occurred with the discovery that the coexpression of Dα6 with Dα5 and ric3 in oocytes could result in nAChRs that were sensitive to nAChR agonists (Fig. 3.7, Watson et al., 2010). While expression was inconsistent, these studies did confirm that this Dα6-containing nAChR was activated by insecticidal spinosyns, including spinosad and spinetoram, and was also activated by general nAChR agonists such as acetylcholine and nicotine (Watson et al., 2010). This finding, when coupled with the loss of spinosyn sensitivity in mutant *D. melanogaster*, provided strong evidence that Dα6 is a relevant insecticidal target site for the spinosyns. Further, all the data derived from the heterologous expression studies were consistent with the early electrophysiological studies that first suggested a significant interaction of the spinosyns with native nAChRs. For example, for both the nAChR expressed in oocytes and the native insect nAChR, insecticidal spinosyns induce currents that are blocked by nAChR antagonists, while spinosyns that are less active as insecticides either fail to induce currents or induce much lower amplitude currents (Salgado and Saar, 2004; Watson et al., 2010).

A recent study (Lansdell et al., 2012) has demonstrated that ACh-sensitive nAChRs can be expressed by the injection of either Dα5 or Dα7 alone (with ric3). Additionally, it was further demonstrated that the coexpression of Dα5, Dα6, and Dα7 (with ric3) yields nAChRs that have significantly higher affinity for ACh than do Dα5 homomeric or Dα5/Dα6 heteromeric nAChRs (Lansdell et al., 2012). As the subunit composition of most native insect nAChRs is not presently known, this leaves open the possibility that nAChRs with all three α7 homologs could represent an nAChR that more closely approximates a native nAChR. Therefore, it is important to determine potential pharmacological differences in spinosyn sensitivity between Dα5/Dα6 nAChRs and Dα5/Dα6/Dα7 nAChRs. However, the spinosyn sensitivity of the Dα5, Dα6, and Dα7 heteromeric nAChR has thus far not been reported.

6.3. Extrapolation of expression studies

The information obtained from the *D. melanogaster* studies has led to the search for spinosyn-sensitive Dα6 homologs in other insect species. For example, recent studies have demonstrated that for *P. xylostella*, field-generated resistance to spinosad is due to a mutation of Pxα6, which has >80% homology to Dα6 (Baxter et al., 2010; Rinkevich et al., 2010). We are not aware of any attempt to express Pxα6 to confirm its function as a molecular target site for spinosad. Nonetheless, the identification of a likely spinosyn target site in

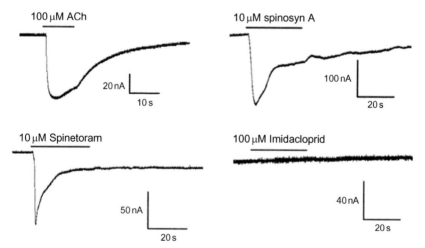

Figure 3.7 Agonist activation of Dα6/Dα5 nAChRs by acetylcholine, spinosyn A, and spinetoram, but not the neonicotinoid imidacloprid. *Reprinted with permission from Watson et al. (2010).*

the diamondback moth underscores the significant research direction that can often be gained from the study of expressed nAChR subunits.

Unfortunately, studies on putative target-site-based spinosyn-resistant housefly (*M. domestica*) have failed to uncover any changes in a Dα6 homolog, or any other nAChR subunit that would be expected to underlie spinosyn resistance (Scott, 2008). This would seem to suggest that there are additional, perhaps nAChR-related, loci for spinosyn resistance that could represent additional target sites for the spinosyns in some species.

6.4. Studies on the spinosyn-binding interaction with nAChRs

The available data support the hypothesis that the spinosyns activate insect nAChRs through a novel mechanism or binding site. However, there is limited information as to the precise nature of the interaction of the spinosyns with nAChRs. Studies on native cockroach nAChRs (Salgado and Saar, 2004; Salgado and Sparks, 2010), combined with the lack of interaction with known nAChR binding sites (Orr et al., 2009), suggest the conclusion that the interaction by spinosyns is allosteric to the nAChR. However, these studies did not provide direct evidence as to whether the interaction occurs directly with the nAChR pentamer or perhaps with an auxiliary protein site. In the studies on expressed nAChR subunits in oocytes (Watson et al., 2010), there are presumably few other potential target proteins with which spinosyns might interact. Therefore, the expression studies provide direct evidence that spinosyns interact directly with nAChRs. This provides the impetus for continued study of the nature of the spinosyn–nAChR interaction.

There are several methods available for refining the nature of this interaction. Ideally, one method of investigation typically involves the study of the interaction of radiolabelled insecticides with their binding site. Indeed, several attempts to use radiolabelled spinosyns and spinosoids have previously been made with limited success (Watson et al., unpublished data). One significant hurdle for these studies is the lipophilic nature of the spinosyns, which often leads to high levels of nonspecific binding of spinosyn-based radioligands (Watson et al., unpublished data). Nonetheless, limited success using this approach has been obtained using an insecticidal analogue of spinosyn A, [^{3}H] 5,6-dihydrospinosyn A, to label Dα6 nAChR subunits expressed with ric3 in *D. melanogaster* S2 cells (Orr et al., 2006). In these studies, dose-dependent displacement of [^{3}H] 5,6-dihydrospinosyn A by spinosyns was observed. However, oocyte expression studies indicate that Dα6 alone does not yield functional nAChRs (Lansdell et al., 2012; Watson et al., 2010). This calls into question the relevance of binding to nAChRs composed only of Dα6 subunits. Unfortunately, coexpression of Dα5 with Dα6 subunits has thus far failed to consistently lead to the expression of a high-affinity binding site for [^{3}H] 5,6-dihydrospinosyn A (Watson et al., unpublished data). Therefore, further research is necessary to advance the understanding of the interaction of spinosyns with nAChRs from studies employing the displacement of radiolabelled spinosyns and spinosoids.

The discovery that molluscan acetylcholine binding proteins (AChBPs) contain high homology with nAChRs (Brejc et al., 2001) has resulted in numerous advances in the understanding of ligand/nAChR interactions. For example, nAChR-acting insecticides such as the neonicotinoids have been crystallized in complex with AChBPs, resulting in highly refined models that predict their binding to nAChRs at the molecular level (Ihara et al., 2008). Further, information derived from receptor binding studies of neonicotinoids to AChBPs has also provided necessary insight into the molecular recognition elements of nAChRs that are important for binding (Casida and Tomizawa, 2008; Talley et al., 2008). Unfortunately, there is no literature yet emanating from studies where these highly powerful molecular tools have been applied to the spinosyns. The spinosyns are quite structurally dissimilar to all other nAChR agonists, including acetylcholine, nicotine, and the neonicotinoids and therefore could bind to a distinct site on nAChRs. Thus, while it seems unlikely that the spinosyns would bind to AChBPs in a manner similar to other nAChR agonists, this hypothesis remains to be tested.

7. NEXT-GENERATION SPINOSYN-BASED INSECTICIDES

Shortly after the discovery of spinosad, investigations began searching for spinosyns and semisynthetic analogues with modifications of the basic structure of spinosyn A to determine the SAR of the spinosyn chemistry. At the same time, a goal was set to discover a next-generation spinosyn-based product that would provide an expanded pest spectrum along with improved insecticidal potency relative to spinosad. Multiple approaches were used to determine the spinosyn SAR. A number of spinosyn factors were isolated from the fermentation broth of *S. spinosa* (Deamicis et al., 1997; Kirst et al., 1992; Sparks et al., 1998). Although many of the spinosyns isolated were active against neonate *H. virescens* larvae, none were more potent than the most abundant and active naturally occurring spinosyn, spinosyn A. From the inception of spinosyn research, chemical modifications were made to the spinosyn structure. However, virtually, all of the initial modifications to the forosamine, tri-O-methyl rhamnose, or the spinosyn tetracycle resulted in decreased insecticidal activity relative to spinosyn A (Crouse and Sparks, 1998; Salgado and Sparks, 2010).

7.1. Nonsynthetic approaches to generating new spinosyns

In addition to synthetic modifications to the spinosyn structure, other approaches were explored to generate new spinosyns and spinosyn analogues. The first was a natural product-based approach, searching for new biologically active plants, microbes, and fungi from which new spinosyns might be discovered. From this programme, a new spinosyn-producing species of *Saccharopolyspora* was identified named *Saccharopolyspora pogona* (Hahn et al., 2006; Lewer et al., 2009). Interestingly, the spinosyns produced by *S. pogona* all contained a butenyl moiety at the C21-position of the spinosyn tetracycle (Fig. 3.8). The C-21 position is typically occupied by an ethyl group in the spinosyns produced by *S. spinosa*. A few of the C-21-butenyl spinosyns exhibited improved activity against certain pest insects compared with spinosyn A (Lewer et al., 2009) (Table 3.3), but, as of yet, none have become commercial products.

Genetic engineering of the biosynthetic pathway for the spinosyn tetracycle has also afforded an opportunity to generate modifications of the spinosyns, primarily for the C21-position, that were not readily accessible by conventional means (Sheehan et al., 2006). Several of the resulting novel C-21 derivatives (e.g., 21-*n*-propyl, 5,6-dihydro-21-cyclobutyl) were far

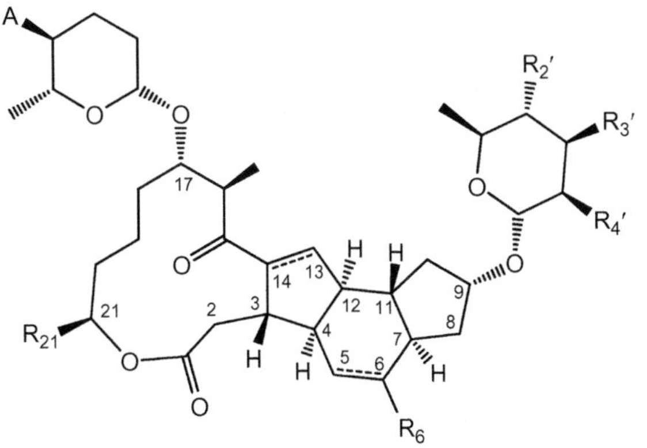

Figure 3.8 General spinosyn structure. For substituents around spinosyn structure— see Table 3.3.

Table 3.3 Structures and biological activity for selected spinosyns and semisynthetic analogues

	Structure	C17	A	R21	R6	R2'	R3'	R4'	5,6	13,14	Other	Hv-n	Hv-t	Se-t	Ag	Tu
1	Spinosyn A	F[a]	NMe2	Et	H	OMe	OMe	OMe	DB	DB	–	0.31	0.03	0.63	18–55	5.3
2	Spinosyn B	F	NHMe	Et	H	OMe	OMe	OMe	DB	DB	–	0.4	–	–	11	0.9
3	Spinosyn C	F	NH2	Et	H	OMe	OMe	OMe	DB	DB	–	0.8	–	–	33	8.4–29
4	Spinosyn D	F	NMe2	Et	Me	OMe	OMe	OMe	DB	DB	–	0.8	–	–	50	3–19
5	Spinosyn E	F	NMe2	Me	H	OMe	OMe	OMe	DB	DB	–	4.6	–	–	–	100
6	Spinosyn H	F	NMe2	Et	H	OH	OMe	OMe	DB	BD	–	5.7	–	–	>50	>50
7	Spinosyn J	F	NMe2	Et	H	OMe	OH	OMe	DB	DB	–	>80	–	–	>50	~63
8	Spinosyn K	F	NMe2	Et	H	OMe	OMe	OH	DB	DB	–	3.5	–	–	12	1.4
9	Spinosyn L	F	NMe2	Et	Me	OMe	OH	OMe	DB	DB	–	6	–	–	>50	48–67
10	Spinosyn M	F	NHMe	Et	H	OMe	OH	OMe	DB	DB	–	22.6	–	–	–	–
11	Spinosyn N	F	NHMe	Et	Me	OMe	OH	OMe	DB	DB	–	40	–	–	–	>50
12	Spinosyn O	F	NMe2	Et	Me	OMe	OMe	OMe	DB	DB	–	1.4	–	–	11	0.8
13	Spinosyn P	F	NMe2	Et	H	OH	OH	OMe	DB	DB	–	>64	–	–	–	–
14	Spinosyn Q	F	NMe2	Et	Me	OH	OMe	OMe	DB	DB	–	0.5	–	–	>50	14
15	Spinosyn R	F	NhMe	Et	H	OH	OMe	OMe	DB	DB	–	14.5	–	–	–	–
16	Spinosyn S	F	NMe2	Me	H	OH	OMe	OMe	DB	DB	–	53	–	–	–	114
17	Spinosyn Y	F	NMe2	Me	H	OMe	OMe	OH	DB	DB	–	20	–	–	–	>50

Forosamine modifications

18	4''-N-Me quat	F	N⁺Me3	Et	H	OMe	OMe	OMe	DB	DB	–	7.1	–	–	–
19	4''-OH	F	OH	Et	H	OMe	OMe	OMe	DB	DB	–	3.7	–	~0	–
20	4''-N-ethyl A	F	NMe–Et	Et	H	OMe	OMe	OMe	DB	DB	–	0.29	–	15.2	0.3
21	4''-N-n-propyl A	F	NMe–nPr	Et	H	OMe	OMe	OMe	DB	DB	–	0.27	–	17.3	0.35
22	4''-N-isopropyl A	F	NMe–isoPr	Et	H	OMe	OMe	OMe	DB	DB	–	4.5	–	25	0.90
23	4''-N-n-butyl A	F	NMe–nBut	Et	H	OMe	OMe	OMe	DB	DB	–	14.1	–	5.3	0.4
24	4''-N,N-diethyl	F	NEt2	Et	H	OMe	OMe	OMe	DB	DB	–	>64	–	4.8	0.3
25	C17 pseudoaglycone A	–	–	Et	H	OMe	OMe	OMe	DB	DB	–	>64	–	–	–
26	C17 O-methyl	OMe	–	Et	H	OMe	OMe	OMe	DB	DB	–	3.0	–	–	–
27	C17 3-OMe glucose	OMG	–	2-Butenyl	H	OMe	OMe	OMe	DB	DB	–	>25	–	17	–
28	C17 D-Amicetose	AMC	–	2-Butenyl	H	OMe	OMe	OMe	DB	DB	–	0.21	–	6	–
29	C17 D-Aamicetose	AMC	–	2-Butenyl	H	OMe	OMe	OMe	DB	DB	8–OH	0.07	–	–	–
30	C17 Me oleanderose	MOD	–	2-Butenyl	H	OMe	OMe	OMe	DB	DB	–	<0.1	–	18	–
31	C17 O-acetate (ACT)	ACT	–	Et	H	OMe	OMe	OMe	DB	DB	–	>64	–	–	47
32	C17 DMAA	DMAA	–	Et	H	OMe	OMe	OMe	DB	DB	–	>64	–	–	–
33	C17 DMAP	DMPA	–	Et	H	OMe	OMe	OMe	DB	DB	–	16	–	–	21
34	C17 NMPz	NMPz	–	Et	H	OMe	Ome	OMe	DB	DB	–	2.5	–	–	–
35	C17 (4-NO2Phe-CH2–C(O)O)	–	–	Et	H	OMe	OMe	OMe	DB	DB	–	4.1	–	–	–

Continued

Table 3.3 Structures and biological activity for selected spinosyns and semisynthetic analogues—cont'd

	Structure	C17	A	R21	R6	R2'	R3'	R4'	5,6	13,14	Other	Hv-n	Hv-t	Se-t	Ag	Tu
Tetracyle modifications																
36	5,6-Dihydro A	F	NMe2	Et	H	OMe	OMe	OMe	**SB**	DB	–	0.46	0.025	1	6.4	0.44
37	5,6-Dihydro B	F	NHMe	Et	H	OMe	OMe	OMe	SN	DB	–	1.78	–	–	–	6.0
38	5,6-α-Epoxide A	F	NMe2	Et	H	OMe	OMe	OMe	SB	DB	5,6-α-Epixide	8	–	–	–	–
39	5,6-β-Epoxide A	F	NMe2	Et	H	OMe	OMe	OMe	SB	DB	5-6-β-Epoxide	0.63	–	–	–	0.50
40	13,14-α-Dihydro A	F	NMe2	Et	H	OMe	OMe	OMe	DB	**α-SB**	–	4.7	–	–	–	–
41	13,14-β-Dihydro A	F	NMe2	Et	H	OMe	OMe	OMe	DB	**β-SB**	–	>80	–	–	–	32
42	13,14-Epoxide A	F	NMe2	Et	H	OMe	OMe	OMe	DB	**SB**	**13,14-epoxide**	1.1	–	–	–	5.6
43	2,3-Double bond A	F	NMe2	Et	H	OMe	OMe	OMe	DB	DB	**2,3-DB**	>50	–	–	–	–
44	3,13-Double bond A	F	NMe2	Et	H	OMe	OMe	OMe	DB	**SB**	**3,13-DB**	>50	–	–	–	–
45	6,7-Double bond A	F	NMe2	Et	Me	OMe	OMe	OMe	**SB**	DB	**6,7-DB**	3.7	–	–	–	–
46	7,8-Double bond A	F	NMe2	Et	Me	OMe	OMe	OMe	DB	DB	**7,8-DB**	0.6	–	–	–	–

No.	Name															
47	7,11-Double bond A	F	NMe2	Et	Me	OMe	OMe	OMe	DB	DB	**7,11-DB**	0.2	–	–	>50	47.4
48	7,11 and12,4 Double bond	F	NMe2	Et	Me	OMe	OMe	OMe	DB	DB	**7,11 and 12,4-DB**	>64	–	–	–	–
49	21-Vinyl, 3'-OEt A	F	NMe2	**Vinyl**	H	OMe	**OEt**	OMe	DB	DB	–	0.5	–	–	–	–
50	21-n-Propyl A	F	NMe2	**n-Propyl**	H	OMe	OMe	OMe	DB	DB	–	0.16	0.025	0.031	27	–
51	21-Isopropyl A	F	NMe2	**Isopropyl**	H	OMe	OMe	OMe	DB	DB	–	–	0.02	0.123	–	–
52	21-Cyclopropyl A	F	NMe2	**Cyclopropyl**	H	OMe	OMe	OMe	DB	DB	–	–	0.125	~1	~46	–
53	21-Cyclobutyl A	F	NMe2	**Cyclobutyl**	H	OMe	OMe	OMe	DB	DB	–	–	0.013	>1	~3	–
54	21-Cyclobutyl D	F	NMe2	**Cyclobutyl**	**Me**	OMe	OMe	OMe	DB	DB	–	–	0.016	0.39	15	–
55	5,6-DH, 21-Cyclobutyl A	F	NMe2	**Cyclobutyl**	H	OMe	OMe	OMe	**SB**	DB	–	–	0.02	0.13	1.7	–
56	21-Butenyl A	F	NMe2	**2-Butenyl**	H	OMe	OMe	OMe	DB	DB	–	0.29	0.012	0.036	5.2	1.4
57	21-Butenyl D	F	NMe2	**2-Butenyl**	**Me**	OMe	OMe	OMe	DB	DB	–	0.3	–	–	2	–
57	21-Butenyl, 3'-OEt A	F	NMe2	**2-Butenyl**	H	OMe	**OEt**	OMe	DB	DB	–	–	0.014	0.004	0.95	–
58	21-Hexenyl, 3'-OEt A	F	NMe2	**2-Hexenyl**	H	OMe	**OEt**	OMe	DB	DB	–	–	–	–	15.2	–

Continued

Table 3.3 Structures and biological activity for selected spinosyns and semisynthetic analogues—cont'd

	Structure	C17	A	R21	R6	R2′	R3′	R4′	5,6	13,14	Other	Hv-n	Hv-t	Se-t	Ag	Tu
59	21-Octenyl, 3′-OEt A	F	NMe2	2-Octenyl	H	OMe	OEt	OMe	DB	DB	–	–	–	–	26	–
60	21-Styryl, 3′-OEt A	F	NMe2	styryl	H	OMe	OEt	OMe	DB	DB	–	–	–	–	16.9	–
	Tri-O-methyl rhamnose modifications															
61	C9 pseudoaglycone A	F	NMe2	Et	H	–	–	–	DB	DB	–	>64	–	–	–	–
62	2′,3′,4′-Tri-O-demethyl A	F	NMe2	Et	H	OH	OH	OH	DB	DB	–	>64	–	–	–	–
63	2′-H A	F	NMe2	Et	H	H	OMe	OMe	DB	DB	–	0.23	–	–	–	134
64	2′-O-acetate A	F	NMe2	Et	H	OAc	OMe	OMe	DB	DB	–	1.2	–	–	–	>100
65	2′-O-vinyl A	F	NMe2	Et	H	Ovinyl	OMe	OMe	DB	DB	–	1.10	–	–	–	–
66	2′-O-ethyl A	F	NMe2	Et	H	OEt	OMe	OMe	DB	DB	–	0.30	–	–	–	3.5
67	2′-O-n-Propyl A	F	NMe2	Et	H	On-Pr	OMe	OMe	DB	DB	–	0.30	–	–	–	–
68	2′-O-n-Pentyl A	F	NMe2	Et	H	On-Pnt	OMe	OMe	DB	DB	–	2.0	–	–	–	–
69	2′-O-Benzyl A	F	NMe2	Et	H	OBnz	OMe	OMe	DB	DB	–	1.8	–	–	–	–

No.	Name															
70	3'-H A	F	NMe2	Et	H	OMe	H	OMe	DB	DB	–	0.36	–	–	–	5->50
71	3'-Methylene A	F	NMe2	Et	H	OMe	Me-thy-lene	OMe	DB	DB	–	0.95	–	–	–	9
72	3'-O-Acetate A	F	NMe2	Et	H	OMe	OAc	OMe	DB	DB	–	33	–	–	>50	22
73	3'-OH-Epi A	F	NMe2	Et	H	OMe	OH-epi	OMe	DB	DB	–	13.2	–	–	55	50
74	3'-O-Methyl-epi A	F	NMe2	Et	H	OMe	OMe-epi	OMe	DB	DB	–	1.86	–	–	>50	>50
75	3'-O-Vinyl A	F	NMe2	Et	H	OMe	Ovi-nyl	OMe	DB	DB	–	<0.06-0.3	–	–	>50	14
76	3'-O-Ethyl A	F	NMe2	Et	H	OMe	OEt	OMe	DB	DB	–	0.03	0.032	0.039	12.7	2.0
77	3'-O-Ethyl B	F	NHMe	Et	H	OMe	OEt	OMe	DB	DB	–	0.35	–	–	>50	1.0
78	3'-O-Ethyl, 5,6-DH A	F	NMe2	Et	H	OMe	OEt	OMe	SB	DB	–	0.05	0.009	0.04	15	0.7
79	3'-O-Ethyl D	F	NMe2	Et	Me	OMe	OEt	OMe	DB	DB	–	0.12	–	–	5-33	0.4
80	3'-OCH$_2$CF$_3$ A	F	NMe2	Et	H	OMe	OCH$_2$CF$_3$	OMe	DB	DB	–	0.33	–	–	25	>50
81	3'-O-n-Propyl A	F	NMe2	Et	H	OMe	OnPr	OMe	DB	DB	–	0.05	0.02	0.04	9.3	2.6
82	3'-O-n-Propyl D	F	NMe2	Et	Me	OMe	OnPr	OMe	DB	DB	–	–	–	–	6	2.4
83	3'-O-n-Propyl, 5,6-DH A	F	NMe2	Et	H	OMe	OnPr	OMe	SB	DB	–	0.04	0.17	0.038	1.7	0.4

Continued

Table 3.3 Structures and biological activity for selected spinosyns and semisynthetic analogues—cont'd

	Structure	C17	A	R21	R6	R2'	R3'	R4'	5,6	13,14	Other	Hv-n	Hv-t	Se-t	Ag	Tu
84	3'-O-Isopropyl A	F	NMe2	Et	H	OMe	**OisoPr**	OMe	DB	DB	–	0.27	–	–	>50	3.6
85	3'-O-Isopropenyl A	F	NMe2	Et	H	OMe	**Oiso-propenyl**	OMe	DB	DB	–	2.85	–	–	>50	>50
86	3'-O-Allyl A	F	NMe2	Et	H	OMe	**Oallyl**	OMe	DB	DB	–	0.06	–	–	>50	11
87	3'-O-propargyl A	F	NMe2	Et	H	OMe	**Opropargyl**	OMe	DB	DB	–	0.05	–	–	>50	10
88	3'-O-n-Butyl A	F	NMe2	Et	H	OMe	**OnBut**	OMe	DB	DB	–	0.38	–	–	2.5	1.5
89	3'-O-n-Butyl, 5,6-DH A	F	NMe2	Et	H	OMe	OnBut	OMe	SB	DB	–		−0.05	0.12	9.2	
90	3'-O-n-Pentyl A	F	NMe2	Et	H	OMe	**OnPnt**	OMe	DB	DB	–		−1	0.19	32	2.5
91	3'-O-phenyl A	F	NMe2	Et	H	OMe	**Ophenyl**	OMe	DB	DB	–	2.8	–	–	–	–
92	4'-H A	F	NMe2	Et	H	OMe	OMe	**H**	DB	DB	–	4.1	–	–	–	–
93	4'-O-acetate A	F	NMe2	Et	H	OMe	OMe	**OAc**	DB	DB	–	1.3	–	–	–	1.8
94	4'-O-ethyl A	F	NMe2	Et	H	OMe	OMe	**OEt**	DB	DB	–	0.24	–	–	>50	4.1
95	2'-H, 3', 4'-di-O-ethyl A	F	NMe2	Et	H	**H**	**OEt**	**OEt**	DB	DB	–	0.12	–	–	–	50
96	2'-H, 3',4'-Di-O-n-propyl A	F	NMe2	Et	H	**H**	**OnPr**	**On-Pr**	DB	DB	–	0.24	–	–	–	>50

No.	Name															
97	2′,3′,4′-Tri-O-ethyl A	F	NMe2	Et	H	**OEt**	**OEt**	OEt	DB	DB	–	0.02	–	–	6.8	1.3
98	2′,3′,4′-Tri-O-ethyl, 5,6-DH	F	NMe2	Et	H	**OEt**	**OEt**	OEt	**SB**	DB	–	0.28	–	–	–	0.3
99	C9-Methyl	F	NMe2	Et	H	–	–		DB	DB	–	>64	–	–	–	–
100	C9-2-Methoxybenzoyl	F	NMe2	Et	H	–	–		DB	DB	–	>64	–	–	–	–
101	C9-3-Methoxybenzoyl	F	NMe2	Et	H	–	–		DB	DB	–	>64	–	–	–	–
102	C9-4-Methoxybenzoyl	F	NMe2	Et	H	–	–		DB	DB	–	>64	–	–	–	–
Standards																
Cypermethrin		–	–	–	–	–	–	–	–	–	–	0.18	0.027	0.008	–	–
Imidacloprid		–	–	–	–	–	–	–	–	–	–	–	–	–	0.06	–
Fenazaquin		–	–	–	–	–	–	–	–	–	–	–	–	–	–	1.0

Bold designates substitutions that are different from spinosyn A.

Hv-n = Heliothis virescens, neonate bioassay; Hv-t = Heliothis virescens, topical bioassay; Se-t = Spodoptera exigua, topical bioassay; Ag = Aphis gossypii, sprayed plant bioassay; Tu = Tetranychus urticae, leaf disk bioassay

[a]F = forosamine; DMAA=dimethylaminoacetate; DMPP=dimethylaminoproprinonate; NMPz=,N-methylpiperazinyl acetate

more active than spinosyn A against many pest insects, but again, for a variety of reasons, none have become commercial products (Sheehan et al., 2006). Other attempts at genetic modification of the macrolide glycosylation pathway aimed at the incorporation of alternative sugars into the spinosyns were less successful (Gaisser et al., 2009). Although these nonsynthesis-based approaches resulted in the discovery of a variety of new spinosyn analogues, improvements in insecticidal activity and spectrum were insufficient, or technically too complex to be feasible for large-scale production, relative to the synthetic modification of the spinosyn sugars and tetracycle.

7.2. Areas of synthetic modification of the spinosyns

7.2.1 Modification to the forosamine

Spinosyn A has two methyl groups on the forosamine nitrogen (Table 3.3, Fig. 3.3). Removal of either one (spinosyn B) or both (spinosyn C) of the forosamine N-methyl groups resulted in slight loss of biological activity relative to spinosyn A (Table 3.3). Replacement of one of the forosamine N-methyl groups with a longer chain alkyl group (ethyl or n-propyl) had minimal impact on insecticidal activity, while a longer chain substituent (n-butyl) significantly reduced the insecticidal activity (Salgado and Sparks, 2010). Removal of the entire forosamine moiety resulted in a total loss of insecticidal activity (Table 3.3) (Salgado and Sparks, 2010). Likewise, replacement of the forosamine with other chemical moieties that partially mimic the forosamine functional groups resulted in losses of insecticidal activity, with the most active analogues being about 5- to 10-fold less insecticidal than spinosyn A (Salgado and Sparks, 2010). Recently, there has been a renewed interest in exploring new substituents as replacements for the alkyl groups on the forosamine nitrogen, and it is yet to be seen whether these efforts will ultimately lead to spinosyn analogues with improved insecticidal activity (Liu et al., 2010).

7.2.2 Modifications to the tetracycle

The spinosyn tetracycle is composed of a C12-macrocyclic ring coupled to a 5-6-5 ring system that, except for the 5,6 and 13,14-double bonds, is totally saturated (Figs. 3.2 and 3.8). The tetracycle is relatively rigid, and any alterations that change its shape resulted in large losses in biological activity. For example, saturation of the 5,6-double bond had little effect on the shape of the tetracycle and consequently only a small effect on the biological activity (Crouse et al., 2001). In contrast, saturation of the 13,14-double bond resulted in a structural shift in the tetracycle (Fig. 3.9) that is associated with a large reduction in insecticidal activity depending on the particular isomer

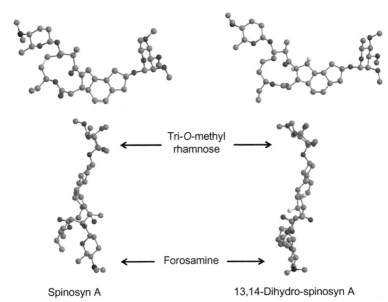

Spinosyn A   13,14-Dihydro-spinosyn A

Figure 3.9 Comparison of predicted shapes for spinosyn A and 13,14-dihydro spinosyn A based on the X-ray crystal structure of spinosyn A and following MOPAC minimization for both spinosyns.

(Crouse and Sparks, 1998; Salgado and Sparks, 2010). The addition of another double bond to the six-member ring of the tetracycle had little effect on activity (Crouse and Sparks, 1998; Crouse et al., 1999). However, aromatization of the six-member ring of the tetracycle resulted in a complete loss of insecticidal activity (Crouse and Sparks, 1998; Crouse et al., 1999).

Given the complexity of the spinosyn tetracycle, it is not surprising that the total synthesis of the tetracycle is a difficult process involving a large number of synthetic steps (Evans and Black, 1993; Mergott et al., 2004; Paquette et al., 1998). As such, fermentation currently remains the only reasonable means to produce the spinosyns. There has been at least one attempt to simplify the structure of the spinosyn tetracycle by incorporating a phenyl ring in place of the cyclopentane ring next to the tri-O-methyl rhamnose (Tietze et al., 2007). To the best of our knowledge, the insecticidal activity of this modification has not been explored.

7.2.3 Modifications to the tri-O-methyl rhamnose

Initially, the synthetic modifications of the spinosyns focused on forosamine, as the amine moiety was amenable to modification and substitution (Salgado and Sparks, 2010). As experience was gained with the chemistry, subsequent synthetic modifications shifted to those on the spinosyn tetracycle and the rhamnose (Crouse and Sparks, 1998; Salgado and Sparks, 2010). As noted above, all of the modifications made in these early explorations resulted in partial, if not total, loss of insecticidal activity relative to spinosyn A (Fig. 3.10). For several years, it appeared that the discovery of spinosyns that were more insecticidally active than spinosyn A might not be possible. One modification that was an exception involved the 2′ and 3′-desmethoxy rhamnose analogues (2′-H or 3′-H). Against a few insect species, these analogues showed insecticidal activity equal to or perhaps slightly better than spinosyn A (Creemer et al., 2000; Salgado and Sparks, 2010). However, an improved spinosyn-based product required improved insecticidal activity across a diverse group of pest species, which these initial analogues did not exhibit. Therefore, a computational approach was taken to determine if new analogues could be identified that might be predicted to have improved insecticidal activity over that of spinosyn A. These initial analyses attempted to define the quantitative structure–activity relationships (QSAR) for insecticidal activity using a variety of techniques, including multiple linear regression (MLR), partial least squares, comparative molecular field analysis, and others. These initial QSAR studies all attempted to identify synthetic directions from analyses of insecticidal potency against *H. virescens* larvae using data from the hundreds of spinosyns and semisynthetic analogues (Sparks et al., 2000, 2001). The results of these initial studies were not immediately successful, in part due to the difficulty of identifying a clear pattern and direction from among all of the widely differing

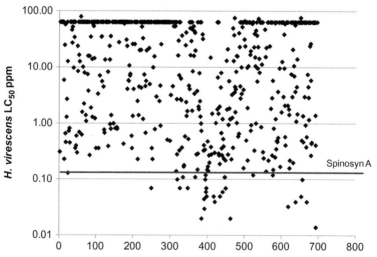

Figure 3.10 Progression of spinosyn analogue efficacy. (See Color Plate 15 at the back of the book.)

analogues, many of which were relatively inactive. Therefore, an alternative approach was investigated that employed a form of artificial intelligence called artificial neural networks (ANN, Salt et al., 1992; Devillers, 1996). The use of ANN-based QSAR was coupled with a simplification of the dataset to a small, selected group of spinosyns and spinosoids with a limited number of structural variations (Sparks et al., 2000). Using an ANN-based QSAR, the resulting models predicted that extension of the alkyl groups on the rhamnose sugar should improve the insecticidal activity (Sparks et al., 2000). When the identified analogues were synthesized, they were indeed more biologically active than spinosyn A (Sparks et al., 2000, 2008). Importantly, the ANN-based analysis identified the 3′-position of the rhamnose as a key site on the spinosyn molecule having the greatest positive effect for improving the biological activity (Sparks et al., 2000, 2001). The coupling of the extended rhamnose alkyl groups, especially in the 3′-position, with saturation of the 5,6-double bond ultimately lead to the discovery of spinetoram, which is composed of two synthetically modified spinosyn factors (Dripps et al., 2008; Sparks et al., 2008). Spinetoram provided improved efficacy, expanded spectrum and longer residual than spinosad, while maintaining the favourable toxicological and environmental profile established with spinosad (Dripps et al., 2008, 2011). Spinetoram was commercially launched in 2007, and similar to spinosad, in 2008 spinetoram also received the United States Environmental Protection Agency Green Chemistry Award (Dripps et al., 2011).

7.3. QSAR of the spinosyns

Given the large molecular weight and complexity of the spinosyn chemical structure, some hesitation about the likelihood of being able to develop useful QSARs with *in vivo* data would be understandable. However, later MLR-based studies on more targeted sets of spinosyns were able to identify useful trends (Sparks et al., 2001). For example, Hansch-style MLR analyses (Kubinyi, 1993) identified equations (e.g., (Eq. 3.1)) that describe relationships between whole molecule physicochemical properties of the spinosyns, and *in vivo* biological activity towards larvae of *H. virescens* (Sparks et al., 2001).

$$H.viresces\log LC_{50} = -6.62\,C\log P + 0.67\,C\log P2 + 3.09\,HOMO \\ + 0.59\,MOPAC\ dipole + 43.05 \qquad [3.1]$$
$$r2 = 0.816, s = 0.475, F < 0.0001, q2 = 0.706, n = 34$$

An equation incorporating the whole molecule properties of calculated log P (C log P), highest occupied molecular orbital (Todeschini and Consonni, 2000) and MOPAC dipole moment can account for much of the observed insecticidal activity for a set of 34 spinosyns and spinosyn analogues against *H. virescens* larvae (Sparks et al., 2001). Other recent MLR-based QSAR studies have again shown that these three parameters are highly correlated with spinosyn insecticidal activity against *H. virescens* larvae (Thomas Sparks, unpublished data). Interestingly, these same three parameters are also well correlated with the insecticidal activity (whole plant assay—% mortality at 50 ppm) of the spinosyns against sap-feeding insects such the cotton aphid (*Aphis gossypii*, (Eq. 3.2)) (Sparks, unpublished data)

$$
\begin{aligned}
&\textit{A.gossypii} \text{ mortality at 50 ppm} = 2.52\,C\log P - 0.28\,C\log P2 \\
&\quad - 0.94\,HOMO - 0.23\,MOPAC \text{ dipole} - 13.35 \qquad\qquad [3.2] \\
&r^2 = 0.772, s = 0.173, F < 0.0001, q2 = 0.650, n = 23
\end{aligned}
$$

Thus, somewhat surprisingly, in spite of the large, complex molecular structure of the spinosyns, it is possible to derive relatively simple equations that provide a good correlation with *in vivo* insecticidal activity and are predictive for the spinosyn chemistry (Sparks, unpublished data). Furthermore, because the spinosyns are not well metabolized by *H. virescens* larvae (Sparks et al., 2001), the QSAR equations (Sparks et al., 2001) for *H. virescens* may, in part, reflect properties for activity at the spinosyn–nAChR target site. Using data for spinosyn efficacy at the target site (Salgado and Sparks, 2010), MLR analysis suggests that a combination of similar types of parameters can be correlated with nAChR target site efficacy. For example, the combination of MR and SPAM can account for much of the observed spinosyn efficacy at the target site (Eq. 3.3), where MR is molecular radius (MR) of the substituent at the 3′-position of the rhamnose and SPAM (average molecular span) is a parameter that is related to the overall size of the whole molecule (Todeschini and Consonni, 2000). Thus, QSAR studies emphasize the likely relevance of the spinosyn target site (nAChR) to the overall efficacy of the spinosyn chemistry.

$$
\begin{aligned}
&\text{Electrophysiology index} = 2.4\,MR \text{ at C3}' + 437.4\,SPAM - 154.9 \\
&\quad r2 = 0.845, s = 2.203, F = 0.00375, q2 = 0.427, n = 9
\end{aligned}
\qquad [3.3]
$$

7.4. Next-generation spinosyns—Summary

Even with two products already developed from the spinosyn class of chemistry, exploration of the spinosyn motif continues to seek the discovery of additional spinosyns with improved insecticidal efficacy and additional pest spectrum. The discovery of the 21-butenyl spinosyns (Lewer et al., 2009) clearly demonstrates that the naturally occurring spinosyn motif can be more varied than might initially have been imagined.

8. RESISTANCE TO THE SPINOSYNS AND RESISTANCE MANAGEMENT

A long-standing problem for the introduction of any new insecticide is the potential for the development of resistance. Presently, there are more than 550 species of insects and mites that have developed resistance to one or more insecticides or acaricides (MSU APRD, 2012). The spinosyns are highly effective as insecticides against a wide range of lepidopteran and dipteran insect pests (Dripps et al., 2011). Many of the insect pests targeted by the spinosyns are some of the most important insect pests that have a history of developing resistance to a wide range of insecticides. Included among this group are *Helicoverpa armigera* (cotton bollworm), *P. xylostella*, and a variety of *Spodoptera* spp. (armyworms; Whalon et al., 2008). As such, prior to its launch, an insecticide resistance management (IRM) programme was put in place for spinosad that included specific recommendations for its use (Thompson et al., 2000; Zhao et al., 2002). However, improper use of any insecticide can nullify such IRM programmes, leading to the development of resistance, and the spinosyns have been no exception. In addition to numerous laboratory resistance studies, there have been several examples where overuse or misuse in the field has resulted in the development of resistance to spinosad.

8.1. Resistance to the spinosyns

In both the laboratory and the field, resistance to the spinosyns has been documented in a number of studies (Table 3.4). The bulk of these studies involve resistance selection in the laboratory or the laboratory selection of strains collected in the field (Fig. 3.11). Resistance to spinosad was initially described in laboratory selection studies with *H. virescens* (Young et al., 2001) and *M. domestica* (Shono and Scott, 2003). The first reports of spinosad resistance in the field were for beet armyworm, *Spodoptera exigua* (Moulton et al., 2000).

P. xylostella has developed resistance to a wide range of insecticides involving more than 450 cases over the past four decades (MSU 2012). This propensity for resistance development is due, in part, to the short generation time of the species and its presence on high value crops in regions where continuous cropping is common (e.g., Southeast Asia, Florida). Spinosad

Table 3.4 Resistance and cross-resistance in spinosad selected insects—representative examples from reports and publications

Species-strain	Select	Assay	Compound	RR[a]	Link[b]	Mech[c]	References
Diptera							
Bactrocera dorsalis (Oriental fruit fly)							
Spinosad-R	L	T	**Spinosad**	>408	–	–	Hsu and Feng (2006)
Spi-Sel	L	T	**Spinosad**	>2000	–	Dα6-target	Hsu et al. (2012)
Spi-Sel	L	T	Imidacloprid	3.0			
Spi-Sel	L	T	Fipronil	1.4			
Bactrocera oleae (wild olive fruit fly)							
Sonoma 1	F	C	**Spinosad**	13.3			Kakani et al. (2009)
Ohlone	F	C	**Spinosad**	11.4			
Sonoma 2	F	C	**Spinosad**	10.9			
Butte	F	C	**Spinosad**	10.1			
Drosophila melanogaster (fruit fly)							
Rsn	L	D	**Spinosyn A**	231	res	Dα6-target	Orr et al. (2006)
Rsn	L	D	**21-Butenyl spinosyn A**	265			
Rsn	L	D	Imidacloprid	4.9			
Rsn	L	D	Abamectin	3.6			

Rsn	L	D	**Spinosad**	370	res	Dα6-target	Watson et al. (2010)
Rsn	L	D	**Spinetoram**	176			
Rsn	L	D	Imidacloprid	0.13			
Rsn	L	D	Acetamiprid	0.88			
Rsn	L	D	Thiamethoxam	0.28			
Rsn	L	D	Avermectin b1a	0.17			
Rsn	L	D	Permethrin	0.18			
Rsn	L	D	Indoxacarb	0.30			
Rsn	L	D	**Spinosad**	483	res	Target	DAS (2003)
Rsn	L	D	**Spinetoram**	566	res	Target	
Dα6	L	D	**Spinosad**	1181	res	Dα6-target[d]	Perry et al. (2007)
Rsn	L	D	**Spinosad**	355	res	Dα6-target	Perry et al. (2012)
Rsn	L	D	Imidacloprid	1.1			
Rsn	L	D	Nitenpyram	0.4			
Rsn	L	D	Acetamiprid	1.2			
Rsn	L	D	Clothianidin	0.8			
Rsn	L	D	Dinotefuran	1.1			

Continued

Table 3.4 Resistance and cross-resistance in spinosad selected insects—representative examples from reports and publications—cont'd

Species-strain	Select	Assay	Compound	RR	Link	Mech	References
Rsn	L	D	Sulfoxaflor	1.1			
Liriomyza trifolii (American serpentine leafminer)							
CA-1[e]	F	S	**Spinosad**	>188	–	–	Ferguson (2004)
CA-1	F	S	Cyromazine	18.1			
CA-1	F	S	Abamectin	22.0			
CA-2[f]	F	S	**Spinosad**	1192	–	–	Ferguson (2004)
CA-2	F	S	Cyromazine	8.2			
CA-2	F	S	Abamectin	1.9			
Musca domestica (house fly)							
rspin	L	T	**Spinosad**	>150	rec	Target?	Shono and Scott (2003)
rspin	L	T	Indoxacarb	0.11			
rspin	L	T	Abamectin	1.2			
rspin	L	T	Fipronil	0.34			
rspin	L	T	Cyfluthrin	0.42			
rspin	L	T	Dimethoate	0.31			
rspin	L	T	Chlorfenapyr	0.77			

rspin	L	T	Methomyl	0.61			
rspin	L	T	**Spinosad**	>150	rec	Target not-α6	Gao et al. (2007)
SpRR	L	T	**Spinosad**	279	icd	>1 gene	Shi et al. (2011)
791spin[h]	FL	D	Spinosad	21	–	CYP6A1	Markussen and Kristensen (2012)
791spin	FL	D	Imidacloprid	2			
791spin	FL	D	Thiamethoxam	6			
791spin	FL	D	Fipronil	2			
Lepidoptera							
Choristoneura rosaceana (obliquebanded leafroller)							
SPIN	L	D	**Spinetoram**	5	–	MFO	Sial et al. (2011)
SPIN-G12	L	D	**Spinetoram**	5.3	–	MFO	Sial and Brunner (2012)
Heliothis virescens (tobacco budworm)							
NCSU spinR	L	LVD	**Spinosad**	>1000	res	Target	Young et al. (2001)
NCSU spinR	L	LVD	Permethrin	0.65			Roe et al. (2010)
NCSU spinR	L	LVD	Profenofos	1.70			
NCSU spinR	L	LVD	Emamectin benz.	1.91			

Continued

Table 3.4 Resistance and cross-resistance in spinosad selected insects—representative examples from reports and publications—cont'd

Species-strain	Select	Assay	Compound	RR	Link	Mech	References
NCSU spinR	L	LVD	Indoxacarb	1.45			
NCSU spinR	L	LVD	Acetamiprid	0.41			
NCSU spinR	L	T	**Spinosad**	177–312	res	Target	DAS (2003)
NCSU spinR	L	T	**3′-O-Ethyl spinosyn J**	241	res	Target	
Spino-SEL	FL	D	Spinosad	213	–	–	Sayyed et al. (2008)
Helicoverpa armigeria (cotton bollworm)							
Guntur	F	T	**Spinosad**	4.0	–	–	Kranthi et al. (2000)
Malisi	F	LD	Fipronil	9.5	–	–	Kranthi et al. (2000)
Malisi	F	LD	Indoxacarb	1.1			
Alipur	F	LD	**Spinosad**	4.7	–	–	Ahmad et al. (2003)
Alipur	F	LD	Fipronil	1.9			
Alipur	F	LD	Chlorfenapyr	3.9			
Alipur	F	LD	Indoxacarb	1.0			
Alipur	F	LD	Abamectin	1.8			
Alipur	F	LD	Emamectin benz.	2.0			
Spinosad–R	L	T	**Spinosad**	24.1		MFO	Wang et al. (2009)

Spinosad-R	L	T	Abamectin	1.4		MFO	
Spinosad-R	L	T	Chlorpyrifos	1.5		MFO	
Spinosad-R	L	T	Methomyl	1.6		MFO	
Spinosad-R	L	T	Fenvalerate	2.4		MFO	
Spinosad-R	L	T	Chlorfenapyr	1.0		MFO	Zhao et al. (2002)
Plutella xylostella (diamondback moth)							
Pearl city	F	LD	**Spinosad**	1080	res	Target?	
Pearl city	F	LD	Emamectin benz.	4			
Pearl city	F	LD	Indoxacarb	1.28			
CH1[g]	F	LD	**Spinosad**	20,600	icd	–	Sayyed et al. (2004)
CH1	F	LD	Abamectin[a]	16,300			
CH1	F	LD	Btk	230			
CH1	F	LD	Bta	60			
CH1	F	LD	Cry1Ac	1680			
CH1	F	LD	Cry1Ca	210			
CH1	F	LD	Fipronil	70			
Camilla GA	F	LD	**Spinosad**	>20,000	–	–	Zhao et al. (2006)

Continued

Table 3.4 Resistance and cross-resistance in spinosad selected insects—representative examples from reports and publications—cont'd

Species-strain	Select	Assay	Compound	RR	Link	Mech	References
Camilla GA	F	LD	Emamectin benz.	60.5			
Camilla GA	F	LD	Indoxacarb	140			
Irwin GA	F	LD	**Spinosad**	9193			Zhao et al. (2006)
Irwin GA	F	LD	Indoxacarb	44			
Irwin GA	F	LD	Emamectin benz.	35.7			
Mitchell GA	F	LD	**Spinosad**	8317	–	–	Zhao et al. (2006)
Mitchell GA	F	LD	Indoxacarb	8	–	–	
Mitchell GA	F	LD	Emamectin benz.	36	–	–	
Oxnard CA	F	LD	Spinosad	15,922	–	–	Zhao et al. (2006)
Oxnard CA	F	LD	Indoxacarb	3.6	–	–	
Oxnard CA	F	LD	Emamectin benz.	3.6	–	–	
Cornell	FL	LD	**Spinosad**	310	res	Target	DAS (2006)★
Cornell	FL	LD	**Spinetoram**	629	res	Target	
Cornell	FL	D	**Spinosad**	3900	res	Target	DAS (2007)★
Cornell	FL	D	**Spinetoram**	10,000	res	Target	
SZ-spin	L	?	**Spinosad**	60.7	ires	?	Zhu et al. (2007)

Strain			Insecticide	Value		Reference
SZ-spin	L	?	Abamectin	18.4		
SZ-spin	L	?	Fipronil	1.3		
SZ-spin	L	?	Monosultap	1.8		
Spino-SEL*	FL	LDK	**Spinosad**	1983	ires MFO	Sayyed et al. (2008)
Spino-SEL	FL	LDK	Indoxacarb	449		
Spino-SEL	FL	LDK	Deltamethrin	152		
Spinoi-SEL	FL	LDK	Abamectin	23		
Spino-SEL	FL	LDK	Acetamiprid	358		
Spino-SEL was derived from MN (Multan—same as Khaliq et al., 2007) strain via 12 gen of selection with spinosad						
Pearl-Sel	FL	–	Spinosad	18,600	Pxα6	Baxter et al. (2010)
Pearl-Sel	FL	–	Spinosad	>18,000	Pxα6	Rinkevich et al. (2010)
Spodoptera exigua (beet armyworm)						
Arizona	F	LD	**Spinosad**	14–20	– –	Moulton et al. (2000)
Thailand	F	LD	**Spinosad**	58–85		Moulton et al. (2000)
Spinosad-R	L	LD	**Spinosad**	345	– –	Wang et al. (2006)
Spinosad-R	L	T	**Spinosad**	29	MFO	Wang et al. (2006)
Spinosad-R	L	T	Fenvalerate	1.1	MFO	

Continued

Table 3.4 Resistance and cross-resistance in spinosad selected insects—representative examples from reports and publications—cont'd

Species-strain	Select	Assay	Compound	RR	Link	Mech	References
Spinosad-R	L	T	Phoxim	0.9	–	MFO	
Spinosad-R	L	T	Methomyl	1.0	–	MFO	
Spinosad-R	L	T	Abamectin	0.7	–	MFO	
Spinosad-R	L	T	Cyfluthrin	0.7	–	MFO	
Spinosad G7	FL	D	**Spinosad**	2.75	–	–	Osorio et al. (2008)
Mehoxy G7	FL	D	Methoxyfenozide	1.25	–	–	
Se-La Floriza	F	D	**Spinosad**	26	–	–	Osorio et al. (2008)
Se-La Floriza	F	D	Methoxyfenozide	3			
Se-Lazareto	F	D	**Spinosad**	37			
Se-Lazareto	F	D	Methoxyfenozide	13			
Se-Bachigualato	F	D	**Spinosad**	29			
Se-Bachigualato	F	D	Methoxyfenozide	5			
Se-Los Agustions	F	D	**Spinosad**	16			
Se-Los Agustions	F	D	Methoxyfenozide	6			
Se-Villa de Arista	F	D	**Spinosad**	2			
Se-Villa de Arista	F	D	Methoxyfenozide	2			

Tuta absoluta (tomato pinworm)

Azapa 1	F	LD	Spinosad	Reduced control	–	metab	Reyes et al. (2012)
Azapa 2	F	LD	Spinosad	Reduced control	–	metab	
Lluta	F	LD	Spinosad	Reduced control	–	metab	
Colin	F	LD	Spinosad	Reduced control	–	metab	

Thysanoptera

Frankliniella occidentalis (Western flower thrips)

Spinosad selected	L	S	**Spinosad**	62.4	–	–	Herron and James (2005)
IL–GH1	F	CF	**Spinosad**	Reduced control	–	–	Loughner et al. (2005)
Field	F	S	**Spinosad**	87	–	–	Herron and James (2007)
Almeria	F	DR	**Spinosad**	13,500	–	Target?	Bielza et al. (2007a)
Murcia	F	DR	**Spinosad**	3682	–	Target?	
R1S	FL	LD	**Spinosad**	570,473	res	Target?	Bielza et al. (2007b)
Almeria	F	DR	**Spinosad**	Reduced control	–	Target?	Bielza et al. (2008)
Almeria	F	DR	Acrinathrin	Reduced control			
Almeria	F	DR	Formeanate	Reduced control			
Almeria	F	DR	Methiocarb	Reduced control			
ICS	L	LD	**Spinosad**	**14**	icd	Target?	Zhang et al. (2008)

Continued

Table 3.4 Resistance and cross-resistance in spinosad selected insects—representative examples from reports and publications—cont'd

Species-strain	Select	Assay	Compound	RR	Link	Mech	References
ICS	L	LD	Prothiophos	2.2			
ICS	L	LD	Chlorphenapyr	2.4			
ICS	L	LD	Thiocyclam	4.3			
R1S	FL	–	Spinosad	>350,000	–	Foa6	Puinean et al. (2012)
PPP	FL	LD	Spinosad	1014	–	–	Guillen and Bielza (2012)
PPP	FL	LD	Thiamethoxam	4.8	–	–	
PPP	FL	LD	Acetamiprid	0.2	–	–	
R1S	FL	LD	Spinosad	13,766	–	–	Guillen and Bielza (2012)
R1S	FL	LD	Thiamethoxam	0.8	–	–	
R1S	FL	LD	Acetamiprid	0.2	–	–	
Hymenoptera							
Cotesia plutellae							
SR-H-L G37	L	C	**Spinosad**	**21.7**	–	–	Liu et al. (2007)

idom, incompletely dominant

Selection: L, lab; F, field; FL, field-lab.

Assay: C, contact; T, topical; D, diet; LD, leaf-dip; LVD, larval dip; LDK, leaf disk; S, spray; DR, drench; C, cup; CV, coated vial; F, feeding; G, glass. Bold designates spinosyn-related.

[a]RR, resistance ratio: reported ratio of LC_{50} of resistant strain/LC_{50} of susceptible strain

[b]link, linkage; rec, recessive; icd, incompletely dominant.

[c]Mech, mechanism; target, target-site–based resistance; target?, target-site–based resistance is likely but not proven: in many of these cases, cross-resistance spectrum, lack of synergism/lack of differences in metabolism, etc., suggest that target-site–based resistance is the most likely basis for the observed resistance.

[d]Knockout strain—missing αD6.

[e]In the 4 months prior to collection, CA-1 strain had multiple applications of cyromazine (4), abamectin (9) and spinosad (4).

[f]Just prior to collection, CA-2 strain had multiple applications of abamectin (2) and spinosad (7); no cyromazine.

[g]CH1 strain also known to have resistance to abamectin and some Bt's prior to use of spinosad.

[h]—multiresistant strain from/derived from the field—known to have resistance to pyrethroids, AChE inhibitors and acylureas

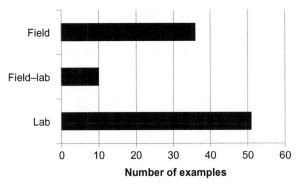

Figure 3.11 Number of documented instances of the development of spinosyn resistance based on the source of the resistant strain. Field, field-isolated resistance; field-lab, field-isolated strain that was further selected for resistance in the laboratory; Lab, resistant strains generated in the laboratory.

resistance in *P. xylostella* was first reported in Hawaii where local practices likely favoured the development of resistance (Zhao et al., 2002). Specifically, continuous cropping, weekly insecticide treatments, limited refugia, the movement of survivors from one sprayed field to adjacent fields, and limited alternative insecticide options due to a history of resistance development all contributed to the development of resistance to spinosad by *P. xylostella*. Failure to follow suggested IRM guidelines resulted in the development of a high level of resistance to spinosad in many of the Hawaiian Islands within a mere 2 years after the introduction of spinosad to the market (Zhao et al., 2002). Approaches taken to address this spinosad resistance problem included the voluntary temporary withdrawal of spinosad from Hawaii (Mao and Gusukuma-Minuto, 2004; Zhao et al., 2006). At the same time, two newly registered insecticides (indoxacarb and emamectin benzoate) became available and an insecticide rotation programme incorporating these two new insecticides was implemented (Mao and Gusukuma-Minuto, 2004; Zhao et al., 2006). This new resistance management programme allowed the reintroduction of spinosad 2 years after its withdrawal (Zhao et al., 2006). However, in some areas of Hawaii, spinosad resistance quickly reappeared, suggesting that the gene frequency for the spinosad resistance mechanisms was still too high to allow for any extended use of spinosad (Zhao et al., 2006). The *P. xylostella* resistance issues were not confined to just spinosad. During this time, in a pattern similar to that of spinosad, resistance to indoxacarb also developed (Zhao et al., 2006). Similar examples of overuse or misuse have also led to spinosad resistance in *P. xylostella* in other areas of the United States (Zhao et al., 2006).

Resistance to the spinosyns in the field accounts for 37% of all of the current examples of resistance to the spinosyns (Table 3.4). However, resistance to the spinosyns that developed in the field involves only six different species, and of these only four species have developed high levels of resistance: *Liriomyza trifolii* (American serpentine miner), *S. exigua*, *P. xylostella*, and *F. occidentalis* (Table 3.4). Therefore, after 15 years of use in the field, the number of species showing field-developed resistance to the spinosyns is restricted to only a small number of species, some of which have typically been the first to develop resistance to any new insecticide. By comparison, in the 15 years following the introduction of the pyrethroid permethrin, more than 25 species had developed resistance in the field (MSU database 2012). Thus, the spinosyns appear to have no greater propensity to develop resistance than other insecticides (see also Sparks et al., 2012).

8.2. Mechanisms of spinosyn resistance

In the 22 studies that have explored the basis for resistance in spinosyn-resistant insects, 14 have identified a target-site-based resistance mechanism, while seven instances of spionosyn resistance were shown to be associated with an enhanced metabolism of spinosad by monooxygenases, and one case was due to a combination of multiple gene mutations (Table 3.4). Thus, where resistance mechanisms have been investigated, alterations in the target site appear to be the predominant mechanism of resistance for the spinosyns. For example, the efficacy of spinosad in a highly resistant (>1000-fold) lab-selected strain of *H. virescens* was not synergized with inhibitors of metabolic enzyme systems, and available data suggest the likely involvement of a target-site mechanism (Roe et al., 2010). Likewise, spinosad-resistant strains of *D. melanogaster* were found to have an altered $D\alpha6$ nAChR subunit (Orr et al., 2006; Perry et al., 2007; Watson et al., 2010). Homologous nAChR subunits, $Px\alpha6$, $Fo\alpha6$ and $Bd\alpha6$, were also found to be associated with spinosad resistance in a field-lab strain of *P. xylostella* (Baxter et al., 2010; Rinkevich et al., 2010 and *F. occidentalis* (Puinean et al., 2012)) and in a lab-selected strain of *B. dorsalis* (Hsu et al., 2012), respectively. A target-site-based resistance mechanism also appears to be the basis for spinosad resistance in *M. domestica* (Scott, 2008). However, unlike *D. melanogaster*, *B. dorsalis*, and *P. xylostella*, $Md\alpha6$ was found to be unchanged in spinosad-resistant *M. domestica*, suggesting that it was not associated with spinosad resistance in this instance (Scott, 2008). Interestingly, $Md\alpha5$ and $Md\beta3$ were also not associated with spinosad resistance in *M. domestica* (Scott, 2008; Shono and Scott, 2003). Thus, the target for this instance of spinosad resistance in *M. domestica* remains unidentified. Based on lack of synergism with a selection of enzyme inhibitors, a target-site-related mechanism may also be involved in spinosad resistance in *F. occidentalis* (Bielza et al., 2007a,b; Zhang et al., 2008).

Several studies provide data suggesting the presence of metabolic-based resistance mechanisms to spinosad (Table 3.4). An abamectin-resistant strain from the field (MN—Pakistan) of *P. xylostella* (spino-SEL) was selected for high levels (ca. 1900-fold) of resistance to spinosad (Sayyed et al., 2008). The spino-SEL strain exhibited cross-resistance to a variety of other insecticides, with synergist studies indicating the presence of enhanced monooxygenase activity (Sayyed et al., 2008). However, the original strain from the field (MN) was known to have developed resistance to a range of insecticides, perhaps including spinosad (Sayyed et al., 2008). It is therefore likely that the spino-SEL strain possessed multiple resistance mechanisms which may explain the spinosad resistance and cross-resistance to other insecticides. A spinosad-resistant strain of *S. exigua* from China was shown to possess enhanced monooxygenase activity (Wang et al., 2006). As there was no cross-resistance to other insecticides in this strain, the monooxygenase(s) associated with resistance may be rather specific for spinosad. A spinosad-selected strain of *H. armigera*, possessing a modest level of resistance (24-fold), was shown to exhibit enhanced levels of *O*-demthylase activity that was associated with the ability of monooxygenase synergists to decrease the level of resistance (Wang et al., 2009). These data suggest the involvement of a monooxygenase-based resistance mechanism for these spinosad-selected *H. armigera* (Wang et al., 2009). Finally, studies with laboratory-selected strains of *Choristoneura rosaceana* (obliquebanded leafroller) that possessed a low level of resistance (ca. fivefold) to spinetoram suggested the involvement of monooxygenases in the observed spinetoram resistance (Sial and Brunner, 2012; Sial et al., 2011). Although not as common as target-site-based resistance, metabolic-based mechanisms do appear to be significant in some instances of resistance to the spinosyns.

8.3. Cross-resistance to other insecticides in spinosyn-resistant strains

One characteristic of target-site-based resistance is a general lack of resistance to insecticides that target different proteins. Where spinosad resistance has been shown to be target-site-based, this has indeed been the case. As might be expected, *D. melanogaster* resistant to spinosad showed cross-resistance to spinetoram and other spinosyns (Orr et al., 2006; Watson et al., 2010). However, no cross-resistance was observed to neonicotinoids, avermectins, sulfoximines, pyrethroids, or oxadiazines (Orr et al., 2006; Perry et al., 2012; Watson et al., 2010). Similar observations have been made for spinosad-resistant *M. domestica* (Shono and Scott, 2003), *B. dorsalis* (Hsu et al., 2012), *H. virescens* (Roe et al., 2010), and *P. xylostella* (Zhao et al., 2002, 2006). As noted above, a spinosad-resistant strain of *S. exigua*, possessing enhanced monooxygenase activity, also showed a general lack of cross-resistance to other classes of insecticides (Wang et al., 2006). However, other strains that exhibit apparent metabolism-based resistance to spinosad have shown varying degrees of cross-resistance to other insecticides (Osorio et al., 2008). As demonstrated by the studies of Sayyed et al. (2008), some field-selected strains of *P. xylostella* resistant to spinosad may have a background of resistance to other insecticides prior to selection with spinosad. Thus, cross-resistance to other insecticides would be anticipated from these multiresistant strains (Sayyed et al., 2008). Other spinosad-resistant strains, where there has been no determination of the resistance mechanism(s), have also shown varying degrees (low to moderate) of cross-resistance to other classes of insecticides (Ahmad et al., 2003; Ferguson, 2004; Sayyed et al., 2004).

Thus, where target-site-based resistance to the spinosyns is present, the likelihood of cross-resistance to other classes of insecticides has been shown to be very low (Table 3.4). Even where metabolism-based resistance to the spinosyns is present, cross-resistance to other classes of insecticides may or may not be present (Table 3.3).

8.4. Cross-resistance to the spinosyns in strains resistant to other classes of insecticides

Numerous studies have examined cross-resistance to the spinosyns in strains that are resistant to one or more insecticides, and the data in Table 3.5 provide a comprehensive view of these studies (Sparks et al., 2012). An analysis of cross-resistance to the spinosyns shows that cross-resistance most often (83% of the cases) is none or low in insects that are resistant to different classes of insecticides (Sparks et al., 2012). The few cases where there has been a substantial level of cross-resistance to the spinosyns tend to be in lepidopteran species where multiple resistance mechanisms were already present prior to exposure to the spinosyns. To a lesser degree, there have also been some cases of spinosyn cross-resistance in species that are resistant to fipronil or abamectin (Sparks et al., 2012). In cases of fipronil- or abamectin-resistant strains exhibiting spinosyn cross-resistance, some of these strains had also previously been selected for resistance to other insecticides in addition to fipronil or abamectin and were also cross-resistant to a

Table 3.5 Cross-resistance to the spinosyns in insect strains resistant to other insecticides—representative examples from reports and publications

Species-strain	Select	Assay	Compound	RR	Mech	Reference
Diptera						
Aedes aegypti (yellow fever mosquito)						
LPH (kdr)	FL	C	**Spinosad**	**0.9**	–	Darriet et al. (2005)
Field	F	C	**Spinosad**	**No cross-R**	kdr and metabolic	Darriet et al. (2010)
Field	F	C	Pyriproxyfen	No cross-R		
Aedes albopictus (Asian tiger mosquito)						
HAmAal	F	C	Permethrin	5		Liu et al. (2004a)
HAmAal	F	C	Deltamethrin	22		
HAmAal	F	C	Resmethrin	0.8		
HAmAal	F	C	Malathion	1		
HAmAal	F	C	Chlorpyrifos	18		
HAmAal	F	C	Propoxur	2		
HAmAal	F	C	Fipronil	3		
HAmAal	F	C	Imidacloprid	0.6		

Continued

Table 3.5 Cross-resistance to the spinosyns in insect strains resistant to other insecticides—representative examples from reports and publications—cont'd

Species-strain	Select	Assay	Compound	RR	Mech	Reference
HAmAal	F	C	**Spinosad**	**0.7**		
HAmAal	F	C	Bti	1		Liu et al. (2004a)
MWTmAal	F	C	Permethrin	2		
MWTmAal	F	C	Deltamethrin	0.6		
MWTmAal	F	C	Resmethrin	0.8		
MWTmAal	F	C	Malathion	1.2		
MWTmAal	F	C	Chlorpyrifos	33		
MWTmAal	F	C	Propoxur	0.6		
MWTmAal	F	C	Fipronil	3.3		
MWTmAal	F	C	Imidacloprid	1.2		
MWTmAal	F	C	**Spinosad**	**1**		
MWTmAal	F	C	Bti	2		
Anopheles gambiae (malaria mosquito)						
VKPR (kdr)	FL	C	**Spinosad**	**1.1**		Darriet et al. (2005)
Bactrocera zonata (fruit fly)						
Multan	F	C	Trichlorfon	32.7		Haider et al. (2011)

Multan	F	C	Profenofos	15.3	
Multan	F	C	Malathion	4.9	
Multan	F	C	Deltamethrin	2.7	
Multan	F	C	*Lambda*-cyhalothrin	3.9	
Multan	F	C	Bifenthrin	4.8	
Multan	F	C	Emamectin benz.	1.5	
Multan	F	C	Imidacloprid	15.2	
Multan	F	C	Indoxacarb	1.8	
Multan	F	C	**Spinosad**	**1.1**	
Culex quinquefasciatus (southern house mosquito)					
VBFmCq	F	C	Permethrin	13	Liu et al. (2004b)
VBFmCq	F	C	Deltamethrin	50	
VBFmCq	F	C	Chlorpyrifos	150	
VBFmCq	F	C	Propoxur	3	
VBFmCq	F	C	Fipronil	10	
VBFmCq	F	C	Imidacloprid	8	
VBFmCq	F	C	**Spinosad**	**3**	

Continued

Table 3.5 Cross-resistance to the spinosyns in insect strains resistant to other insecticides—representative examples from reports and publications—cont'd

Species-strain	Select	Assay	Compound	RR	Mech	Reference
VBFrmCq	F	C	Bti	1		
HAmCq	F	C	Permethrin	100		Liu et al. (2004b)
HAmCq	F	C	Deltamethrin	100		
HAmCq	F	C	Resmethrin	200		
HAmCq	F	C	Malathion	4		
HAmCq	F	C	Chlorpyrifos	33		
HAmCq	F	C	Propoxur	1		
HAmCq	F	C	Fipronil	5		
HAmCq	F	C	Imidacloprid	5		
HAmCq	F	C	**Spinosad**	**0.7**		
HAmCq	F	C	Bti	0.4		
MAmCq	F	C	Permethrin	940		Liu et al. (2004b)
MAmCq	F	C	Deltamethrin	300		
MAmCq	F	C	Resmethrin	830		
MAmCq	F	C	Malathion	70		
MAmCq	F	C	Chlorpyrifos	720		

MAmCq	F	C	Propoxur	5		
MAmCq	F	C	Fipronil	15		
MAmCq	F	C	Imidacloprid	10		
MAmCq	F	C	**Spinosad**	**3**		
MAmCq	F	C	Bti	0.8		
Ace.1^R	FL	C	**Spinosad**	**1.3**		Darriet et al. (2005)

Ceratitis capitata (Mediterranean fruit fly)

W	F	F	Malathion	79	Ace, Est	Magana et al. (2007)
W	F	F	Fenthion	9		
W	F	F	**Spinosad**	**2**		
W–4Km	FL	F	Malathion	178	–	Couso-Ferrer et al. (2011)(JEE)
W–4Km	FL	F	Trichlorphos	7		
W–4Km	FL	F	Diazinon	16		
W–4Km	FL	F	Phosmet	>15		
W–4Km	FL	F	Chlorpyrifos-methyl	7		
W–4Km	FL	F	Carbaryl	>4		
W–4Km	FL	F	*Lambda*-cyhalothrin	3		

Continued

Table 3.5 Cross-resistance to the spinosyns in insect strains resistant to other insecticides—representative examples from reports and publications—cont'd

Species-strain	Select	Assay	Compound	RR	Mech	Reference
W–4Km	FL	F	**Spinosad**	**1.5**		
W–4Km	FL	F	Lufenuron	6		
Drosophila melanogaster (fruit fly)						
Dα1	L	D	Nitenpyram	13.3	target	Perry et al. (2012)
Dα1	L	D	Imidacloprid	20.8		
Dα1	L	D	Acetamprid	23.4		
Dα1	L	D	Clothianidin	15.5		
Dα1	L	D	Dinotefuran	2.1		
Dα1	L	D	Sulfoxalfor	1.5		
Dα1	L	D	Spinosad	0.13		
Dβ2	L	D	Nitenpyram	70.7	target	Perry et al. (2012)
Dβ2	L	D	Imidacloprid	28.9		
Dβ2	L	D	Acetamprid	27.4		
Dβ2	L	D	Clothianidin	33.4		
Dβ2	L	D	Dinotefuran	24.6		
Dβ2	L	D	Sulfoxalfor	2.8		

Dβ2	L	D	Spinosad	0.67	
4A4D	L	D	Nitenpyram	35.4	target	Perry et al. (2012)
4A4D	L	D	Imidacloprid	27.0	
4A4D	L	D	Acetamprid	27.4	
4A4D	L	D	Clothianidin	11.9	
4A4D	L	D	Dinotefuran	36.7	
4A4D	L	D	Sulfoxalfor	2.6	
4A4D	L	D	Spinosad	0.73	

Liriomyza trifolii (American serpentine leafminer)

GA-1	F	S	Abamectin	30.6	–	Ferguson (2004)
GA-1	F	S	**Spinosad**	**1.9**	
GA-1	F	S	Cyromazine	5.4	

Musca domestica (house fly)

AVER	??	T	Abamectin	>1000	–	Scott (1998)
AVER		T	**Spinosad**	**1.9**	
LPR (Multi-R)	??	T	**Spinosad**	**4.3**	
OCR (cyclodiene)	??	T	Cyclodiene	High	

Continued

Table 3.5 Cross-resistance to the spinosyns in insect strains resistant to other insecticides—representative examples from reports and publications—cont'd

Species-strain	Select	Assay	Compound	RR	Mech	Reference
OCR (cyclodiene)	??	T	**Spinosad**	**0.4**		
Cornell-R		T	**Spinosad**	0.9		
R12 (CYP6D1)		T	**Spinosad**	1.8		
R3 (kdr & pen)		T	**Spinosad**	1.4		
ALHF (multi R)	F	T	Permethrin	1800	–	Liu and Yue (2000)
ALHF (multi R)	F	T	Cypermethrin	4200		
ALHF (multi R)	F	T	**Spinosad**	**1.4**		
Several	F	G	Permethrin	High	–	Scott et al. (2000)
Several	F	G	Cyfluthrin	High		
Several	F	F	**Spinosad**	**Low**		
Several	F	G	Fipronil	Very low		
Several	F	G	Dimethoate	Low		
Several	F	F	Methomyl	Low–high		
17e	L	T	Lindane	14,000	–	Kirstensen and Jespersen (2004)
17e	L	T	**Spinosad**	**4.1**		
39m2b	L	T	Tetrachlorvinphos	>3000		

39m$_2$b	L	T	**Spinosad**	**3.3**		
381zb	L	T	Permethrin	600		
381zb	L	T	Dimethoate	86		
381zb	L	T	**Spinosad**	**1.5**		
690ab	L	T	Methomyl	32		
690ab	L	T	**Spinosad**	**3.5**		
698ab	L	T	Ddt	>750		
698ab	L	T	**Spinosad**	**5.5**		
NYINDR	L	T	Indoxacarb	>118	–	Shono et al. (2004)
NYINDR	L	T	Dcjw	3.2		
NYINDR	L	T	**Spinosad**	**4.1**		
NYINDR	L	T	Frpronil	0.5		
NYINDR	L	T	Cyfluthrin	4.4		
NYINDR	L	T	Chlorfenapyr	0.7		
NYINDR	L	T	Dimethoate	1.0		
NYINDR	L	T	Methomyl	1.0		
NYINDR	L	T	Abamectin	2.1		

Continued

Table 3.5 Cross-resistance to the spinosyns in insect strains resistant to other insecticides—representative examples from reports and publications—cont'd

Species-strain	Select	Assay	Compound	RR	Mech	Reference
NYINDR	L	T	Dieldrin	2.5		
791a[b]	F	D	Thiamethoxam	23	MFO	Markussen and Kristensen (2012)
791a	F	D	Imidacloprid	20		
791a	F	D	Spinosad	27		
791tmx[b]	FL	D	Thiamethoxam	22	MFO	Markussen and Kristensen (2012)
791tmx	FL	D	Imidacloprid	23		
791tmx	FL	D	Spinosad	13		
791imi[b]	FL	D	Thiamethoxam	24	MFO	Markussen and Kristensen (2012)
791imi	FL	D	Imidacloprid	150		
791imi	FL	D	Spinosad	52		
791fip[b]	FL	D	Spinosad	21	MFO	Markussen and Kristensen (2012)
766b[b]	F	D	Thiamethoxam	28	MFO	Markussen and Kristensen (2012)

| 766b | F | D | Imidacloprid | 140 | | |
| 766b | F | D | Spinosad | 5 | | |

Lepidoptera

Choristoneura rosaceana (obliquebanded leafroller)

Strain			Compound	Value		Reference
Brown 97 (OP-R)	F	LD	Tebufenozide	12.8	–	Waldstein and Reissig (2000)
Brown-96 (OP-R)	F	LD	**Spinosad**	**2.0**	–	
Berrien	F	D	Azinphos-methyl	27	–	Ahmad et al. (2002)
Berrien	F	D	Chlorpyrifos	25		
Berrien	F	D	**Spinosad**	**0.83**		
Berrien	F	D	Cypermethrin	8		
Berrien	F	D	Methoxyfenozide	3.1		
Berrien	F	D	Indoxacarb	705		
Site 2	F	LDK	Azinphosmethyl	15.5	–	Smirle et al. (2003)
Site 2	F	LDK	**Spinosad**	**1.9**	–	
MA1	F	LD	Azinphosmethyl	10	–	Dunley et al. (2006)
MA1	F	LD	Chlorpyrifos	0.9		
MA1	F	LD	Tebufenozide	2.7		

Continued

Table 3.5 Cross-resistance to the spinosyns in insect strains resistant to other insecticides—representative examples from reports and publications—cont'd

Species-strain	Select	Assay	Compound	RR	Mech	Reference
MA1	F	LD	Methoxyfenozide	26		
MA1	F	LD	**Spinosad**	**0.5**		Dunley et al. (2006)
MF2	F	LD	Azinphosmethyl	25	–	
MF2	F	LD	Methoxyfenozide	8.2		
MF2	F	LD	**Spinosad**	**1.0**		Dunley et al. (2006)
AR1	F	LD	Azinphosmethyl	6.2	–	
AR1	F	LD	Methoxyfenozide	5.6		
AR1	F	LD	**Spinosad**	**0.7**		
AR1	F	LD	Indoxacarb	4440		
WEB	F	D	Azinphosmethyl	39.7	–	Sial et al. (2010)
WEB	F	D	Chloranthraniliprole	5.03		
WEB	F	D	**Spinosad**	**0.95**		
WEB	F	D	**Spinetoram**	**0.4**		
GRF	F	D	Azinphosmethyl	28.8	–	Sial et al. (2010)
GRF	F	D	Chloranthraniliprole	5.3		
GRF	F	D	**Spinosad**	**3.6**		
GRF	F	D	**Spinetoram**	**3.0**		

Cydia pomonella (codling moth)

Calderwood	F	D	Azinphos-methyl	5	–	Mota-Sanchez et al. (2008)
Calderwood	F	D	Phosmet	8		
Calderwood	F	D	*Lambda*-cyhalothrin	6		
Calderwood	F	D	Methoxyfenozide	14		
Calderwood	F	D	Acetamiprid	1		
Calderwood	F	D	Indoxacarb	1		
Calderwood	F	D	**Spinosad**	**2**		

Other field strains displayed little resistance

Raz	L*	?	Azinphosmethyl	9.5	–	Reyes and Sauphanor (2008)
Raz	L	?	Chlorpyriphos	2.0		
Raz	L	?	Emamectin benz	2.3		
Raz	L	?	**Spinosad**	**1.3**		
Raz	L	?	Thiacloprid	8.7		
Raz	L**	?	Azinphosmethyl	5.7	–	Reyes and Sauphanor (2008)
Raz	L	?	Chlorpyriphos	1.4		
Raz	L	?	Emamectin benz	0.54		

Continued

Table 3.5 Cross-resistance to the spinosyns in insect strains resistant to other insecticides—representative examples from reports and publications—cont'd

Species-strain	Select	Assay	Compound	RR	Mech	Reference
Raz	L	?	**Spinosad**	**12.2**		
Raz	L	?	Thiacloprid	1.6		
★Neonate larvae						
★★Diapausing larvae						
Earias vittella (spotted bollworm)						
Multan 2007	F	??	Cypermethrin	520	–	Ahamad and Arif (2009)
Multan 2007	F	??	Deltamethrin	62		
Multan 2007	F	??	*Zeta*-cypermetrin	24		
Multan 2007	F	??	Esfenvalerate	115		
Multan 2007	F	??	Bifenthrin	474		
Multan 2007	F	??	*Lambda*-cyhalothrin	407		
Multan 2007	F	??	**Spinosad**	**5.3**		
Multan 2007	F	??	Indoxacarb	53		
Multan 2006	F	??	Cypermethrin	120	–	Ahamad and Arif (2009)
Multan 2006	F	??	Deltamethrin	46		

Multan 2006	F	??	Zeta-cypermetrin	23		
Multan 2006	F	??	Esfenvalerate	98		
Multan 2006	F	??	Bifenthrin	388		
Multan 2006	F	??	Lambda-cyhalothrin	175		
Multan 2006	F	??	Chlorpyrifos	1.1		
Multan 2006	F	??	Profenofos	3.3		
Multan 2006	F	??	Triazophos	3.1		
Multan 2006	F	??	**Spinosad**	**2.4**		
Multan 2006	F	??	Abamectin	13		
Multan 2006	F	??	Emamectin benz.	4.2		
Multan 2006	F	??	Methoxyfenozide	3.1		
Multan 2006	F	??	Indoxacarb	36		
Multan 2006	F	??	Chlorfenapyr	74		
Helicoverpa armigera (cotton bollworm)						
Mansa (Oct)★	F	T	Cypermethrin	139	–	Gill and Dhawan (2006)
Mansa (Oct)	F	T	**Spinosad**	**1.5**		
Mansa (Oct)	F	T	Quinalphos	3.4		
Mansa (Oct)	F	T	Chlorpyriphos	4.2		

Continued

Table 3.5 Cross-resistance to the spinosyns in insect strains resistant to other insecticides—representative examples from reports and publications—cont'd

Species-strain	Select	Assay	Compound	RR	Mech	Reference
Mansa (Oct)	F	T	Endosulfan	8.9		
★Representative of several collections						
GS-RR06	L	T	Cypermethrin	1529	–	Achaleke et al. (2009)
GS-RR06	L	T	Deltamethrin	2972		
GC-RR06	L	T	Etofenprox	2		
GC-RR06	L	T	DDT	1.0		
GC-RR06	L	T	Endosulfan	0.7		
GC-RR06	L	T	Indoxacarb	1.0		
GC-RR06	L	T	Profenofos	3.9		
GC-RR06	L	T	**Spinosad**	**1.6**		
Pitoa–Maize	F	T	Cypermethrin	510	–	Achaleke et al. (2009)
Pitoa–Maize	F	T	Deltamethrin	590		
Pitoa–Maize	F	T	Etofenprox	5		
Pitoa–Maize	F	T	Ddt	1.9		
Pitoa–Maize	F	T	Endosulfan	1.0		
Pitoa–Maize	F	T	Indoxacarb	1.1		

Pitoa-Maize	F	T	Profenofos	4.2		
Pitoa-Maize	L	T	**Spinosad**	**1.0**		
Heliothis virescens (tobacco budworm)						
PYR (pyreth–R)	L	D	Cypermethrin	26	–	Polizzi and Payne (2004)
PYR (pyreth–R)	L	D	**Spinosad**	**1.1**		
Tif03	F	D	Cypermethrin	23		
Tif03	F	D	**Spinosad**	**1.4**		
Field	F	D	Deltamethrin	25	–	Sayyed et al. (2008)
Field	F	D	Indoxacarb	40		
Field	F	D	**Spinosad**	**86**		
Spino-SEL	FL	D	**Spinosad**	**213**	–	Sayyed et al. (2008)
Delta-SEL	FL	D	Deltamethrin	65	–	Sayyed et al. (2008)
Indoxa-SEL	FL	D	Indoxacarb	55	–	Sayyed et al. (2008)
Pandemis pyrusana (pandemis leafroller)						
TF1	F	LD	Azinphosmethyl	13.9	–	Dunley et al. (2006)
TF1	F	LD	**Spinosad**	**1.2**		
TF1	F	LD	Indoxacarb	1.0		

Continued

Table 3.5 Cross-resistance to the spinosyns in insect strains resistant to other insecticides—representative examples from reports and publications—cont'd

Species-strain	Select	Assay	Compound	RR	Mech	Reference
WV	F	LD	Azinphosmethyl	17.8	–	Dunley et al. (2006)
WV	F	LD	**Spinosad**	**2.5**		
Plutella xylostella (diamondback moth)						
Fenvalerate-R	L	T	Cypermethrin	1520	–	Sun 1991★
Fenvalerate-R	L	T	**Spinosad**	**3.2**		
Tebufenozide-R	L	T	Cypermethrin	27	–	Sun 1991★
Tebufenozide-R	L	T	**Spinosad**	**0.9**		
Abamectin-R	L	T	Cypermethrin	64	–	Sun 1991★
Abamectin-R	L	T	**Spinosad**	**2.4**		
Ocean cliff	F	LD	Permethrin	126	–	Shelton et al. (2000)
Ocean cliff	F	LD	**Spinosad**	**14.5**		
Ocean cliff	F	LD	Emamectin benz	13.0		
Ocean cliff	F	LD	Chlorfenapyr	1.0		
Ocean cliff	F	LD	Methomyl	7.1		
Oxnard	F	LD	Permethrin	206	–	Shelton et al. (2000)
Oxnard	F	LD	**Spinosad**	**12.8**		

Oxnard	F	LD	Emamectin benz	6.0	
Oxnard	F	LD	Methomyl	6.5	
His–hu 2001	F	LD	Abamectin	2497	–
His–hu 2001	F	LD	Emamectin benz	305	
His–hu 2001	F	LD	Fipronil	65	
His–hu 2001	F	LD	Chlorfeapyr	9	
His–hu 2001	F	LD	**Spinosad**	**65**	
His–hu 2001	F	LD	Azadirachtin	>1000	
Lu–chu 2001	F	LD	Abamectin	4988	–	Kao and Cherg (2001)
Lu–chu 2001	F	LD	Emamectin benz	153	
Lu–chu 2001	F	LD	Fipronil	104	
Lu–chu 2001	F	LD	Chlorfeapyr	10	
Lu–chu 2001	F	LD	**Spinosad**	**59**	
Lu–chu 2001	F	LD	Azadirachtin	>1000	
Fip–Sel[a]	FL	LD	Fipronil	769	–	Sayyed and Wright (2004)
Fip–Sel	FL	LD	**Spinosad**	**1170**	
Fip–Sel	FL	LD	Indoxacarb	165	

Continued

Table 3.5 Cross-resistance to the spinosyns in insect strains resistant to other insecticides—representative examples from reports and publications—cont'd

Species-strain	Select	Assay	Compound	RR	Mech	Reference
Fip-Sel	FL	LD	Cry1Ac	53		
Fip-Sel	FL	LD	Cry1Ca	53		
Multan	F	LD	Deltamethrin	507	—	Sayyed et al. (2005)
Multan	F	LD	**Spinosad**	**10**		
Multan	F	LD	Fipronil	1		
Multan	F	LD	Indoxycarb	10		
Multan	F	LD	Abamectin	4		
Multan	F	LD	Btk	<1		
Multan	F	LD	Bta	<1		
Multan	F	LD	Cry1Ac	2		
Delta-Sel	FL	LD	Deltamethrin	6736	—	Sayyed et al. (2005)
Delta-Sel	FL	LD	**Spinosad**	**50**		
Delta-Sel	FL	LD	Indoxacarb	0.3		
Delta-Sel	FL	LD	Fipronil	2		
February	F	LD	Cypermethrin	47,600	—	Sayyed et al. (2005b)
February	F	LD	Deltamethrin	872,000		

February	F	LD	Bifenthrin	30,500		
February	F	LD	Λ-cyhalothrin	223,000		
February	F	LD	Chlorpyriphos	598,000		
February	F	LD	Triazophos	98,800		
February	F	LD	Profenophos	36,500		
February	F	LD	Emamectin benz.	500		
February	F	LD	**Spinosad**	**9**		
February	F	LD	Indoxacarb	464		
Field	F	LD	Chlorpyrifos	331,150	–	Attique et al. (2006)
Field	F	LD	Bifenthrin	45,200		
Field	F	LD	Emamectin	1800		
Field	F	LD	**Spinosad**	**11**		
UNSEL G_2	F	LDK	Indoxacarb	813	–	Sayyed and Wright (2006)
UNSEL G_2	F	LDK	Fipronil	79		
UNSEL G_2	F	LDK	Spinosad	171		
UNSEL G_2	F	LDK	Deltamethrin	498		
UNSEL G_2	F	LDK	Cry1Ac	1285		

Continued

Table 3.5 Cross-resistance to the spinosyns in insect strains resistant to other insecticides—representative examples from reports and publications—cont'd

Species-strain	Select	Assay	Compound	RR	Mech	Reference
Indoxa-SEL G$_8$	F	LDK	Indoxacarb	90	–	Sayyed and Wright (2006)
Indoxa-SEL G$_8$	F	LDK	Fipronil	9		
Indoxa-SEL G$_8$	F	LDK	Spinosad	5		
Indoxa-SEL G$_8$	F	LDK	Deltamethrin	11		
Indoxa-SEL G$_8$	F	LDK	Cry1Ac	3		
Field	F	LD	Esfenvalerate	1875	–	Cordero and Kuhar (2007)
Field	F	LD	Acetamiprid	139		
Field	F	LD	Methomyl	32		
Field	F	LD	Acephate	8		
Field	F	LD	Methoxyfenozide	22		
Field	F	LD	Indoxacarb	19		
Field	F	LD	Bt	3.2*		
Field	F	LD	Novaluron	1.8*		
Field	F	LD	Azadiractin	2.3*		
Field	F	LD	Emamectin benz.	2.0*		
Field	F	LD	**Spinosad**	**9.7***		

*Not statistically difference from susceptible strain.

Lahore–2	F	LD	Cypermethrin	54	–	Khaliq et al. (2007)
Lahore–2	F	LD	Deltamethrin	6		
Lahore–2	F	LD	*Lambda*-cyhalothrin	24		
Lahore–2	F	LD	Bifenthrin	20		
Lahore–2	F	LD	Chlorpyrifos	24		
Lahore–2	F	LD	Triazophos	12		
Lahore–2	F	LD	Profenophos	16		
Lahore–2	F	LD	Indoxacarb	38		
Lahore–2	F	LD	**Spinosad**	**5**		
Lahore–2	F	LD	Emamectin benz.	6		
Multan–6	F	LD	Cypermethrin	22	–	Khaliq et al. (2007)
Multan–6	F	LD	Deltamethrin	14		
Multan–6	F	LD	*Lambda*-cyhalothrin	35		
Multan–6	F	LD	Bifenthrin	26		
Multan–6	F	LD	Chlorpyrifos	91		
Multan–6	F	LD	Triazophos	43		

Continued

Table 3.5 Cross-resistance to the spinosyns in insect strains resistant to other insecticides—representative examples from reports and publications—cont'd

Species-strain	Select	Assay	Compound	RR	Mech	Reference
Multan-6	F	LD	Profenophos	43		
Multan-6	F	LD	Indoxacarb	11		
Multan-6	F	LD	**Spinosad**	**1**		
Multan-6	F	LD	Emamectin benz.	<1		
Multan-7	F	LD	Cypermethrin	21	–	Khaliq et al. (2007)
Multan-7	F	LD	Deltamethrin	37		
Multan-7	F	LD	*Lambda*-cyhalothrin	32		
Multan-7	F	LD	Bifenthrin	18		
Multan-7	F	LD	Chlorpyrifos	110		
Multan-7	F	LD	Triazophos	83		
Multan-7	F	LD	Profenophos	31		
Multan-7	F	LD	Indoxacarb	6		
Multan-7	F	LD	**Spinosad**	**12**		
Multan-7	F	LD	Emamectin benz.	5		
Islamabad-2	F	LD	Cypermethrin	67	–	Khaliq et al. (2007)
Islamabad-2	F	LD	Deltamethrin	27		

Islamabad-2	F	LD	*Lambda*-cyhalothrin	23	
Islamabad-2	F	LD	Bifenthrin	21	
Islamabad-2	F	LD	Chlorpyrifos	141	
Islamabad-2	F	LD	Triazophos	50	
Islamabad-2	F	LD	Profenophos	28	
Islamabad-2	F	LD	Indoxacarb	6	
Islamabad-2	F	LD	**Spinosad**	**10**	
Islamabad-2	F	LD	Emamectin benz.	2	
CrC	F	T	Esfenvalerate	300	–	Eziah et al. (2008)
CrC	F	T	Methamidophos	21.2	
CrC	F	T	Chlorpyrifos	27.8	
CrC	F	T	Permethrin	65	
CrC	F	T	**Spinosad**	**2.4**	
CrC	F	T	Indoxacarb	28.1	
WiC	F	T	Esfenvalerate	490	–	Eziah et al. (2008)
WiC	F	T	Methamidophos	21.6	
WiC	F	T	Chlorpyrifos	36.2	

Continued

Table 3.5 Cross-resistance to the spinosyns in insect strains resistant to other insecticides—representative examples from reports and publications—cont'd

Species-strain	Select	Assay	Compound	RR	Mech	Reference
WiC	F	T	Permethrin	125		
WiC	F	T	**Spinosad**	**2.6**		
WiC	F	T	Indoxacarb	25.2		
MN	F	LDK	**Spinosad**	**7**	–	Sayyed et al. (2008)
MN	F	LDK	Indoxacarb	36		
MN	F	LDK	Deltamethrin	265		
MN	F	LDK	Abamectin	2557		
MN	F	LDK	Acetamiprid	5		
TH (F2)	F	LD	Abamectin	312	–	Pu et al. (2010)
TH (F2)	F	LD	Emamectin benz.	3390		
TH (F2)	F	LD	Fipronil	913		
TH (F2)	F	LD	**Spinosad**	**17**		
TH (F2)	F	LD	Indoxacarb	44		
TH (F2)	F	LD	Chlorfenapyr	10		
TH (F2)	F	LD	Tebufenozide	2		
TH (F2)	F	LD	Chlorfluazuron	0.5		

TH-Abm (F14)	FL	LD	Abamectin	9840	–	Pu et al. (2010)
TH-Abm (F14)	FL	LD	Emamectin benz.	273,730		
TH-Abm (F14)	FL	LD	Fipronil	282		
TH-Abm (F14)	FL	LD	**Spinosad**	**150**		
TH-Abm (F14)	FL	LD	Indoxacarb	32		
TH-Abm (F14)	FL	LD	Chlorfenapyr	2.2		
TH-Abm (F14)	FL	LD	Tebufenozide	4.2		
TH-Abm (F14)	FL	LD	Chlorfluazuron	0.9		
Teb-R	L	–	Tebufenozide	185.5	–	Yin et al. (2010)
Teb-R	L	–	Abamectin	41.0		
Teb-R	L	–	Indoxacarb	11.4		
Teb-R	L	–	Chlofenapyr	5.3		
Teb-R	L	–	**Spinosad**	**1.7**		
Teb-R	L	–	Chloranthranilliprole	1.4		
Aba-R★	L	–	Tebufenozide	28.0	-	Yin et al. (2010)
Aba-R	L	–	Abamectin	593.8		
Aba-R	L	–	Indoxacarb	12.3		

Continued

Table 3.5 Cross-resistance to the spinosyns in insect strains resistant to other insecticides—representative examples from reports and publications—cont'd

Species-strain	Select	Assay	Compound	RR	Mech	Reference
Aba-R	L	–	Chlofenapyr	2.7		
Aba-R	L	–	**Spinosad**	**7.9**		
Aba-R	L	–	Chloranthranilliprole	0.9		
★Aba-R, Teb-R strain selected for 39 gen with abamectin						
Pseudoplusia includens (soybean looper)						
Jen95	F	D	Permethrin	13.6	–	Mscarenhas and Boethel (1997)
Jen95	F	D	**Spinosad**	**0.9**		
Jen95	F	D	Btk	7.9		
Jen95	F	D	Thidicarb	1.7		
Jen95	F	D	Chlorfenapyr	1.0		
Jen95	F	D	Emamectin benz.	1.9		
Win95	F	D	Permethrin	38.2	–	Mscarenhas and Boethel (1997)
Win95	F	D	**Spinosad**	**1.8**		
Win95	F	D	Btk	1.9		

Win95	F	D	Thiodicarb	2.7	
Win95	F	D	Chlorfenapyr	0.9	
Win95	F	D	Emamectin benz.	6.1	
GA95	F	D	Permethrin	28.8	–	Msacarenhas and Boethel (1997)
GA95	F	D	**Spinosad**	**8.8**	
GA95	F	D	Btk	6.1	
GA95	F	D	Thiodicarb	1.5	
GA95	F	D	Chlorfenapyr	1.2	
GA95	F	D	Emamectin benz.	4.4	
Spodoptera exigua (beet armyworm)					
SPODEX E-98	F	S	Deltamethrin	708	–	Nauen et al. (2007)
SPODEX E-98	F	S	Etofenprox	45	
SPODEX E-98	F	S	Triazophos	29	
SPODEX E-98	F	S	Methamidophos	>5.5	
SPODEX E-98	F	S	Enodsulfan	>5.4	
SPODEX E-98	F	S	Triflumuron	>46	
SPODEX E-98	F	S	Indoxacarb	1.0	

Continued

Table 3.5 Cross-resistance to the spinosyns in insect strains resistant to other insecticides—representative examples from reports and publications—cont'd

Species-strain	Select	Assay	Compound	RR	Mech	Reference
SPODEX E-98	F	S	Emamectin benz.	1.0		
SPODEX E-98	F	S	Chlorfenapyr	1.8		
SPODEX E-98	F	S	Methoxyfenozide	2.2		
SPODEX E-98	F	S	**Spinosad**	**1.8**		
SPODEX E-98	F	S	Flubendiamide	0.8		
Multan	F	LD	Deltamethrin	5	–	Ishitiaq and Saleem (2011) (JEE)
Multan	F	LD	Cypermethrin	6		
Multan	F	LD	Chlorpyrifos	8		
Multan	F	LD	Profenofos	86		
Multan	F	LD	**Spinosad**	**33**		
Multan	F	LD	Ememectin benz.	4		
Multan	F	LD	Abamectin	50		
Multan	F	LD	Indoxacarb	58		
Multan	F	LD	Lufenuron	16		
Multan	F	LD	Methoxyfenozide	31		

DG Khan	F	LD	Deltamethrin	33	–	Ishitiaq and Saleem (2011) (JEE)
DG Khan	F	LD	Cypermethrin	132		
DG Khan	F	LD	Chlorpyrifos	62		
DG Khan	F	LD	Profenofos	126		
DG Khan	F	LD	**Spinosad**	**158**		
DG Khan	F	LD	Ememectin benz.	8		
DG Khan	F	LD	Abamectin	104		
DG Khan	F	LD	Indoxacarb	42		
DG Khan	F	LD	Lufenuron	9		
DG Khan	F	LD	Methoxyfenozide	25		
Multan (cotton)	F	LD	Deltamethrin	9	–	Ishitiaq et al. (2012) (Crop protect)
Multan (cotton)	F	LD	Cypermethrin	12		
Multan (cotton)	F	LD	Profenofos	74		
Multan (cotton)	F	LD	**Spinosad**	**30**		
Multan (cotton)	F	LD	Ememectin benz.	12		
Multan (cotton)	F	LD	Abamectin	75		

Continued

Table 3.5 Cross-resistance to the spinosyns in insect strains resistant to other insecticides—representative examples from reports and publications—cont'd

Species-strain	Select	Assay	Compound	RR	Mech	Reference
Multan (cotton)	F	LD	Indoxacarb	27		
Multan (cotton)	F	LD	Lufenuron	31		
Multan (cotton)	F	LD	Methoxyfenozide	3.5		
Khanewal (cotton)	F	LD	Deltamethrin	17	–	Ishitiaq et al. (2012) (Crop protect)
Khanewal (cotton)	F	LD	Cypermethrin	42		
Khanewal (cotton)	F	LD	Chlorpyrifos	103		
Khanewal (cotton)	F	LD	Profenofos	97		
Khanewal (cotton)	F	LD	**Spinosad**	**3**		
Khanewal (cotton)	F	LD	Ememectin benz.	75		
Khanewal (cotton)	F	LD	Abamectin	34		
Khanewal (cotton)	F	LD	Indoxacarb	47		
Khanewal (cotton)	F	LD	Lufenuron	59		
Khanewal (cotton)	F	LD	Methoxyfenozide	5		
Shujabad (trap)	F	LD	Deltamethrin	51	–	Ishitiaq et al. (2012) (Crop protect)
Shujabad (trap)	F	LD	Cypermethrin	51		

Shujabad (trap)	F	LD	Chlorpyrifos	68		
Shujabad (trap)	F	LD	Profenofos	42		
Shujabad (trap)	F	LD	**Spinosad**	**73**		
Shujabad (trap)	F	LD	Abamectin	11		
Shujabad (trap)	F	LD	Indoxacarb	40		
Shujabad (trap)	F	LD	Lufenuron	5.3		
Shujabad (trap)	F	LD	Methoxyfenozide	14		
Taian 2008	F	LD	β-cypermethrin	95.3	–	Zhou et al. (2011)
Taian 2008	F	LD	Chlorpyrifos	40.6		
Taian 2008	F	LD	Methomyl	2.6		
Taian 2008	F	LD	Chlorfluazuron	36.9		
Taian 2008	F	LD	Tebufenozide	30.1		
Taian 2008	F	LD	Methoxyfenozide	191.9		
Taian 2008	F	LD	Emamectin benz.	4.6		
Taian 2008	F	LD	Chlorfenapyr	2.0		
Taian 2008	F	LD	Indoxacarb	15.5		
Taian 2008	F	LD	**Spinosad**	**5.1**		

Continued

Table 3.5 Cross-resistance to the spinosyns in insect strains resistant to other insecticides—representative examples from reports and publications—cont'd

Species-strain	Select	Assay	Compound	RR	Mech	Reference
Taian 2010	F	LD	β-cypermethrin	669	–	Zhou et al. (2011)
Taian 2010	F	LD	Chlorpyrifos	176		
Taian 2010	F	LD	Methomyl	4.8		
Taian 2010	F	LD	Chlorfluazuron	2783		
Taian 2010	F	LD	Tebufenozide	85.6		
Taian 2010	F	LD	Methoxyfenozide	442		
Taian 2010	F	LD	Emamectin benz.	10.2		
Taian 2010	F	LD	Chlorfenapyr	1.4		
Taian 2010	F	LD	Indoxacarb	17.3		
Taian 2010	F	LD	**Spinosad**	**44.6**		
Spodoptera litura (cutworm)						
Delta-SEL★	FL	LDK	Deltamethrin	63	indom	Ahmad et al. (2007)
Delta-SEL	FL	LDK	Cypermethrin	83		
Delta-SEL	FL	LDK	DDT	187		
Delta-SEL	FL	LDK	Profenofos	97		
Delta-SEL	FL	LDK	Chlorpyrifos	104		

Delta-SEL	FL	LDK	Triazofos	86		
Delta-SEL	FL	LDK	Phoxim	2		
Delta-SEL	FL	LDK	Endosulfan	86		
Delta-SEL	FL	LDK	**Spinosad**	**2**		
Delta-SEL	FL	LDK	Fipronil	2		
Delta-SEL	FL	LDK	Chlorfenapyr	2		
Sah–1 2005	F	LDK	**Spinosad**	**68.0**	–	Ahmad et al. (2008)
Sah–1 2005	F	LDK	Indoxacarb	94.8		
Sah–1 2005	F	LDK	Abamectin	31.8		
Sah–1 2005	F	LDK	Emamectin	77		
Sah–1 2005	F	LDK	Fipronil	151		
Sah–1 2005	F	LDK	Lufenuron	65.6		
Sah–1 2005	F	LDK	Diflubenzuron	56.3		
Sah–1 2005	F	LDK	Methoxyfenozide	104		
Dgk–1 2004	F	LDK	**Spinosad**	**122**	–	Ahmad et al. (2008)
Dgk–1 2004	F	LDK	Indoxacarb	45.8		
Dgk–1 2004	F	LDK	Abamectin	60.8		

Continued

Table 3.5 Cross-resistance to the spinosyns in insect strains resistant to other insecticides—representative examples from reports and publications—cont'd

Species-strain	Select	Assay	Compound	RR	Mech	Reference
Dgk-1 2004	F	LDK	Emamectin	37		
Dgk-1 2004	F	LDK	Lufenuron	22.3		
Dgk-1 2004	F	LDK	Methoxyfenozide	11.4		
Dgk-1 2005	F	LDK	**Spinosad**	**16.5**	–	Ahmad et al. (2008)
Dgk-1 2005	F	LDK	Indoxacarb	7.5		
Dgk-1 2005	F	LDK	Abamectin	53.6		
Dgk-1 2005	F	LDK	Emamectin	6		
Dgk-1 2005	F	LDK	Lufenuron	20.6		
Dgk-1 2005	F	LDK	Methoxyfenozide	31.7		
Field	F	LDK	Fipronil	94	–	Sayyed et al. (2008)
Field	F	LDK	Indoxacarb	15		
Field	F	LDK	**Spinosad**	**23**		
Field	F	LDK	Abamectin	37		
Field	F	LDK	Emamectin benz.	16		
Indoxa-SEL (G9)**	FL	LDK	Indoxacarb	44	indom	Sayyed et al. (2008)
Indoxa-SEL (G4)	FL	LDK	**Spinosad**	**68**		

Indoxa-SEL (G4)	FL	LDK	Abamectin	32		
Indoxa-SEL (G4)	FL	LDK	Emamectin benz.	77		
Indoxa-SEL (G4)	FL	LDK	Fipronil	151		

*Delta-SEL, field strain selected with deltamethrin.

**Indoxa-SEL, field strain selected with indoxacarb.

Spodoptera littoralis (cotton leafworm)

LET	F	LDK	Cypermethrin	369	–	Miles and Lysandrou (2002)
LET	F	LDK	Chlorpyrifos	9.2		
LET	F	LDK	**Spinosad**	**0.51**		
MNF (2003)	F	LD	Profeofos	3.8	–	Abo-Elghar et al. (2005)
MNF (2003)	F	LD	Chlorpyrifos	25.4		
MNF (2003)	F	LD	Methomyl	39.9		
MNF (2003)	F	LD	Thiodicarb	18.0		
MNF (2003)	F	LD	Carbaryl	26.8		
MNF (2003)	F	LD	Esfenvalerate	25.9		
MNF (2003)	F	LD	Fenpropathrin	9.4		
MNF (2003)	F	LD	Cypermethrin	98.5		

Continued

Table 3.5 Cross-resistance to the spinosyns in insect strains resistant to other insecticides—representative examples from reports and publications—cont'd

Species-strain	Select	Assay	Compound	RR	Mech	Reference
MNF (2003)	F	LD	Hexaflumuron	2.1		
MNF (2003)	F	LD	Flufenoxuron	4.5		
MNF (2003)	F	LD	**Spinosad**	**0.3**		
KFR (2003)	F	LD	Profeofos	5.5	–	Abo-Elghar et al. (2005)
KFR (2003)	F	LD	Chlorpyrifos	23.4		
KFR (2003)	F	LD	Methomyl	37.2		
KFR (2003)	F	LD	Carbaryl	15.6		
KFR (2003)	F	LD	Cypermethrin	129		
KFR (2003)	F	LD	Flufenoxuron	2.0		
KFR (2003)	F	LD	**Spinosad**	**0.3**		
Tuta absoluta (tomato pinworm)						
Camocim	F	LD	Bifenthrin	11.4		Silva et al. (2011)
Camocim	F	LD	Permethrin	12.5		
Camocim	F	LD	Deltmethrin + triazophos	1.6		
Camocim	F	LD	Abamectin	1.1		

Camocim	F	LD	Spinosad	3.1	
Camocim	F	LD	Indoxacarb	4.1	
Camocim	F	LD	Diflubenzuron	57.4	
Camocim	F	LD	Teflubenzuron	3.0	
Camocim	F	LD	Triflumuron	4.1	
Camocim	F	LD	Bt	1.3	
Camocim	F	LD	Bifenthrin	11.4	Silva et al. (2011)
Camocim	F	LD	Permethrin	12.5	
Camocim	F	LD	Deltmethrin + triazophos	1.6	
Camocim	F	LD	Abamectin	1.1	
Camocim	F	LD	Spinosad	3.1	
Camocim	F	LD	Indoxacarb	4.1	
Camocim	F	LD	Diflubenzuron	57.4	
Camocim	F	LD	Teflubenzuron	3.0	
Camocim	F	LD	Triflumuron	4.1	
Camocim	F	LD	Bt	1.3	

Continued

Table 3.5 Cross-resistance to the spinosyns in insect strains resistant to other insecticides—representative examples from reports and publications—cont'd

Species-strain	Select	Assay	Compound	RR	Mech	Reference
Thysanoptera						
Frankliniella occidentalis (Western flower thrips)						
Field (2000–2001)	F	S	Bifenthrin	41		Herron and James (2005)
Field (2000–2001)	F	S	Abamectin	2.0–2.3		
Field (2000–2001)	F	S	Chlorpyrifos	2.6–6.3		
Field (2000–2001)	F	S	Diclorvos	0.9–2.5		
Field (2000–2001)	F	S	Malathion	2.3		
Field (2000–2001)	F	S	Methamidophos	0.7–1.5		
Field (2000–2001)	F	S	Methidathion	1.1–2.0		
Field (2000–2001)	F	S	Endosulfan	0.6–1.1		
Field (2000–2001)	F	S	Pyrazophos	0.7		
Field (2000–2001)	F	S	Fipronil	0.2–0.9		
Field (2000–E2001)	F	S	Methomyl	1.6–2.5		
Field (2000–2001)	F	S	**Spinosad**	**1.2–1.8**		
ABA-R	L	LD	Abamectin	45.5		Chen et al. (2011)
ABA-R	L	LD	**Spinosad**	**2.0**		
ABA-R	L	LD	Chlorpyrifos	11.4		

ABA-R	L	LD	*Lambda*-cyhalothrin	4.0	
ABA-R	L	LD	Acetamiprid	1.5	
ABA-R	L	LD	Chlorfenapyr	0.3	Wang et al. (2011)
Haidain 2010	F	TS	Cyhalothrin	39.7	
Haidain 2010	F	TS	Deltamethrin	4.55	
Haidain 2010	F	TS	Cypermethrin	1.62	
Haidain 2010	F	TS	Esfenvalerate	0.72	
Haidain 2010	F	TS	Abamectin	1.01	
Haidain 2010	F	TS	Chlorpyrifos	0.33	
Haidain 2010	F	TS	Methomyl	1.53	
Haidain 2010	F	TS	Emamectin benz.	1.47	
Haidain 2010	F	TS	Chlorfenapyr	1.17	
Haidain 2010	F	TS	**Spinosad**	**2.89**	
Haidain 2010	F	TS	Imidacloprid	2.85	
Haidain 2010	F	TS	Indoxacarb	2.38	

Continued

Table 3.5 Cross-resistance to the spinosyns in insect strains resistant to other insecticides—representative examples from reports and publications—cont'd

Species-strain	Select	Assay	Compound	RR	Mech	Reference
Coleoptera						
Leptionotarsa decemlineata (Colorado potato beetle)						
R	F	T	Imidacliprid	310		Mota-Sanchez et al. (2006)
R	F	T	Dinotefuran	59		
R	F	T	Clothianidin	33		
R	F	T	Acetamiprid	29		
R	F	T	Thiacloprid	25		
R	F	T	Thiamethoxam	15		
R	F	T	Nitenpyram	10		
R	F	T	Nicotine	1.7		
R	F	T	**Spinosad**	**7.6**		
Tribolium castaneum (red flour beetle)						
PAK	L	C	Malathion	40		Hussain et al. (2005)
PAK	L	C	Abamectin	0.68		
PAK	L	C	**Spinosad**	**0.81**		
PAK	L	C	Indoxacarb	1.89		

PAK	L	C	Azadirachtrin	1.22	
PAK	L	C	Polycl hydrocarbon	1.58	
PAK	L	C	Buprofezin	0.75	
PAK	L	C	Malathion	56	Hussain et al. (2009)
PAK	L	C	**Spinosad**	**0.85**	

Homoptera

Bemisia tabaci (sweetpotato whitefly)

Sj	F	CV	Methamidophos	29.3	–
Sj	F	CV	Chlorpyrifos	21.8	Kang et al. (2006)
Sj	F	CV	Phoxim	2.4	
Sj	F	CV	Fenvalerate	72.4	
Sj	F	CV	Avermectin	9.4	
Sj	F	CV	Emamectin benz.	5.5	
Sj	F	CV	Fipronil	11.6	
Sj	F	CV	**Spinosad**	**1.8**	
Sj	F	CV	Imidacloprid	8.0	
NJ-IMI-G30	L	LD	Imidacloprid	490	Wang et al. (2009)

Continued

Table 3.5 Cross-resistance to the spinosyns in insect strains resistant to other insecticides—representative examples from reports and publications—cont'd

Species-strain	Select	Assay	Compound	RR	Mech	Reference
NJ-IMI-G30	L	LD	Acetamiprid	143		
NJ-IMI-G30	L	LD	Thiamethoxam	243		
NJ-IMI-G30	L	LD	Thiacloprid	37		
NJ-IMI-G30	L	LD	Nitenpyram	94		
NJ-IMI-G30	L	LD	**Spinosad**	**8**		
NJ-IMI-G30	L	LD	Cartap	11		
NJ-IMI-G30	L	LD	Monsultap	7		
NJ-IMI-G30	L	LD	Cypermethrin	1		
NJ-IMI-G30	L	LD	Abamectin	1		Wang et al. (2010)
GZ–B	F	LD	Imidacloprid	320		
GZ–B	F	LD	Thiamethoxam	240		
GZ–B	F	LD	**Spinosad**	**5.7**		
GZ–B	F	LD	Abamectin	2.3		
GZ–B	F	LD	Fipronil	2.3		
GZ–B	F	LD	α–cypermethrin	460		
LH–Q	F	LD	Imidacloprid	54		Wang et al. (2010)

LH-Q	F	LD	Thiamethoxam	78	
LH-Q	F	LD	**Spinosad**	**1.8**	
LH-Q	F	LD	Abamectin	1	
LH-Q	F	LD	Fipronil	3.2	
LH-Q	F	LD	α-cypermethrin	89	Wang et al. (2010)
YC-Q	F	LD	Imidacloprid	1900	
YC-Q	F	LD	Thiamethoxam	1200	
YC-Q	F	LD	**Spinosad**	**6**	
YC-Q	F	LD	Abamectin	1	
YC-Q	F	LD	Fipronil	26	
YC-Q	F	LD	α-cypermethrin	22	Wang et al. (2010)
HZ-QB	F	LD	Imidacloprid	1100	
HZ-QB	F	LD	Thiamethoxam	520	
HZ-QB	F	LD	**Spinosad**	**1.8**	
HZ-QB	F	LD	Abamectin	2.8	
HZ-QB	F	LD	Fipronil	8.3	
HZ-QB	F	LD	α-cypermethrin	610	

Continued

Table 3.5 Cross-resistance to the spinosyns in insect strains resistant to other insecticides—representative examples from reports and publications—cont'd

Species-strain	Select	Assay	Compound	RR	Mech	Reference
These data are representative of the many other strains tested						
Hemiptera						
Diaphorina citri (Asian citrus psyllid)						
La Belle	F	T	Abamectin	0.8		Tiwari et al. (2011)
La Belle	F	T	Acetamiprid	1.0		
La Belle	F	T	Imidacloprid	35.0		
La Belle	F	T	Thiamethoxam	13.0		
La Belle	F	T	Aldicarb	0.8		
La Belle	F	T	Carbaryl	0.65		
La Belle	F	T	Bifenthrin	0.33		
La Belle	F	T	Cypermethrin	0.80		
La Belle	F	T	Fenpropathrin	1.23		
La Belle	F	T	Chlorpyrifos	6.92		
La Belle	F	T	Dimethoate	0.51		
La Belle	F	T	Malathion	1.18		
La Belle	F	T	**Spinetoram**	**1.31**		
Winter garden	F	LD	Imidacloprid	2.32		Tiwari et al. (2011)

Winter garden	F	LD	Carbaryl	2.88	
Winter garden	F	LD	Fenpropathrin	3.73	
Winter garden	F	LD	Chlorpyrifos	2.07	
Winter garden	F	LD	**Spinetoram**	**5.92**	Wei et al. (2001)

Blattodea

Blatella germanica (German cockroach)

Apyr-R	F	T	Permethrin	97	–
Apyr-R	F	T	Deltamethrin	480	
Apyr-R	F	T	**Spinosad**	**1.3**	Nasirian et al. (2011)
Boostan7	F	T	Permethrin	High	
Boostan7	F	T	Cypermethrin	High	
Boostan7	F	T	Cyfluthrin	High	
Boostan7	F	T	DDT	High	
Boostan7	F	T	**Spinosad**	**1.31**	Nasirian et al. (2011)
A22	F	T	Permethrin	High	
A22	F	T	Cypermethrin	High	
A22	F	T	Cyfluthrin	High	
A22	F	T	DDT	High	
A22	F	T	**Spinosad**	**1.07**	Nasirian et al. (2011)
Saman	F	T	Permethrin	11.9	

Continued

Table 3.5 Cross-resistance to the spinosyns in insect strains resistant to other insecticides—representative examples from reports and publications—cont'd

Species-strain	Select	Assay	Compound	RR	Mech	Reference
Saman	F	T	Cypermethrin	5.4		
Saman	F	T	Cyfluthrin	2.9		
Saman	F	T	DDT	High		
Saman	F	T	**Spinosad**	**1.65**		
Saman	F	T	Fipronil	2.4		
Acari—Ixodidae						
Rhipicephalus microplus (southern cattle tick)						
San Alfonso	F	TS	**Spinosad**	**0.30**	MR	Miller et al. (2011)
Santa Luiza	F	TS	**Spinosad**	**1.41**	MR	
B&H Ranch	F	TS	**Spinosad**	**0.74**	py-R	
San Roman	F	TS	**Spinosad**	**1.73**	MR	
Pesqueria	F	TS	**Spinosad**	**1.18**	MR	

Selection: L, lab; F, field; FL, field-lab

Assay: T, topical; D, diet; LVD, larval dip; LD, leaf-dip; LDK, leaf disk; S, spray; DR, drench; C, cup.

Bold designates spinosyn-related.

[a]Fip-Sel—strain derived from CH1—known to have spinosad and fipronil resistance prior to further selection with fipronil.

[b]___multiresistant strain from/derived from the field—known to have resistance to pyrethroids, AChE inhibitors and acylureas.

variety of insecticides. There have been no reported cases of cross-resistance to the spinosyns in strains selected for resistance to the cyclodienes (Table 3.5, Fig. 3.12). Thus, there does not appear to be a general trend for cross-resistance between insecticides acting on GABA-gated chloride channels (fiproles, avermectins, cyclodienes, and lindane) and the spinosyns.

Early studies noted a lack of spinosyn cross-resistance in lepidopteran species that were resistant to pyrethroid or organophosphorus insecticides, through kdr or enhanced monooxygenase-based resistance mechanisms (Salgado and Sparks, 2010; Sparks et al., 1995). This trend has continued, with more recent studies in a variety of pyrethroid and organophosphorate insecticide-resistant insects failing to demonstrate meaningful cross-resistance to the spinosyns (Table 3.5, Fig. 3.12) (Sparks et al., 2012). Similarly, there is a broad lack of cross-resistance to the spinosyns in insect strains exhibiting resistance to carbamates, ecdysone agonists (moult accelerating compounds, MACs), and acylureas (Table 3.5, Fig. 3.12). Cross-resistance to the spinosyns in insect strains resistant to other classes of insecticides, including oxadiazines (e.g., indoxacarb) and neonicotinoids (e.g., imidacloprid), is highly variable ranging from none to moderate (Table 3.5, Fig. 3.12). Thus far, there does not appear to be any direct relationship between a specific mode

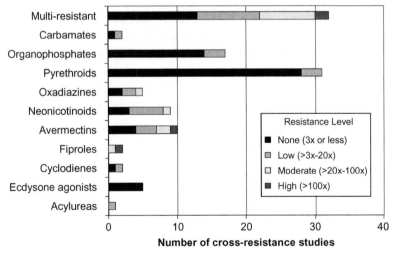

Figure 3.12 Distribution of examples of cross-resistance to the spinosyns in insect species resistant to other insecticides. Level of cross-resistance to the spinosyns—none (3 × or less);low (>3×–20×); moderate (>20–100); high (>100). Multiresistant = likely presence of more than resistance mechanism. Total number of studies = 116. (See Color plate 16 at the back of the book.)

of action or chemistry and cross-resistance to the spinosyn class of insecticides (Sparks et al., 2012).

8.5. Resistance management

The development of any new pesticide is increasingly expensive with current estimates in the range of $250 million total cost to develop a new pesticidal product (CropLife, 2012). Thus, it is critical that a new insecticide product has as long a useful life as possible, and that the product life not be cut short by the development of resistance. As noted above, many of the insect pests targeted by the spinosyns also have a history of developing resistance to the available insecticides. Therefore, IRM was an important consideration for spinosad during its commercial development process, and similar guidelines were also put in place with the launch of the second spinosyn insecticide product, spinetoram.

In IRM programmes, the spinosyns possess a favourable set of characteristics. They have a unique mode of action that distinguishes them from all other current insecticides. The IRAC has classified the spinosyns as a separate group (Group 5) that enables rotation with all other insecticides, regardless of mode of action (Nauen et al., 2012). In addition, the spinosyns show favourable selectivity for many beneficial insects making them useful tools in many Integrated Pest Management (IPM) programmes (Dripps et al., 2011). This, in turn, aids in the reduction of selection for resistance by maintaining a biotic selection component against the target pest species. Target site resistance to the spinosyns has, in general, been recessive in nature and often associated with a genetic instability or fitness cost (Dripps et al., 2011; Sparks et al., 2012). In light of these attributes, in some instances, it has been possible to re-introduce spinosad into an area where it had limited utility because of resistance at an earlier date (Dripps et al., 2011).

As noted above, cases of resistance to the spinosyns in the field are limited to some local populations of just a few species such as *P. xylostella*, *Spodoptera* spp. *L. trifolii*, and *F. occidentalis*. Thus, following 15 years of world-wide use, the spinosyns remain highly useful tools in IPM and IRM programmes.

9. CONCLUSION

Discovered and characterized a little more than 20 years ago, the spinosyns have come to epitomize the natural product-as-insecticide concept. During this time, no other insecticides have been discovered that share the spinosyn mode of action. *D. melanogaster* mutants with significantly reduced susceptibility to spinosyns were instrumental in the identification of

the Dα6 nAChR subunit as a significant target of spinosyns. The demonstra-
tion that spinosyn-resistant Dα6 mutants do not show cross-resistance to
neonicotinoids and that *D. melanogaster* strains possessing target site resistance
to the neonicotinoids do not show cross-resistance to the spinosyns indicates
that spinosyns and neonicotinoids are acting upon different subtypes of
nAChRs. Thus far, only the naturally derived spinosyns and their semisynthetic analogues appear to target Dα6-like nAChR subunits (Dα6, Pxα6,
Bdα6, etc.). Due, in part, to their inherent selectivity for Dα6-like nAChR
subunits, the spinosyns fit well into many IPM and IRM programmes. To
date, field-selected resistance to the spinosyns has occurred in only a limited
number of insect pest species. As such, the spinosyns remain highly effica-
cious and valuable insect management tools for use in a broad range of IPM
and IRM programmes.

REFERENCES

Abo Elghar, G.E., Elbermawy, Z.A., Yousef, A.G., Abd Elhady, H.K., 2005. Monitoring
and characterization of insecticide-resistance in cotton leafworm, Spodoptera littoralis
(Boisd) (Lepidoptera: Noctuidae). J. Asia-Pac. Entomol. 8, 397–410.

Achaleke, J., Martin, T., Ghogomu, R.T., Vaissayre, M., Brevault, T., 2009. Esterase-
mediated resistance to pyrethroids in field populations of Helicoverpa armigera (Lepidoptera: Noctuidae) from Central Africa. Pest Manag. Sci. 65, 1147–1154.

Ahmad, M., Arif, M.I., 2009. Resistance of Pakistani field populations of spotted bollworm
Earias vittella (Lepidoptera: Noctuidae) to pyrethroid, organophosphorus and new
chemical insecticides. Pest Manag. Sci. 65, 433–439.

Ahmad, M., Arif, M.I., Ahmad, Z., 2003. Susceptibility of Helicoverpa armigera (Lepidoptera: Noctuidae) to new chemistries in Pakistan. Crop Prot. 22, 539–544.

Ahmad, M., Hollingworth, R.M., Wise, J.C., 2002. Broad-spectrum insecticide resistance in
obliquebanded leafroller Choristoneura rosaceana (Lepidoptera: Tortricidae) from Michigan. Pest Manag. Sci. 58, 834–838.

Ahmad, M., Sayyed, A.H., Saleem, M.A., Ahmad, M., 2008. Evidence for field evolved resistance to newer insecticides in Spodoptera litura (Lepidoptera: Noctuidae) from Pakistan.
Crop Prot. 27, 1367–1372.

Ahrnad, M., Sayyed, A.H., Crickmore, N., Saleem, M.A., 2007. Genetics and mechanism of
resistance to deltamethrin in a field population of Spodoptera litura (Lepidoptera:
Noctuidae). Pest Manag. Sci. 63, 1002–1010.

Attique, M.N.R., Khaliq, A., Sayyed, A.H., 2006. Could resistance to insecticides in Plutella
xylostella (Lep., Plutellidae) be overcome by insecticide mixtures? J. Appl. Entomol. 130,
122–127.

Baxter, S.W., Chen, M., Dawson, A., Zhao, J.Z., Vogel, H., Shelton, A.M., Heckel, D.G.,
Jiggins, C.D., 2010. Mis-spliced transcripts of nicotinic acetylcholine receptor alpha6 are
associated with field evolved spinosad resistance in Plutella xylostella (L.). PLoS Genet. 6,
e1000802.

Bertrand, D., Ballivet, M., Gomez, M., Bertrand, S., Phannavong, B., Gundelfinger, E.D.,
1994. Physiological properties of neuronal nicotinic receptors reconstituted from the
vertebrate beta 2 subunit and Drosophila alpha subunits. Eur. J. Neurosci. 6, 869–875.

Bielza, P., 2008. Insecticide resistance management strategies against the western flower
thrips, Frankliniella occidentalis. Pest Manag. Sci. 64, 1131–1138.

Bielza, P., Quinto, V., Contreras, J., Torne, M., Martin, A., Espinosa, P.J., 2007a. Resistance to spinosad in the western flower thrips, Frankliniella occidentalis (Pergande), in greenhouses of south-eastern Spain. Pest Manag. Sci. 63, 682–687.

Bielza, P., Quinto, V., Fernandez, E., Gravalos, C., Contreras, J., 2007b. Genetics of spinosad resistance in Frankliniella occidentalis (Thysanoptera: Thripidae). J. Econ. Entomol. 100, 916–920.

Brejc, K., Van Dijk, W.J., Klaassen, R.V., Schuurmans, M., Van Der Oost, J., Smit, A.B., Sixma, T.K., 2001. Crystal structure of an ACh-binding protein reveals the ligand-binding domain of nicotinic receptors. Nature 411, 269–276.

Casida, J.E., Tomizawa, M., 2008. Insecticide interactions with gamma-aminobutyric acid and nicotinic receptors: predictive aspects of structural models. J. Pestic. Sci. 33, 4–8.

Changeux, J., Edelstein, S.J., 2005. Nictinic Acetylcholine Receptors. Odile Jacob Publishing Corporation, New York.

Chen, X.L., Yuan, L.Z., Du, Y.Z., Zhang, Y.J., Wang, J.J., 2011. Cross-resistance and biochemical mechanisms of abamectin resistance in the western flower thrips, Frankliniella occidentalis. Pestic. Biochem. Physiol. 101, 34–38.

Cordero, R.J., Kuhar, T.P., Speese, J.I., 2005. Susceptibility of diamondback moth and its parasitoids to commercial insecticides in Virginia. In: Entomological Society of America, Annual Meeting, Ft Lauderdale, FL, USA.

Couso-Ferrer, F., Arouri, R., Beroiz, B., Perera, N., Cervera, A., Navarro-Llopis, V., Castanera, P., Hernandez-Crespo, P., Ortegoa, F., 2011. Cross-resistance to insecticides in a malathion-resistant strain of Ceratitis capitata (Diptera: Tephritidae). J. Econ. Entomol. 104, 1349–1356.

Creemer, L.C., Kirst, H.A., Paschal, J.W., Worden, T.V., 2000. Synthesis and insecticidal activity of spinosyn analogs functionally altered at the 2'-, 3'- and 4'-positions of the rhamnose moiety. J. Antibiot. 53, 171–178.

Crouse, G.D., Sparks, T.C., 1998. Naturally derived materials as products and leads for insect control: the spinosyns. Rev. Toxicol. 2, 133–146.

Crouse, G.D., Sparks, T.C., Deamicis, C.V., Kirst, H.A., Martynow, J.G., Creemer, L.C., Worden, T.V., Anzeveno, P.B., 1999. Chemistry and insecticidal activity of the spinosyns. In: Brooks, G.T., Roberts, T.R. (Eds.), Pesticide Chemistry and Bioscience: The Food-Environment Challenge. Royal Society of Chemistry, Cambridge, UK, pp. 155–166.

Crouse, G.D., Sparks, T.C., Schoonover, J., Gifford, J., Dripps, J., Bruce, T., Larson, L.L., Garlich, J., Hatton, C., Hill, R.L., Worden, T.V., Martynow, J.G., 2001. Recent advances in the chemistry of spinosyns. Pest Manag. Sci. 57, 177–185.

Darriet, F., Duchon, S., Hougard, J.M., 2005. Spinosad: a new larvicide against insecticide-resistant mosquito larvae. J. Am. Mosq. Control Assoc. 21, 495–496.

Darriet, F., Marcombe, S., Etienne, M., Yebakima, A., Agnew, P., Yp-Tcha, M.M., Corbel, V., 2010. Field evaluation of pyriproxyfen and spinosad mixture for the control of insecticide resistant Aedes aegypti in Martinique (French West Indies). Parasit. Vectors 3, 88.

Deamicis, C.V., Dripps, J.E., Hatton, C.J., Karr, L.L., 1997. Physical and biological properties of the spinosyns: Novel macrolide pest-control agents from fermentation. In: Hedin, P.A., Hollingworth, R.M., Masler, E.P., Miyamoto, J., Thompson, D.G. (Eds.), Phytochemicals for Pest Control. American Chemical Society, Washington, pp. 144–154 Symposium Series, no.658.

Devillers, J.E., 1996. Neural Networks in QSAR and Drug Design. Academic Press, New York.

Dripps, J.E., Deamicis, C.V., Sparks, T.C., Crouse, G.D., 2008. AGRO 186-Spinosad to spinetoram: evolution of the spinosyns. Abstracts of Papers of the American Chemical Society, 236.

Dripps, J.E., Boucher, R.E., Chloridis, A., Cleveland, C.B., Deamicis, C.V., Gomez, L.E., Paroonagian, D.L., Pavan, L.A., Sparks, T.C., Watson, G.B., 2011. The spinosyn insecticides. In: Lopez, O., Fernandez-Bolanos, J.G. (Eds.), Green Trends in Insect Control. Royal Society of Chemistry, Cambridge, UK, pp. 163–212.

Dunley, J.E., Brunner, J.F., Doerr, M.D., Beers, E.H., 2006. Resistance and cross-resistance in populations of the leafrollers, Choristoneura rosaceana and Pandemis pyrusana, in Washington apples. J. Insect Sci. 6, 1 7.

Evans, D.A., Black, W.C., 1993. Total synthesis of (+)-A83543A [(+)-lepicidin A. J. Am. Chem. Soc. 115, 4497–4513.

Eziah, V.Y., Rose, H.A., Clift, A.D., Mansfield, S., 2008. Susceptibility of four field populations of the diamondback moth Plutella xylostella L. (Lepidoptera: Yponomeutidae) to six insecticides in the Sydney region, New South Wales, Australia. Aust. J. Entomol. 47, 355–360.

Ferguson, J.S., 2004. Development and stability of insecticide resistance in the leafminer Liriomyza trifolii (Diptera: Agromyzidae) to cyromazine, abamectin, and spinosad. J. Econ. Entomol. 97, 112–119.

Forsayeth, J.R., Franco, A., Rossi, A.B., Lansman, J.B., Hall, Z.W., 1990. Expression of functional-mouse muscle acetylcholine-receptors in chinese hamster ovary cells. J. Neurosci. 10, 2771–2779.

Gaisser, S., Carletti, I., Schell, U., Graupner, P.R., Sparks, T.C., Martin, C.J., Wilkinson, B., 2009. Glycosylation engineering of spinosyn analogues containing an L-olivose moiety. Org. Biomol. Chem. 7, 1705–1708.

Gao, J.R., Deacutis, J.M., Scott, J.G., 2007. The nicotinic acetylcholine receptor subunits Mdalpha5 and Mdbeta3 on autosome 1 of Musca domestica are not involved in spinosad resistance. Insect Mol. Biol. 16, 691–701.

Gill, H.K., Dhawan, A.K., 2006. Monitoring of insecticide resistance to Helicoverpa armigeria (Hubner) in cotton growing areas of Punjab. Pest. Res. J. 18, 150–153.

Grauso, M., Reenan, R.A., Culetto, E., Sattelle, D.B., 2002. Novel putative nicotinic acetylcholine receptor subunit genes, Dalpha5, Dalpha6 and Dalpha7, in Drosophila melanogaster identify a new and highly conserved target of adenosine deaminase acting on RNA-mediated A-to-I pre-mRNA editing. Genetics 160, 1519–1533.

Guillen, J., Bielza, P., 2012. Thiamethoxam acts as a target-site synergist of spinosad in resistant strains of Frankliniella occidentalis. Pest Manag. Sci. http://dx.doi.org/10.1002/ps.3372.

Hahn, D.R., Gustafson, G., Waldron, C., Bullard, B., Jackson, J.D., Mitchell, J., 2006. Butenyl-spinosyns, a natural example of genetic engineering of antibiotic biosynthetic genes. J. Ind. Microbiol. Biotechnol. 33, 94–104.

Haider, H., Ahmed, S., Khan, R.R., 2011. Determination of level of insecticide resistance in fruit fly, Bactrocera zonata (Saunders) (Diptera: Tephritidae) by bait bioassay. Int. J. Agric. Biol. 13, 815–818.

Halevi, S., Mckay, J., Palfreyman, M., Yassin, L., Eshel, M., Jorgensen, E., Treinin, M., 2002. The C-elegans ric-3 gene is required for maturation of nicotinic acetylcholine receptors. EMBO J. 21, 1012–1020.

Herron, G.A., James, T.M., 2005. Monitoring insecticide resistance in Australian Frankliniella occidentalis Pergande (Thysanoptera: Thripidae) detects fipronil and spinosad resistance. Aust. J. Entomol. 44, 299–303.

Hsu, J.C., Feng, H.T., 2006. Development of resistance to spinosad in oriental fruit fly (Diptera: Tephritidae) in laboratory selection and cross-resistance. J. Econ. Entomol. 99, 931–936.

Hsu, J.C., Feng, H.T., Wu, W.J., Geib, S.M., Mao, C.H., Vontas, J., 2012. Truncated transcripts of nicotinic acetylcholine subunit gene Bd alpha 6 are associated with spinosad resistance in Bactrocera dorsalis. Insect Biochem. Mol. Biol. 42, 806–815.

Hussain, R., Ashfaq, M., Saleem, M.A., Ahmed, S., 2005. Toxicity of some insecticides with novel mode of action against malathion-resistant organophosphate-susceptible strains of *Tribolium castaneum* larvae. J. Agric. Biol. 7, 768–772.

Hussain, R., Ashfaq, M., Saleem, M.A., 2009. Biochemical abnormalities produced by spinosad in *Tribolium castaneum* adult beetles. Int. J. Agric. Biol. 11, 241–244.

Ihara, M., Matsuda, K., Shimomura, M., Sattelle, D.B., Komai, K., 2004. Super agonist actions of clothianidin and related compounds on the SAD beta 2 nicotinic acetylcholine receptor expressed in Xenopus laevis oocytes. Biosci. Biotechnol. Biochem. 68, 761–763.

Ihara, M., Okajima, T., Yamashita, A., Oda, T., Hirata, K., Nishiwaki, H., Morimoto, T., Akamatsu, M., Ashikawa, Y., Kuroda, S., Mega, R., Kuramitsu, S., Sattelle, D.B., Matsuda, K., 2008. Crystal structures of Lymnaea stagnalis AChBP in complex with neonicotinoid insecticides imidacloprid and clothianidin. Invert. Neurosci. 8, 71–81.

Ishtiaq, M., Saleem, M.A., 2011. Generating susceptible strain and resistance status of field populations of *Spodoptera exigua* (Lepidoptera: Noctuidae) against some conventional and new chemistry insecticides in Pakistan. J. Econ. Entomol. 104, 1343–1348.

Ishtiaq, M., Salem, M.A., Razaq, M., 2012. Monitoring of resistance in *Spodoptera exigua* (Lepidoptera: Noctuidae) from four districts of the Southern Punjab, Pakistan to four conventional and six new chemistry insecticides. Crop Prot. 33, 13–20.

Jeschke, P., Nauen, R., 2008. Neonicotinoids—from zero to hero in insecticide chemistry. Pest Manag. Sci. 64, 1084–1098.

Jeschke, P., Nauen, R., 2012. Nicotinic Acetylcholine Receptor Agonists: Target and Selectivity Aspects. Weinheim, Germany, Wiley-VCH.

Jones, A.K., Sattelle, D.B., 2010. Diversity of insect nicotinic acetylcholine receptor subunits. Adv. Exp. Med. Biol. 683, 25–43.

Kakani, E.G., Zygouridis, N.E., Tsoumani, K.T., Seraphides, N., Zalom, F.G., Mathiopoulos, K.D., 2010. Spinosad resistance development in wild olive fruit fly *Bactrocera oleae* (Diptera: Tephritidae) populations in California. Pest Manag. Sci. 66, 447–453.

Kang, C.Y., Wu, G., Miyata, T., 2006. Synergism of enzyme inhibitors and mechanisms of insecticide resistance in *Bemisia tabaci* (Gennadius) (Hom, Aleyrodidae). J. Appl. Entomol. 130, 377–385.

Kao, C.H., Cheng, E.Y., 2001. Insecticide resistance in *Plutella xylostella* (L.) XI. Resistance to new introduced insecticides in Taiwan (1990–2001). J. Agric. Res. China 50, 80–89.

Karlin, A., 2002. Emerging structure of the nicotinic acetylcholine receptors. Nat. Rev. Neurosci. 3, 102–114.

Khaliq, A., Attique, M.N.R., Sayyed, A.H., 2007. Evidence for resistance to pyrethroids and organophosphates in Plutella xylostella (Lepidoptera: Plutellidae) from Pakistan. Bull. Entomol. Res. 97, 191–200.

Kirst, H.A., 2010. The spinosyn family of insecticides: realizing the potential of natural products research. J. Antibiot. (Tokyo) 63, 101–111.

Kirst, H.A., Michel, K.H., Mynderase, J.S., Chio, E.H., Yao, R.C., Nakasukasa, W.M., Boeck, L.D., Occlowitz, J.L., Paschal, J.W., Deeter, J.B., Thompson, G.D., 1992. Discovery, isolation and structure elucidation of a family of the structurally unique, fermentation derived tetracyclic macrolides. In: Baker, D.R., Fenyes, J.G., Steffens, J.J. (Eds.), Synthesis and Chemistry of Agrochemicals III, 214–225 ACS Symposium Series, Vol. 504.

Kranthi, K.R., Kranthi, S., Ali, S., Banerjee, S.K., 2000. Resistance to CrylAc deltaendotoxin of Bacillus thuringiensis' in a laboratory selected strain of *Helicoverpa armigera* (Hubner). Curr. Sci. 78, 1001–1004.

Kristensen, M., Jespersen, J.B., 2004. Susceptibility of spinosad in *Musca domestica* (Diptera: Muscidae) field populations. J. Econ. Entomol. 97, 1042–1048.

Kubinyi, H., 1993. QSAR: Hansch Analysis and Related Approaches. VCH, New York.

Lansdell, S.J., Millar, N.S., 2000. The influence of nicotinic receptor subunit composition upon agonist, alpha-bungarotoxin and insecticide (imidacloprid) binding affinity. Neuropharmacology 39, 671–679.

Lansdell, S.J., Collins, T., Goodchild, J., Millar, N.S., 2012. The Drosophila nicotinic acetylcholine receptor subunits Dalpha5 and Dalpha7 form functional hemomeric and heteromeric channels. BMC Neurosci. 13, 73. http://dx.doi.org/10.1186/1471-2202 13 73.

Lewer, P., Hahn, D.R., Karr, L.L., Duebelbeis, D.O., Gilbert, J.R., Crouse, G.D., Worden, T., Sparks, T.C., Edwards, P.M., Graupner, P.R., 2009. Discovery of the butenyl-spinosyn insecticides: novel macrolides from the new bacterial strain Saccharopolyspora pogona. Bioorg. Med. Chem. 17, 4185–4196.

Liu, N.N., Yue, X., 2000. Insecticide resistance and cross-resistance in the house fly (Diptera: Muscidae). J. Econ. Entomol. 93, 1269–1275.

Liu, H., Cupp, E.W., Guo, A.G., Liu, N.N., 2004a. Insecticide resistance in Alabama and Florida mosquito strains of Aedes albopictus. J. Med. Entomol. 41, 946–952.

Liu, H.Q., Cupp, E.W., Micher, K.M., Guo, A.G., Liu, N.N., 2004b. Insecticide resistance and cross-resistance in Alabama and Florida strains of Culex quinquefaciatus. J. Med. Entomol. 41, 408–413.

Liu, S.S., Li, Z.M., Liu, Y.Q., Feng, M.G., Tang, Z.H., 2007. Promoting selection of resistance to spinosad in the parasitoid Cotesia plutellae by integrating resistance of hosts to the insecticide into the selection process. Biol. Control 41, 246–255.

Liu, S., Li, M., Ou, X., Pei, H., Liu, L., Wang, X., 2010. Spinosad derivatives, their preparation method and application as insecticide. China Patent Application

Loughner, R.L., Warnock, D.F., Cloyd, R.A., 2005. Resistance of greenhouse, laboratory, and native populations of western flower thrips to spinosad. HortScience 40, 146–149.

Magana, C., Hernandez-Crespo, P., Ortego, F., Castanera, P., 2007. Resistance to malathion in field populations of Ceratitis capitata. J. Econ. Entomol. 100, 1836–1843.

Mao, F.L.R., Gusukuma-Minuto, L., 2004. Diamondback moth, Plutella xylostella (L.), resistance management in Hawaii. In: Endersby, N.M., Ridland, P.M. (Eds.), The Management of Diamondback Moth and Other Crucifer Pests: Proceedings of the Fourth International Workshop, Melbourne, Australia.

Markussen, M.D.K., Kristensen, M., 2012. Spinosad resistance in female Musca domestica L. from a field-derived population. Pest Manag. Sci. 68, 75–82.

Mascarenhas, R.N., Boethel, D.J., 1997. Responses of field-collected strains of soybean looper (Lepidoptera: Noctuidae) to selected insecticides using an artificial diet overlay bioassay. J. Econ. Entomol. 90, 1117–1124.

Matsuda, K., Buckingham, S.D., Freeman, J.C., Squire, M.D., Baylis, H.A., Sattelle, D.B., 1998. Effects of the alpha subunit on imidacloprid sensitivity of recombinant nicotinic acetylcholine receptors. Br. J. Pharmacol. 123, 518–524.

Mergott, D.J., Frank, S.A., Roush, W.R., 2004. Total synthesis of (-)- spinosyn A. Proc. Natl. Acad. Sci. U.S.A. 101, 11955–11959.

Mertz, F.P., Yao, R.C., 1990. Saccharopolyspora spinosa sp. nov. isolated from soil collected in a sugar mill rum still. Int. J. Syst. Bacteriol. 40, 34–39.

Meunier, J.C., Boquet, P., Changeux, J.P., Fromageo, P., Menez, A., Olsen, R.W., 1972. Studies on cholinergic receptor protein of electrophorus-electricus. 2. Some physical properties of cholinergic receptor protein from electrophorus-electricus revealed by a tritiated alpha-toxin from Naja nigricollis venom. Biochemistry 11, 1200–1210.

Miles, M., Lysandrou, M., 2002. Evidence for negative cross resistance to insecticides in field collected Spodoptera littoralis (Boisd.) from Lebanon in laboratory bioassays. Meded. Rijksuniv. Gent Fak. Landbouwkd. Toegep. Biol. Wet. 67, 665–669.

Millar, N.S., 2009. A review of experimental techniques used for the heterologous expression of nicotinic acetylcholine receptors. Biochem. Pharmacol. 78, 766–776.

Millar, N.S., Denholm, I., 2007. Nicotinic acetylcholine receptors: targets for commercially important insecticides. Invert. Neurosci. 7, 53–66.

Miller, R.J., White, W.H., Davey, R.B., George, J.E., Perez De Leon, A., 2011. Efficacy of spinosad against acaricide-resistant and -susceptible *Rhipicephalus* (Boophilus) *microplus* and acaricide-susceptible *Amblyomma americanum* and *Dermacentor variabilis*. J. Med. Entomol. 48, 358–365.

Mishina, M., Kurosaki, T., Tobimatsu, T., Morimoto, Y., Noda, M., Yamamoto, T., Terao, M., Lindstrom, J., Takahashi, T., Kuno, M., Numa, S., 1984. Expression of functional acetylcholine-receptor from cloned cDNAs. Nature 307, 604–608.

Mota-Sanchez, D., Hollingworth, R.M., Grafius, E.J., Moyer, D.D., 2006. Resistance and cross-resistance to neonicotinoid insecticides and spinosad in the Colorado potato beetle, *Leptinotarsa decemlineata* (Say) (Coleoptera: Chrysomelidae). Pest Manag. Sci. 62, 30–37.

Mota-Sanchez, D., Wise, J.C., Poppen, R.V., Gut, L.J., Hollingworth, R.M., 2008. Resistance of codling moth, *Cydia pomonella* (L.) (Lepidoptera: Tortricidae), larvae in Michigan to insecticides with different modes of action and the impact on field residual activity. Pest Manag. Sci. 64, 881–890.

Moulton, J.K., Pepper, D.A., Dennehy, T.J., 2000. Beet armyworm (Spodoptera exigua) resistance to spinosad. Pest Manag. Sci. 56, 842–848.

MSU, 2012. Arthropod Pesticide Resistance Database. Michigan State University. www.pesticideresistance.org.

Nasirian, H., Ladonni, H., Aboulhassani, M., Limoee, M., 2011. Susceptibility of field populations of *Blattella germanica* (Blattaria: Blattellidae) to spinosad. Pak. J. Biol. Sci. 14, 862–868.

Nauen, R., Elbert, A., Mccaffery, A., Slater, R., Sparks, T.C., 2012. IRAC: Insecticide resistance, and mode of action classification of insecticides In: *Modern Crop Protection Compounds*. Kramer, W., Schirmer, U., Jeschke, P. and Witschel, M. (Eds.). 2 ed. Weinheim, Germany: Wiley-VCH.

Nauen, R., Konanz, S., Hirooka, H., Nishimatsu, T., Kodama, H., 2007. Flubendiamide: a unique tool in resistance management tactics for pest Lapidoptera difficult to control. Pflanzenschutz Nachr. Bayer 60, 247–262.

Noda, M., Furutani, Y., Takahashi, H., Toyosato, M., Tanabe, T., Shimizu, S., Kikyotani, S., Kayano, T., Hirose, T., Inayama, S., Numa, S., 1983. Cloning and sequence-analysis of calf cDNA and human genomic DNA encoding alpha-subunit precursor of muscle acetylcholine-receptor. Nature 305, 818–823.

Numa, S., Noda, M., Takahashi, H., Tanabe, T., Toyosato, M., Furutani, Y., Kikyotani, S., 1983. Molecular-structure of the nicotinic acetylcholine-receptor. Cold Spring Harb. Symp. Quant. Biol. 48, 57–69.

Orr, N., Hasler, J., Watson, G.B., Mitchell, J., Gustafson, G., Gifford, J., Geng, C., Chouinard, S.W., Cook, K.D., 2006. Spinosad: from nature to green chemistry to novel mode of action. In: 11th IUPAC International Congress of Pesticide Chemistry Abstracts.

Orr, N., Shaffner, A.J., Richey, K., Crouse, G.D., 2009. Novel mode of action of spinosad: receptor binding studies demonstrating lack of interaction with known insecticidal target sites. Pestic. Biochem. Physiol. 95, 1–5.

Osorio, A., Martinez, A.M., Schneider, M.I., Diaz, O., Corrales, J.L., Aviles, M.C., Smagghe, G., Pineda, S., 2008. Monitoring of beet armyworm resistance to spinosad and methoxyfenozide in Mexico. Pest Manag. Sci. 64, 1001–1007.

Paquette, L.A., Collado, I., Purdie, M., 1998. Total synthesis of spinosyn A. 2. Degradation studies involving the pure factor and its complete reconstitution. J. Am. Chem. Soc. 120, 2553–2562.

Perry, T., Mckenzie, J.A., Batterham, P., 2007. A Dalpha6 knockout strain of *Drosophila melanogaster* confers a high level of resistance to spinosad. Insect Biochem. Mol. Biol. 37, 184–188.

Perry, T., Chan, J.Q., Batterham, P., Watson, G.B., Geng, C.X., Sparks, T.C., 2012. Effects of mutations in Drosophila nicotinic acetylcholine receptor subunits on sensitivity to insecticides targeting nicotinic acetylcholine receptors. Pestic. Biochem. Physiol. 102, 56–60.

Pollizzi, S., Payne, G., 2004. Laboratory evaluation of selected insecticides on field-collected populations on bollworm and tobacco budworm larvae. In: Beltwide Cotton Conference, San Antonio, Texas, 1832–1835.

Pu, X., Yang, Y.H., Wu, S.W., Wu, Y.D., 2010. Characterisation of abamectin resistance in a field-evolved multiresistant population of *Plutella xylostella*. Pest Manag. Sci. 66, 371–378.

Puinean, A.M., Lansdell, S.J., Collins, T., Bielza, P., Millar, N.S., 2012. A nicotinic acetylcholine receptor transmembrane point mutation (G275E) associated with resistance to spinosad in *Frankliniella occidentalis*. J. Neurochem. http://dx.doi.org/10.1111/jnc.12029.

Raftery, M.A., Hunkapiller, M.W., Strader, C.D., Hood, L.E., 1980. Acetylcholine-receptor—complex of homologous subunits. Science 208, 1454–1457.

Raymond Delpech, V., Ihara, M., Coddou, C., Matsuda, K., Sattelle, D.B., 2003. Action of nereistoxin on recombinant neuronal nicotinic acetylcholine receptors expressed in *Xenopus laevis* oocytes. Invert. Neurosci. 5, 29–35.

Reiter, M.J., Prives, J.M., Cowburn, D.A., Karlin, A., 1972. Affinity labeling of acetylcholine receptor in electroplax—electrophoretic separation in sodium dodecyl sulfate. Proc. Natl. Acad. Sci. U.S.A. 69, 1168–1172.

Reyes, M., Sauphanor, B., 2008. Resistance monitoring in codling moth: a need for standardization. Pest Manag. Sci. 64, 945–953.

Reyes, M., Rocha, K., Alarcon, L., Siegwart, M., Sauphanor, B., 2012. Metabolic mechanisms involved in the resistance of field populations of *Tuta absoluta* (Meyrick) (Lepidoptera: Gelechiidae) to spinosad. Pestic. Biochem. Physiol. 102, 45–50.

Rinkevich, F.D., Chen, M., Shelton, A.M., Scott, J.G., 2010. Transcripts of the nicotinic acetylcholine receptor subunit gene Pxyl alpha 6 with premature stop codons are associated with spinosad resistance in diamondback moth, *Plutella xylostella*. Invert. Neurosci. 10, 25–33.

Roe, R.M., Young, H.P., Iwasa, T., Wyss, C.F., Stumpf, C.F., Sparks, T.C., Watson, G.B., Sheets, J.J., Thompson, G.D., 2010. Mechanism of resistance to spinosyn in the tobacco budworm, *Heliothis virescens*. Pestic. Biochem. Physiol. 96, 8–13.

Salgado, V.L., 1997. The modes of action of spinosad and other insect control products. Down to Earth 52, 35–43.

Salgado, V.L., 1998. Studies on the mode of action of spinosad: insect symptoms and physiological correlates. Pestic. Biochem. Physiol. 60, 91–102.

Salgado, V.L., Saar, R., 2004. Desensitizing and non-desensitizing subtypes of alpha-bungarotoxin-sensitive nicotinic acetylcholine receptors in cockroach neurons. J. Insect Physiol. 50, 867–879.

Salgado, V.L., Sparks, T.C., 2010. The spinosyns: chemistry, biochemistry, mode of action, and resistance. In: Gilbert, L.I., Gill, Sarjeet S. (Eds.), Insect Control: Biological and Synthetic Agents. Academic Press, New York, pp. 207–243.

Salgado, V.L., Sheets, J.J., Watson, G.B., Schmidt, A.L., 1998. Studies on the mode of action of spinosad: the internal effective concentration and the concentration dependence of neural excitation. Pestic. Biochem. Physiol. 60, 103–110.

Salgado, V.L., Watson, G.B., Sheets, J.J., 1997. Studies on the mode of action of spinosad, the active ingredient in Tracer insect control. In: Proceedings—Beltwide Cotton Conferences, vol. 2, 1082–1084.

Salt, D.W., Yildiz, N., Livingstone, D.J., Tinsley, C.J., 1992. The use of artificial neural networks in QSAR. Pestic. Sci. 36, 161–170.

Sattelle, D.B., 2009. Invertebrate nicotinic acetylcholine receptors-targets for chemicals and drugs important in agriculture, veterinary medicine and human health. J. Pestic. Sci. 34, 233–240.

Sattelle, D.B., Jones, A.K., Sattelle, B.M., Matsuda, K., Reenan, R., Biggin, P.C., 2005. Edit, cut and paste in the nicotinic acetylcholine receptor gene family of Drosophila melanogaster. Bioessays 27, 366–376.

Sayyed, A.H., Omar, D., Wright, D.J., 2004. Genetics of spinosad resistance in a multiresistant field-selected population of Plutella xylostella. Pest Manag. Sci. 60, 827–832.

Sayyed, A.H., Ahmad, M., Saleem, M.A., 2008. Cross-resistance and genetics of resistance to indoxacarb in Spodoptera litura (Lepidoptera: Noctuidae). J. Econ. Entomol. 101, 472–479.

Sayyed, A.H., Wright, D.J., 2004. Fipronil resistance in the diamondback moth (Lepidoptera: Plutellidae): inheritance and number of genes involved. J. Econ. Entomol. 97, 2043–2050.

Sayyed, A.H., Attique, M.N.R., Khaliq, A., 2005a. Stability of field-selected resistance to insecticides in Plutella xylostella (Lep., Plutellidae) from Pakistan. J. Appl. Entomol. 129, 542–547.

Sayyed, A.H., Attique, M.N.R., Khaliq, A., Wright, D.J., 2005b. Inheritance of resistance and cross-resistance to deltamethrin in Plutella xylostella (Lepidoptera: Plutellidae) from Pakistan. Pest Manag. Sci. 61, 636–642.

Scott, J.G., 1998. Toxicity of spinosad to susceptible and resistant strains of house flies, Musca domestica. Pestic. Sci. 54, 131–133.

Scott, J.G., 2008. Unraveling the mystery of spinosad resistance in insects. J. Pestic. Sci. (Tokyo, Japan) 33, 221–227.

Sheehan, L.S., Lill, R.E., Wilkinson, B., Sheridan, R.M., Vousden, W.A., Kaja, A.L., Crouse, G.D., Gifford, J., Graupner, P.R., Karr, L., Lewer, P., Sparks, T.C., Leadlay, P.F., Waldron, C., Martin, C.J., 2006. Engineering of the spinosyn PKS: directing starter unit incorporation. J. Nat. Prod. 69, 1702–1710.

Shelton, A.M., Sances, F.V., Hawley, J., Tang, J.D., Boune, M., Jungers, D., Collins, H.L., Farias, J., 2000. Assessment of insecticide resistance after the outbreak of diamondback moth (Lepidoptera: Plutellidae) in California in 1997. J. Econ. Entomol. 93, 931–936.

Shi, J., Zhang, L., Gao, X.W., 2011. Characterisation of spinosad resistance in the housefly Musca domestica (Diptera: Muscidae). Pest Manag. Sci. 67, 335–340.

Shono, T., Scott, J.G., 2003. Spinosad resistance in the housefly, Musca domestica, is due to a recessive factor on autosome 1. Pestic. Biochem. Physiol. 75, 1–7.

Shono, T., Zhang, L., Scott, J.G., 2004. Indoxacarb resistance in the house fly, Musca domestica. Pestic. Biochem. Physiol. 80, 106–112.

Sial, A.A., Brunner, J.F., 2012. Selection for resistance, reversion towards susceptibility and synergism of chlorantraniliprole and spinetoram in obliquebanded leafroller, Choristoneura rosaceana (Lepidoptera: Tortricidae). Pest Manag. Sci. 68, 462–468.

Sial, A.A., Brunner, J.F., Doerr, M.D., 2010. Susceptibility of Choristoneura rosaceana (Lepidoptera: Tortricidae) to two new reduced-risk insecticides. J. Econ. Entomol. 103, 140–146.

Sial, A.A., Brunner, J.F., Garczynski, S.F., 2011. Biochemical characterization of chlorantraniliprole and spinetoram resistance in laboratory-selected obliquebanded leafroller, Choristoneura rosaceana (Harris) (Lepidoptera: Tortricidae). Pestic. Biochem. Physiol. 99, 274–279.

Silva, G.A., Picanco, M.C., Bacci, L., Crespo, A.L., Rosado, J.F., Guedes, R.N., 2011. Control failure likelihood and spatial dependence of insecticide resistance in the tomato pinworm, Tuta absoluta. Pest Manag. Sci. 67, 913–920.

Smirle, M.J., Lowery, D.T., Zurowski, C.L., 2003. Susceptibility of leafrollers (Lepidoptera: Tortricidae) from organic and conventional orchards to azinphosmethyl, spinosad, and *Bacillus thuringiensis*. J. Econ. Entomol. 96, 879–884.

Sparks, T.C., Thompson, G.D., Larson, L.L., Kirst, H.A., Jantz, O.K., Worden, T.V., Hertlein, M.B., Busacca, J.D., 1995. Biological characteristics of the spinosyns: a new class of naturally derived insect control agents. In: Proceedings of the 1995 Beltwide Cotton Production Conference. Beltwide Cotton Production Conference. National Cotton Council, Memphis, TN.

Sparks, T.C., Thompson, G.D., Kirst, H.A., Hertlein, M.B., Larson, L.L., Worden, T.V., Thibault, S.T., 1998. Biological activity of the spinosyns, new fermentation derived insect control agents, on tobacco budworm (Lepidoptera : Noctuidae) larvae. J. Econ. Entomol. 91, 1277–1283.

Sparks, T.C., Thompson, G.D., Kirst, H.A., Hertlein, M.B., Mynderse, J.S., Turner, J.R., Worden, T.V., 1999. Fermentation-detived insect control agents: the spinosyns. In: Hall, R., Menn, J.J. (Eds.), Biopesticides: Use and Delivery. Humana Press, Totowa, New Jersey, pp. 171–188.

Sparks, T.C., Anzeveno, P.B., Martynow, J.G., Gifford, J.M., Hertlein, M.B., Worden, T.V., Kirst, H.A., 2000. The application of artificial neural networks to the identification of new spinosoids with improved biological activity toward larvae of Heliothis virescens. Pestic. Biochem. Physiol. 67, 187–197.

Sparks, T.C., Crouse, G.D., Durst, G., 2001. Natural products as insecticides: the biology, biochemistry and quantitative structure-activity relationships of spinosyns and spinosoids. Pest Manag. Sci. 57, 896–905.

Sparks, T.C., Crouse, G.D., Dripps, J.E., Anzeveno, P., Martynow, J., Deamicis, C.V., Gifford, J., 2008. Neural network-based QSAR and insecticide discovery: spinetoram. J. Comput. Aided Mol. Des. 22, 393–401.

Sparks, T.C., Dripps, J.E., Watson, G.B., Paroonagian, D., 2012. Resistance and cross-resistance to the spinosyns - A review and analysis. Pestic. Biochem. Physiol. 102, 1–10.

Talley, T.T., Harel, M., Hibbs, R.E., Radic, Z., Tomizawa, M., Casida, J.E., Taylor, P., 2008. Atomic interactions of neonicotinoid agonists with AChBP: molecular recognition of the distinctive electronegative pharmacophore. Proc. Natl. Acad. Sci. U.S.A. 105, 7606–7611.

Thompson, G.D., Sparks, T.C., 2002. Spinosad: a green natural product for insect control. In: Lankey, R.L., Anastas, P.T. (Eds.), Advancing Sustainability through Green Chemistry and Engineering. American Chemical Society, Washington, D.C, pp. 61–73.

Thompson, G.D., Dutton, R., Sparks, T.C., 2000. Spinosad – a case study: an example from a natural products discovery programme. Pest Manag. Sci. 56, 696–702.

Tietze, L.E., Brasche, G., Grube, A., Bohnke, N., Stadler, C., 2007. Synthesis of novel spinosyn A analogues by Pd-mediated transformations. Chemistry 13, 8543–8563.

Tiwari, S., Mann, R.S., Rogers, M.E., Stelinski, L.L., 2011. Insecticide resistance in field populations of Asian citrus psyllid in Florida. Pest Manag. Sci. 67, 1258–1268.

Todeschini, R., Consonni, V., 2000. Handbook of Molecular Descriptors. Wiley-VCH, New York.

Waldstein, D.E., Reissig, W.H., 2000. Synergism of tebufenozide in resistant and susceptible strains of obliquebanded leafroller (Lepidoptera: Tortricidae) and resistance to new insecticides. J. Econ. Entomol. 93, 1768–1772.

Wang, W., Mo, J.C., Cheng, J.A., Zhuang, P.J., Tang, Z.H., 2006. Selection and characterization of spinosad resistance in *Spodoptera exigua* (Hubner) (Lepidoptera: Noctuidae). Pestic. Biochem. Physiol. 84, 180–187.

Wang, D., Qiu, X.H., Ren, X.X., Niu, F., Wang, K.Y., 2009. Resistance selection and biochemical characterization of spinosad resistance in *Helicoverpa armigera* (Hubner) (Lepidoptera: Noctuidae). Pestic. Biochem. Physiol. 95, 90–94.

Wang, G.L., Huang, X.L., Wei, H.Y., Fadamiro, H.Y., 2011. Sublethal effects of larval exposure to indoxacarb on reproductive activities of the diamondback moth, *Plutella xylostella* (L.) (Lepidoptera: Plutellidae). Pestic. Biochem. Physiol. 101, 227–231.

Wang, Z.Y., Yan, H.F., Yang, Y.H., Wu, Y.D., 2010. Biotype and insecticide resistance status of the whitefly Bemisia tabaci from China. Pest Manag. Sci. 66, 1360–1366.

Watson, G.B., 2001. Actions of insecticidal spinosyns on gamma-aminobutyric acid responses from small-diameter cockroach neurons. Pestic. Biochem. Physiol. 71, 20–28.

Watson, G.B., Chouinard, S.W., Cook, K.R., Geng, C., Gifford, J.M., Gustafson, G.D., Hasler, J.M., Larrinua, I.M., Letherer, T.J., Mitchell, J.C., Pak, W.L., Salgado, V.L., Sparks, T.C., Stilwell, G.E., 2010. A spinosyn-sensitive Drosophila melanogaster nicotinic acetylcholine receptor identified through chemically induced target site resistance, resistance gene identification, and heterologous expression. Insect Biochem. Mol. Biol. 40, 376–384.

Watson, G.B., Loso, M.R., Babcock, J.M., Hasler, J.M., Letherer, T.J., Young, C.D., Zhu, Y., Casida, J.E., Sparks, T.C., 2011. Novel nicotinic action of the sulfoximine insecticide sulfoxaflor. Insect Biochem. Mol. Biol. 41, 432–439.

Wei, Y.P., Appel, A.G., Moar, W.J., Liu, N.N., 2001. Pyrethroid resistance and crossresistance in the German cockroach, *Blattella germanica* (L). Pest Manag. Sci. 57, 1055–1059.

Weill, C.L., Mcnamee, M.G., Karlin, A., 1974. Affinity-labeling of purified acetylcholine receptor from *Torpedo californica*. Biochem. Biophys. Res. Commun. 61, 997–1003.

Whalon, M.E., Mota-Snachez, D., Hollingworth, R.M., 2008. Analysis of global pesticide resistance in arthropods. In: Whalon, M.E., Mota-Snachez, D., Hollingworth, R.M. (Eds.), Global Pesticide Resistance in Arthropods. CAB International, Cambridge, MA, pp. 5–31.

Yin, Q., Qian, L., Cao, G., Han, Z., 2010. Cross-resistance to several new pesticides exhibited by abamectin- and tebufenozide-resistant *Plutella xylostella* (L.). J. Nanjing Agric. Univ. 33, 60–64.

Young, H.P., Bailey, W.D., Roe, R.M., Iwasa, T., Sparks, T.C., Thompson, G.D., Watson, G.B., 2001. Mechanism of resistance and cross-resistance in a laboratory, spinosad-selected strain of the tobacco budworm and resistance in laboratory-selected cotton bollworms. In: Proceedings of the 2001 Beltwide Cotton Production Conference. National Cotton Council, Memphis TN.

Zhang, S.Y., Kono, S., Murai, T., Miyata, T., 2008. Mechanisms of resistance to spinosad in the western flower thrip, *Frankliniella occidentalis* (Pergande) (Thysanoptera: Thripidae). Insect Sci. 15, 125–132.

Zhao, J.Z., Li, Y.X., Collins, H.L., Gusukuma-Minuto, L., Mau, R.F.L., Thompson, G.D., Shelton, A.M., 2002. Monitoring and characterization of diamondback moth (Lepidoptera: Plutellidae) resistance to spinosad. J. Econ. Entomol. 95, 430–436.

Zhao, J.Z., Collins, H.L., Li, Y.X., Mau, R.F.L., Thompson, G.D., Hertlein, M., Andaloro, J.T., Boykin, R., Shelton, A.M., 2006. Monitoring of diamondback moth (Lepidoptera: Plutellidae) resistance to spinosad, indoxacarb, and emamectin benzoate. J. Econ. Entomol. 99, 176–181.

Zhu, Y.M., Loso, M.R., Watson, G.B., Sparks, T.C., Rogers, R.B., Huang, J.X., Gerwick, B.C., Babcock, J.M., Kelley, D., Hegde, V.B., Nugent, B.M., Renga, J.M., Denholm, I., Gorman, K., Deboer, G.J., Hasler, J., Meade, T., Thomas, J.D., 2011. Discovery and characterization of sulfoxaflor, a novel insecticide targeting sap-feeding pests. J. Agric. Food Chem. 59, 2950–2957.

Zhu, J.X., Wu, S.W., Yang, Y.H., Wu, Y.D., 2007. Laboratory selection and inheritance mode of resistance to spinosad in the diamondback moth, *Plutella xylostella* L. J. Nanjing Agric. Univ. 30, 61–65.

γ-Aminobutyrate- and Glutamate-gated Chloride Channels as Targets of Insecticides

Yoshihisa Ozoe

Department of Life Science and Biotechnology, Faculty of Life and Environmental Science, Shimane University, Matsue, Shimane, Japan

Contents

Abstract

γ-Aminobutyrate (GABA)- and glutamate-gated chloride channels (GABACls and GluCls) are members of the Cys-loop receptor channel family. Both channels are widely distributed in the nervous system of invertebrates and function as neurotransmitter receptors to mediate inhibitory synaptic transmission. These receptor channels are important targets for widely used insecticides and parasiticides. Diverse noncompetitive antagonists,

Advances in Insect Physiology, Volume 44
ISBN 978-0-12-394389-7
http://dx.doi.org/10.1016/B978-0-12-394389-7.00004-1

including the phenylpyrazole insecticide fipronil, inhibit inhibitory neurotransmission by binding to a site deep within the pentameric channel. Macrocyclic lactones, such as the insecticide/parasiticide avermectin, activate GluCls by irreversibly binding to the subunit interfaces. Recent genome sequencing and molecular biology studies have revealed a diversity of subunit isoforms arising from the duplication, alternative splicing, and RNA editing of the genes encoding these channels in various insect species, which may lead to physiological and pharmacological diversification of the receptor channels. Fipronil resistance due to target-site insensitivity in GABACls is a growing concern in several insect pest species. Nevertheless, identification of new chemistry and multiple potential binding sites for agonists, antagonists, and modulators in these channels suggests potential for developing novel pest control chemicals.

ABBREVIATIONS

3-APMPA 3-aminopropyl(methyl)phosphinic acid
3-APPA 3-aminopropylphosphinic acid
BIDN 3,3-bis(trifluoromethyl)bicyclo[2.2.1]heptane-2,2-dicarbonitrile
CACA *cis*-aminocrotonic acid
CaM kinase calcium/calmodulin-dependent protein kinase
DUM neuron dorsal unpaired median neuron
EBOB 4′-ethynyl-4-*n*-propylbicycloorthobenzoate
GABA γ-aminobutyrate
GABACl γ-aminobutyrate-gated chloride channel
GluCl glutamate-gated chloride channel
GPCR G protein-coupled receptor
IBIPPS 4-isobutyl-3-isopropylbicyclophosphorothionate
IR immunoreactivity
LGIC ligand-gated chloride channel
NA nodulisporic acid
nAChR nicotinic acetylcholine receptor
NCA noncompetitive antagonist
RDL, GRD, and LCCH3 insect GABA receptor subunits
rdl, lcch3, grd, glucl, and gaba$_b$r genes encoding the corresponding proteins
RNAi RNA interference
RT-PCR reverse transcription polymerase chain reaction
SAR structure–activity relationship
TACA *trans*-aminocrotonic acid
TBPS *tert*-butylbicyclophosphorothionate
TM transmembrane domain

1. INTRODUCTION

Recent molecular cloning of cDNAs has revealed the presence of multiple amino acid- and amine-gated chloride channels in invertebrates, including γ-aminobutyrate (GABA)-, glutamate-, acetylcholine-, serotonin-, histamine-, tyramine-, and dopamine-gated chloride channels (Cully et al., 1994, 1996; ffrench-Constant et al., 1991; Gengs et al., 2002; Gisselmann et al., 2002; Pirri et al., 2009; Putrenko et al., 2005; Ranganathan et al., 2000; Rao et al., 2009, 2010; Ringstad et al., 2009; Witte et al., 2002; Zheng et al., 2002). These channels are members of the Cys-loop ligand-gated ion channel (LGIC) family, which include nicotinic acetylcholine receptors (nAChRs) and glycine receptors, and mediate inhibitory neurotransmission in the nervous system. Members of this family have several characteristic conserved motifs such as a Cys-loop in the N-terminal domain. Of these members, the GABA- and glutamate-gated chloride channels (GABACls and GluCls), also known as GABA receptors and inhibitory glutamate receptors, respectively, are considered the most important receptor channels from a pesticide science perspective because they are validated targets for currently used insecticides and parasiticides such as fipronil, ethiprole, avermectins, and milbemycins (Ozoe et al., 2009; Raymond-Delpech et al., 2005). As these receptor channels have been shown to have multiple sites for a variety of ligands, including naturally occurring and synthetic compounds, they remain the promising targets for the future development of novel pest control chemicals.

The GABA receptors include metabotropic receptors (G protein–coupled receptors, GPCRs) and ionotropic, inhibitory receptors (ligand-gated chloride channels) (Bowery et al., 2002; Olsen and Sieghart, 2008), and the glutamate receptors comprise metabotropic receptors, ionotropic, excitatory receptors (ligand-gated cation channels), and ionotropic, inhibitory receptors (ligand-gated chloride channels) (Ferraguti et al., 2008; Niswener and Conn, 2010; Raymond and Sattelle, 2002; Traynelis et al., 2010). Inhibitory glutamate receptors are found only in invertebrates. This review focuses on insect GABACls and GluCls as important targets of insecticides.

Studies on the distribution of receptors will aid in elucidating their physiological roles. The agonistic, antagonistic, and modulatory actions of a variety of substances affecting native and recombinant receptors will form the basis of receptor pharmacology. Ligand structure–activity relationships (SARs) will aid the identification of their modes and sites of action. Identification of the location of the ligand-binding site (site of action) will be crucial for a full understanding of the molecular mechanism of any given ligand. In many cases, binding-site insensitivity is a mechanism of insecticide resistance. Advances in the field of molecular and structural biology will contribute to the clarification of the molecular structures of receptors and channels. Functional genomics, proteomics, and bioinformatics will provide a wealth of information on target biomolecules of insect pests, which is necessary for rational agrochemical discovery. With these in mind, this chapter presents an overview of recent advances in the physiology and pharmacology of insect GABA and inhibitory glutamate receptors. For important classical literature that was not included in this review, the reader is referred to additional excellent reviews (Bloomquist, 2001; Buckingham and Sattelle, 2004; Sattelle, 1990; Usherwood, 1994). For further information on the pharmacology and physiology of GABA and inhibitory glutamate receptors, other excellent reviews may be consulted (Buckingham and Sattelle, 2010; Gill, 2010; Raymond and Sattelle, 2002).

2. GABA AND GLUTAMATE

GABA was first identified in the brain more than half a century ago (Awapara et al., 1950; Roberts and Frankel, 1950; Udenfriend, 1950). A beef brain extract that inhibited spontaneous discharges of crayfish neurons was identified as GABA (Bazemore et al., 1957). Determination of the GABA and glutamate contents of the cell bodies of excitatory and inhibitory neurons of lobster ganglia revealed that the GABA content of the inhibitory cell bodies was considerably higher than that of the excitatory cell bodies (Otsuka et al., 1966, 1967). GABA is released by the stimulation of the inhibitory nerve innervating lobster muscles, but not the excitatory nerve. A recent analysis on the heads of the housefly and the diamondback moth, using the capillary electrophoresis/laser-induced fluorescence detection method, demonstrated that their brains contain high and low nanogram levels of GABA, respectively (Shi et al., 2012). Thus, GABA was established as the major inhibitory neurotransmitter in numerous studies, including mammalian studies (Krnjević, 2010).

Glutamate is present in both types of excitatory and inhibitory neurons. Glutamate was originally shown to exhibit excitatory action on the opener muscle of the crayfish, and this action was prevented by GABA (Robbins, 1959). Glutamate was detected in the perfusate of the coxa of the leg of the American cockroach *Periplaneta americana* by the stimulation of the thoracic ganglion (Kerkut et al., 1965). Glutamate also functions as an inhibitory transmitter in invertebrates (Cleland, 1996; Cull-Candy, 1976; Cull-Candy and Usherwood, 1973; Lea and Usherwood, 1973), although it has been mostly studied as an excitatory neurotransmitter.

3. GABA RECEPTORS

The inhibitory neurotransmitter GABA is released into synapses from the presynaptic neuron by depolarization triggered by the action potential. The released GABA traverses the synaptic cleft and binds to ionotropic receptors localized in the postsynaptic membrane to provoke an electrical change in the postsynaptic neuron (Hille, 2001). GABA also activates GPCRs to elicit intracellular signal transduction (Bowery et al., 2002). GABA receptors are well-established targets for drugs and pesticides. Functional and structural differences between insect and vertebrate receptors, which are revealed by molecular cloning, would provide opportunities to design selective pest control agents. This section compares vertebrate GABA receptors with insect GABA receptors.

3.1. Vertebrate GABA receptors

Vertebrate GABA receptors are classified as types A and B (Bettler et al., 2004; Hevers and Lüddens, 1998). The GABA receptors are prevalent in the central nervous system, but many of their subunits are also expressed in peripheral organs (Akinci and Schofield, 1999). The $GABA_A$ receptors belong to the Cys-loop LGIC family, which is named for the cysteine disulfide loop that is present in members of this family. The Cys-loop LGIC includes nAChRs, glycine receptors, and serotonin type-3 receptors. The $GABA_A$ receptors are therefore described as ionotropic GABA receptors or GABACls. This type of receptor mediates fast synaptic transmission. The opening of the channels that pass cations results in the depolarization of the postsynaptic neurons and triggers the generation of action potentials, whereas the opening of the channels that pass anions (like $GABA_A$ receptors) normally leads to membrane hyperpolarization and inhibits nerve excitation elicited by membrane depolarization.

$GABA_A$ receptors are composed of five homologous subunits. In mammals, the genes encoding 19 distinct subunits ($\alpha1-6$, $\beta1-3$, $\gamma1-3$, δ, ϵ, π, θ, and $\rho1-3$) are known from their cloned and expressed cDNAs and the complete sequence of the genome (Enna, 2007). The major receptor in the brain has a combination of two $\alpha1$ subunits, two $\beta2$ subunits, and one $\gamma2$ subunit. The ρ subunit-containing receptors are classified as $GABA_C$ receptors (Zhang et al., 2001), but this type of receptors is considered a specialized set of $GABA_A$ receptors (Barnard et al., 1998). $GABA_A$ and $GABA_C$ receptors have different pharmacological properties. A large N-terminal extracellular domain, four transmembrane domains (TM1–TM4), and a long, variable intracellular loop connecting the TM3 and TM4 segments are the hallmarks of every subunit (Fig. 4.1). The N-terminal extracellular domain contains a cysteine disulfide loop. Five subunits assemble to form an integral chloride channel at the center, with the TM2 domain arranged toward the channel pore. The GABA binding site resides in the extracellular interface between the N-terminal domains of two adjacent subunits; this extracellular interface is called the orthosteric site.

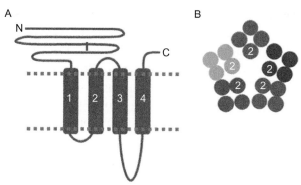

Figure 4.1 Schematic presentation of an LGIC subunit. (A) Side view. (B) Top view showing arrangement of subunits and transmembrane domains. (See Color Plate 17 at the back of the book.)

The $GABA_A$ receptors are inhibited by the competitive antagonist bicuculline and the noncompetitive antagonist (NCA) picrotoxinin. Because the $GABA_A$ receptor is an allosteric protein, the receptor is regulated by a variety of positive modulators such as barbiturates, benzodiazepines, anesthetics, avermectins, and neurosteroids (Sieghart, 1995). In this respect, NCAs can be rephrased as negative allosteric modulators.

The $GABA_B$ receptors are heterodimeric GPCRs. This type of receptor transduces signals via the activation of G proteins specifically coupled to effector proteins such as adenylate cyclase, phospholipase C-β, potassium channels, and calcium channels. $GABA_B$ receptors are also referred to as metabotropic GABA receptors. $GABA_B$ receptors are distributed widely throughout the nervous system, and some cells coexpress $GABA_A$ receptors (Waldvogel et al., 2004). These receptors are found in both the presynaptic and postsynaptic membranes. On the presynaptic membranes, the activation of $GABA_B$ receptors induces potassium conductance and reduces calcium conductance via an autoreceptor- or heteroreceptor-like feedback mechanism and thereby inhibits synaptic release of GABA or glutamate (Bonanno et al., 1998; Thompson and Gähwiler, 1992). On the postsynaptic membranes, the receptor signal is mediated by pertussis-toxin-sensitive G proteins that cause slow hyperpolarization by potassium conductance. $GABA_B$ receptors share a high amino acid sequence similarity with the family of metabotropic glutamate receptors relative to other GPCRs (Kaupmann et al., 1997). The molecular structures and physiological functions of $GABA_B$ receptors have been recently reviewed (Bettler et al., 2004; Bowery et al., 2002).

3.2. Insect GABA receptors

Both ionotropic and metabotropic GABA receptors are present in insects (Buckingham and Sattelle, 2004). Insect GABA receptors are found in both the central nervous system and the neuromuscular junction. For instance, the pharmacology of peripheral GABA receptors has been studied with the extensor tibiae muscle fibers of the desert locust *Schistocerca gregaria* (Murphy and Wann, 1988; Scott and Duce, 1987). The extensor tibiae muscle of the metathoracic leg of *S. gregaria* expresses ionotropic GABA receptors together with inhibitory ionotropic glutamate receptors (Cull-Candy and Miledi, 1981). In *P. americana*, ionotropic GABA receptors inhibited pre- and postsynaptic excitation by increasing chloride permeability through the membrane of the sixth abdominal ganglion neuron (Hue et al., 1979). Dorsal unpaired neurons of the sixth abdominal ganglion of *P. americana* express ionotropic GABA receptors on the soma and the neurite arborization (Dubreil et al., 1994). The pharmacological properties of insect ionotropic GABA receptors are distinct from those of vertebrate $GABA_A$ receptors, although they have similar functions and structures (Buckingham et al., 2005). Insect GABA receptors are activated not only by the $GABA_A$ receptor agonist muscimol but also by the $GABA_C$ receptor agonists *cis*-aminocrotonic acid (CACA) and *trans*-aminocrotonic acid (TACA). Insect GABA receptors are sensitive to picrotoxinin, but not bicuculline.

3.2.1 RDL subunit

The first gene encoding a GABA receptor subunit was cloned from the fruit fly *Drosophila melanogaster* and named *rdl* because the mutation of this gene confers resistance to the cyclodiene insecticide dieldrin (ffrench-Constant et al., 1991). The RDL protein encoded by the *rdl* gene is a common subunit of insect ionotropic GABA receptors. Orthologous genes have been identified from many other insect species, such as Acari (*Rhipicephalus microplus*), Diptera (*Anopheles funestus, Anopheles gambiae, Culex quinquefaciatus, Drosophila simulans, Aedes aegypti, Lucilia cuprina*, and *Musca domestica*), Lepidoptera (*Heliothis virescens, Spodoptera litura, Spodoptera exigua, Pultella xylostella*, and *Bombyx mori*), Coleoptera (*Tribolium castaneum*), and Hemiptera (*Ladelphax striatellus* and *Sogatella furcifera*) (http://www.ncbi.nlm.nih.gov/).

Recently, the cloning of full-length cDNAs encoding RDL subunits from *P. xylostella, S. exigua, B. mori, S. furcifera, L. striatellus*, and *Oulema oryzae* has been reported (Nakao et al., 2010, 2012a,b; Narusuye et al., 2007; Shang et al., 2009; Yu et al., 2010; Yuan et al., 2010; Zhou et al., 2008) in addition to those of *D. melanogaster* (Chen et al., 1994; ffrench–Constant et al., 1991, 1993a), *A. aegypti* (Shotkoski et al., 1994; Thompson et al., 1993b), and *H. virescens* (Wolff and Wingate, 1998). A cDNA clone encoding an RDL subunit composed of 487 amino acids was isolated from the small brown planthopper *L. striatellus* (Narusuye et al., 2007). The percent identity in the mature protein sequence between the *L. striatellus* GABA receptor subunit and the *D. melanogaster* bd-type splice variant is 70.5%. *S. exigua* RDL shares a high sequence similarity with the *H. virescens* α2 subunit, and both contain a serine at the $2'$-position (see Section 9.2.1 for index numbering), as was identified in dieldrin-resistant *Drosophila* (Shang et al., 2009). Three orthologous genes of the RDL subunits were cloned from the susceptible strain of the diamondback moth *P. xylostella* (Yuan et al., 2010; Zhou et al., 2008). The α1 and α2 subunits contain an alanine and a serine at the $2'$-position, respectively. The *P. xylostella* α1 and α2 subunits contain two variants generated by alternative splicing at exon 3 of the encoding gene. Three *rdl* cDNA clones that share high (77–91%) sequence similarity were isolated from the silkworm *B. mori* (Yu et al., 2010). Two of the three *B. mori rdl* clones are alternatively spliced at exons 3 and 6, and the three types of RDL have different amino acids, alanine, serine and glutamine, at the $2'$-position within TM2. Three variants of exon 3 and two variants of exon 6 are present in *S. furcifera rdl* (Nakao et al., 2012a). Two variants of exon 3 were found in the rice leaf beetle *O. oryzae rdl* cDNA (Nakao et al., 2012b).

GABA receptors that contain the RDL subunit were shown to be involved in mediating synaptic inhibition in *D. melanogaster* neuronal circuits (Lee et al., 2003). Embryonic *Drosophila* neurons have the capacity to form functional interactions with each other in cultures and thus provide a unique opportunity to examine the electrophysiological characteristics of synaptic transmission mediated by acetylcholine and GABA. GABA-induced currents in embryonic *Drosophila* neurons were blocked by 1 μM picrotoxin. Glutamate-induced currents were also recorded in the same neurons but were reduced slightly by 10 μM picrotoxin. Blockade of the GABA-induced currents with picrotoxin resulted in an increase in the spontaneous action potential frequency in neurons, indicating GABAergic synaptic inhibition in the neurons. The decay kinetics of miniature inhibitory postsynaptic currents varied in cells, indicating functional heterogeneity of the receptors. The functional heterogeneity of the receptors may suggest the expression of multiple receptor subtypes, given that embryos most likely contain neurons with the potential to function in the various nervous systems. Neurons from *Drosophila Rdl[MD-RR]*embryos containing the *rdl* homozygous A2'S mutation showed a marked reduction in the sensitivity to blockade by picrotoxin, supporting the direct involvement of RDL channels in mediating synaptic inhibition in neuronal circuits formed in cultures.

Several systems, such as native neurons, cell lines, and the oocytes of the African clawed frog *Xenopus laevis* expressing ligand-gated chloride channels, can be used to study the action mechanisms of insecticidal compounds acting at ligand-gated chloride channels or to screen those compounds (Belelli et al., 1996; Chen et al., 1994; Eguchi et al., 2006; ffrench-Constant et al., 1993a; Hosie and Sattelle, 1996a,b; Lee et al., 1993; Shotkoski et al., 1994; Wolff and Wingate, 1998; Zhang, et al., 1995). It is helpful to generate and use cell lines that permanently express the receptor channels for these studies (Buckingham et al., 1996; Millar et al., 1994; Narusuye et al., 2007). To facilitate the identification of stable cell lines expressing GABA receptors, the cDNA encoding the V5 epitope tagged *L. striatellus* GABA receptor subunit was subcloned into the expression vector pAc5.1-lac-Hygro, which was transfected into a *Drosophila* cell line (Narusuye et al., 2007). Cells resistant to hygromycin B were isolated and grown into clonal cell lines. The expression of receptors in clonal cell lines was examined by immunostaining with an anti-V5 antibody, followed by confocal microscopy. A cell line expressing high levels of GABA receptors was selected, and GABA-induced currents in the cell line were recorded with the whole-cell patch–clamp method.

The fast inward currents induced by the activation of *L. striatellus* GABA receptors expressed in the cell line reversed in polarity at a membrane potential of $+0.2$ mV, which was close to the calculated chloride equilibrium potential of -0.5 mV. This finding indicates that GABA-induced currents are carried by chloride ions. The EC_{50} value and the Hill coefficient for GABA-induced peak currents were estimated to be 29 μM and 1.7, respectively. *L. striatellus* GABA receptors were sensitive to both vertebrate $GABA_A$ and $GABA_C$ receptor agonists such as GABA, muscimol, isoguvacine, CACA, or 5-(4-piperidyl)-3-isoxazolol (4-PIOL).

The blocking action of the phenylpyrazole insecticide fipronil on GABA-induced currents in *L. striatellus* GABA receptors was also investigated. Inward currents were induced by 100 μM GABA at a holding potential of -60 mV with an interval of 90–120 s. Fipronil was then applied in the bath for 5 min. The peak amplitude of the inward currents was gradually inhibited and reached nearly complete inhibition 5 min after the bath application of 250 nM fipronil. The current amplitude partially recovered after washout, indicating that fipronil trapped within the channel was slowly released from the channel. The NCAs 4'-ethynyl-4-*n*-propylbicycloorthobenzoate (EBOB) and dieldrin showed inhibitory activity. The competitive antagonist SR95531, but not bicuculline, suppressed GABA-induced currents. It is thus advantageous to use stable cell lines; once established, these cells can be proliferated and used for other systems such as high-throughput screening.

Expression of ligand-gated chloride channels in the oocytes of *X. laevis* is a convenient means to study the functions of LGICs. The genes encoding subunits of two LGICs, GABACls (RDL channels) and GluCls, were cloned from the housefly (*M. domestica*). These genes were expressed in *Xenopus* oocytes by injecting their cRNAs to investigate the pharmacology of these LGICs using two-electrode voltage clamp electrophysiology (Eguchi et al., 2006). GABA activated the RDL homo-oligomers with an EC_{50} value of 101 μM, while glutamate activated the homo-oligomeric GluCls with an EC_{50} value of 30 μM. Both channels were chloride ion-permeable, and the RDL channel was more sensitive to chloride channel blockers, such as γ-BHC, fipronil and picrotoxinin, than the GluCl. The expression of GluCls in oocytes required only 1–2 days of incubation after cRNA injection, whereas 4–7 days of incubation were necessary to achieve RDL channel expression. However, injection of the cRNA of GluCl at a dose of 1% (w/w) 1 day after the injection of the cRNA of RDL resulted in a significant increase in the current amplitude of responses to GABA, and the incubation period necessary for RDL channel expression became shorter. These results suggest that the GluCl or its RNA may enhance the expression of RDL channels when the two are coexpressed.

3.2.2 LCCH3 and GRD subunits

Two cDNAs encoding GABA receptor-like subunits, LCCH3 (Henderson et al., 1993, 1994) and GRD (Harvey et al., 1994), were first cloned from *D. melanogaster*. Shortly after this cloning, LCCH3 was shown to form a heteropentamer with RDL when expressed in Sf21 cells (Zhang et al., 1995). However, this coassembly is unlikely in native neurons or muscles, for the following reasons. First, the heteropentamer showed distinct pharmacological properties (i.e., picrotoxin insensitive and bicuculline sensitive) that are not found in native receptors. Second, the subunits are not co-localized in the *Drosophila* body (Aronstein et al., 1996). However, an argument for the presence of the heteropentamers still remains viable. Immunocytochemical and pharmacological analysis indicated that GABA plays an important role in neurons in the antennal lobe of the honeybee brain. Single-cell RT-PCR demonstrated the expression of *rdl* and *lcch3* but not *grd* in antennal lobe neurons (Dupuis et al., 2010). Dieldrin and fipronil completely blocked GABA-induced currents, whereas picrotoxin failed to produce complete inhibition, even at a concentration as high as 1 mM. These findings may favor the possible presence of minor picrotoxin-insensitive heteropentamers in addition to major picrotoxin-sensitive homopentamers.

LCCH3 and GRD form a GABA-gated cation channel that was blocked by picrotoxin but not bicuculline, dieldrin, or γ-BHC, when expressed in *Xenopus* oocytes (Gisselmann et al., 2004). This finding is consistent with the absence of the PAR motif upstream of TM2, which is found in anion channels, in LCCH3 and GRD. However, cation-permeable GABA receptor channels are not known to exist in insects. Because cation channels produce neuronal excitation, RDL is the only subunit isolated to date that constitutes an inhibitory GABA receptor.

cDNA clones of *lcch3* and *grd* were also isolated from *B. mori* (Yu et al., 2010). Orthologues of these genes were also identified in *Apis mellifera*, *T. cantaneum*, *Nasonia vitripennis*, and the two-spotted spider mite *Tetranychus urticae* (Dermauw et al., 2012; Jones and Sattelle, 2006, 2007; Jones et al., 2010).

3.2.3 Heterogeneity of ionotropic receptors

Heterogeneity of GABA receptors has been reported in several insect species. The dorsal unpaired median (DUM) neurons of *P. americana* express at least two subtypes of chloride-dependent GABA receptors, one of which is regulated by calcium/calmodulin-dependent protein kinase II (CaM kinase II; Alix et al., 2002). The current–voltage (I–V) plots for GABA-induced currents, as determined by patch–clamp recordings, displayed the equilibrium potential for chloride and a biphasic relationship that can be fitted by two linear regressions. The I–V plots became uniphasic in the presence of cadmium or a CaM kinase II inhibitor, indicating that the currents are composed of two components.

Alternative splicing and RNA editing are the two mechanisms for producing protein heterogeneity. Alternative splicing generates variant isoforms by splicing of pre-mRNA. RNA editing involves the enzymatic conversion of adenosine to inosine at specific positions of the pre-mRNA; inosine is interpreted as guanosine by the ribosome. This processing of *rdl* creates the diversity of insect GABA receptor functions and pharmacology (Buckingham et al., 2005). Although in *D. melanogaster*, the RDL subunit is encoded by the single gene *rdl*, the gene is subject to alternative splicing to generate four RDL subunit variants. Alternative splicing of *rdl* generates mRNAs with different nucleotide sequences of equal length at positions of exons 3 (alternative exons termed a and b) and 6 (alternative exons termed c and d), producing four subunit isoforms with partially different amino acid sequences (ffrench-Constant and Rocheleau, 1993). Exon 3 encodes 23 amino acids with 2 amino acid substitutions in the N-terminal region of the RDL subunit, while exon 6 encodes 46 amino acids with 10 amino acid substitutions in the N-terminal region. Transcripts with the b/c combination were the most abundant in *Drosophila* embryos, while those with the a/d combination were the least abundant. As the region encoded by exons 3 and 6 lies in the N-terminal region of the RDL subunit, where the agonist binding site is located, alternative splicing affects the sensitivity of the GABA receptors containing the RDL subunit to agonists (Hosie and Sattelle, 1996b; Hosie et al., 2001).

Drosophila RDL apparently undergoes RNA editing at four sites to generate four amino acid changes: R122G in the N-terminal region, I283V in TM1, N294D in the TM1–TM2 linker, and M360V in the TM3–TM4 linker. The R122G editing affected the sensitivity of the receptor to fipronil (Es-Salah et al., 2008). Recently, the effects of RNA editing and alternative splicing on the sensitivity of *Drosophila* RDL channels to GABA were investigated in detail (Jones et al., 2009). RNA editing varied with developmental stage and between splicing variants and primarily occurred in the adult nervous system. The most abundant variant in the adult nervous system was the bd variant, followed by the ad variant, in contrast to *Drosophila* embryos (ffrench-Constant and Rocheleau, 1993). Three combinations, R122G/I283V/M360V, R122G/I283V, and I283/N294D, accounted for 78% of the editing in the bd variant. cRNAs of the 32 ac and bd variants containing amino acids substituted by RNA editing were injected into *Xenopus* oocytes for two-electrode voltage clamp studies of the GABA sensitivity of homo–oligomeric variant receptors. The EC_{50} values of GABA for the 32 variant channels ranged from 3.2 to 192.5 μM. The results indicate that the combination of editing and splicing of *rdl* permits precise regulation of rapid synaptic inhibition.

In the German cockroach *Blattella germanica*, the presence of six *rdl* variants in the RT-PCR products of the TM was reported (Kaku and Matsumura, 1994). As described in Section 3.2.1, three types of *rdl* were identified in *P. xylostella* and *B. mori*. The partial sequences of three paralogous genes of *H. virescens rdl* are found in the GenBank database. Thus, gene duplication generates the diversification of GABA receptors in lepidopteran insects. Furthermore, two of the three *B. mori rdl* clones are alternatively spliced at exons 3 and 6 to produce a total of eight transcripts, while the other clone has only one version of both exons (Yu et al., 2010). The alternative splicing also results in truncated transcripts lacking an exon or having two copies of exon 6. Interestingly, exons of each type from different insect species (*B. mori, D. melanogaster, An. gambiae, T. castaneum,* and *A. mellifera*) are clustered together in a phylogenetic tree, indicating that exons of one type from different insect species share a higher similarity than exons of different types from the same insect species. Amino acid changes by RNA editing were also identified in two sites of one type of the clones. The three types of the RDL subunit have different 2′ amino acids, alanine, serine, and glutamine; the amino acid residues at this position are involved in forming the NCA binding site.

Bioinformatic analysis of data from insect genome sequencing projects has permitted the identification of the genes encoding the Cys-loop LGIC subunits. The analysis data from several insect species show that this superfamily consists of over 20 Cys-loop LGIC subunit genes. The genome of *A. mellifera* contains 21 candidate subunit genes (Jones and Sattelle, 2006). The *A. mellifera* RDL subunit contains two variants of exon 3 and two variants of exon 6. Additional variants with short and long intracellular TM3–TM4 loops were also identified in the *A. mellifera* genome by RT-PCR analysis. Twenty-four candidate Cys-loop LGIC genes were identified in the *T. castaneum* genome, which contains a sequence for the RDL subunit (Jones and Sattelle, 2007). Exons 3 and 6 are alternatively spliced with two and three variants, respectively. Twenty-six Cys-loop LGIC subunit genes were identified in the genome of the parasitoid wasp *N. vitripennis* (Jones et al., 2010). One RDL subunit was identified with two variants each of exons 3 and 6.

More recently, annotation of the Cys-loop LGIC family in *T. urticae* was performed after complete genome sequencing (Dermauw et al., 2012). In this analysis, three *rdl* orthologues were identified in the mite genome. No splice variant was detected in *T. urticae rdl*, in contrast to other insect species in which alternative splicing may contribute to variations in GABA receptors. With respect to the NCA resistance-associated mutation, the three *T. urticae* RDL orthologues did not contain 2′Ala, which was found in the wild-type *Drosophila* RDL subunit, but two orthologues contained a serine and the other one contained a histidine at this position (see Section 10). Furthermore, two important substitutions, T6′I and T321A in TM3, were detected in one of the RDL orthologues. Mutations at the equivalent positions are associated with dieldrin resistance in cattle ticks (*R. microplus*) and laboratory-selected fipronil resistance in *Drosophila* (see Section 10; Hope et al., 2010; Le Goff et al., 2005). It remains to be investigated whether these mutations are related to the low sensitivity of spider mites to fipronil.

3.2.4 Insect GABA_B receptors

Picrotoxinin-insensitive $GABA_B$-like responses were elicited by GABA in the motor neurons and cercal-nerve giant-interneuron (GI) synapses of *P. americana*, in which a similar response was induced by the vertebrate $GABA_B$ receptor agonist 3-aminopropylphosphinic acid (3-APPA; Bai and Sattelle, 1995; Hue, 1991). Picrotoxin-sensitive and -resistant receptors in cockroach dorsal paired median neurons are located on the surface of the terminal abdominal ganglion and deeper within the ganglion, respectively (Amat and Hue, 1997). Picrotoxin-resistant GABA receptors modulate the rhythmic firing of neurons, and this modulation was reproduced by 3-APPA. Three cDNAs encoding $GABA_B$ receptors, two with high sequence identity to their mammalian counterparts and one insect-specific cDNA, were cloned from *D. melanogaster* (Mezler et al., 2001). Two of the receptors are expressed in similar regions of the embryonic nervous system, suggesting coexpression as a heterodimer. The two receptors heterologously expressed in HEK293 cells are coupled to G_i protein, and when expressed with inwardly rectifying potassium channels in *Xenopus* oocytes, the receptors induced inwardly rectifying currents. The pharmacological properties of the two receptors are different from those of their mammalian counterparts. *Drosophila* $GABA_B$ receptors are involved in the behavior-impairing actions of alcohol and γ-hydroxybutyric acid (Dimitrijevic et al., 2005; Dzitoyeva et al., 2003), modulation of circadian modulation neurons (Hamasaka et al., 2005), and normal development (Dzitoyeva et al., 2005).

4. TISSUE DISTRIBUTIONS OF INSECT GABA RECEPTORS

An antibody raised against a TM2–TM3 loop peptide fused with maltose-binding protein was used to investigate the localization of GABA receptors in *Drosophila* (Aronstein and ffrench-Constant, 1995). RDL subunits were highly expressed throughout the developing central nervous system of embryos. Consistent with anti-GABA antibody staining, strong RDL-like immunohistochemical reactivity (IR) was detected in the optic lobes, ellipsoid body, fan-shaped body, ventrolateral protocerebrum, and the glomeruli of the antennal lobes in the brain.

Application of picrotoxin to the locust antennal lobe selectively blocked the fast inhibitory synapse between local and projection neurons and abolished their oscillatory synchronization (MacLeod and Laurent, 1996). Direct injection of picrotoxin into the antennal lobe resulted in the loss of the ability of honeybees to perceive odor mixtures of configural stimulant, indicating that GABA receptors are involved in the formation of configural olfactory perception (Choudhary et al., 2012).

In situ hybridization to whole-mount *Drosophila* embryos clearly demonstrated the presence of *rdl* transcripts in ventral sensory neuron clusters (Stilwell and ffrench-Constant, 1998). With an antibody raised against a C-terminal peptide of the *Drosophila* RDL protein, strong RDL–IR was detected in several layers of neurons in the optic lobe, the ellipsoid body in the central complex, and all lobes of the mushroom body of *Drosophila* (Harrison et al., 1996). The presence of functional GABA receptors in Kenyon cells isolated from the mushroom body of the house cricket *Acheta domesticus* was reported using fluorescence calcium imaging with Fura 2-AM and whole-cell patch-clamp recordings (Cayre et al., 1999). Application of GABA, muscimol, and 3-APPA evoked a decrease in intracellular calcium levels, although the mechanism is unknown. Fast synaptic currents mediated by picrotoxin-sensitive GABA receptors were detected in *Drosophila* pupae Kenyon cells identified using the GAL4–UAS system (Su and O'Dowd, 2003). The lamina in the optic lobe showed no RDL–IR reactivity (Harrison et al., 1996). However, it was recently shown using an *rdl*-GAL4 line that the RDL subunit may be expressed in the L4 monolayer neurons and the tangential neuron in the lamina (Kolodziejczyk et al., 2008). Using GAL4–UAS system and immunohistochemistry, RDL receptors were demonstrated to be localized in the tips of the dendrites of the lobula plate tangential cells, which are an important visual navigation processing center in optic flow (Raghu et al., 2007).

LCCH3–IR was examined to determine if RDL co-localizes with LCCH3 (Aronstein et al., 1996), revealing a distinct pattern of its distribution. LCCH3 was expressed in the developing neuroblasts before neuronal differentiation and was primarily expressed in neuronal cell bodies at a later embryonic stage, whereas RDL–IR was detected only in the neuropil in all developmental stages. The results indicate that the two subunits may not coassemble in native tissues. Recently, however, Okada et al. (2009) found that many neurons in the antennal lobe expressed GABA$_B$ receptor, RDL, and LCCH3 subunits. The number of cells labeled with hybridization probes for each subunit was quantified with double labeling with *in situ* hybridization and a cell-specific GAL4 driver. Cells expressing *gaba$_b$r*, *lcch3*, and *rdl* accounted for more than 90% of the expression of each gene in inner antennocerebral tracts, suggesting that the three receptors may be often coexpressed in the same cells.

The localization of *Drosophila* RDL receptors was also compared with that of GABA$_B$2 receptors (Enell et al., 2007). Both RDL–IR and GABA$_B$2 receptor–IR were distributed in the calyx, protocerebral bridge, ellipsoid body, antennal glomeruli, and optic lobe neuropils. RDL–IR was detected throughout the ellipsoid body, whereas GABA$_B$ receptor–IR was observed only in the peripheral parts. The RDL–IR was detected in the peduncles, β and γ lobes, whereas GABA$_B$ receptor–IR was not observed in these regions. The distribution patterns of the RDL and GABA$_B$ receptor differed in the lamina and medulla. In the larval central nervous system, RDL–IR was observed in the cell bodies of Kenyon cells, while GABA$_B$ receptor–IR occurred in neuropils.

In the blowfly *Calliphora erythrocephala*, strong RDL–IR was detected in the mushroom body with an antibody against *Drosophila* RDL, although the reactivity was not high in *Drosophila* (Brotz et al., 1997). The staining pattern suggested that GABA receptors are located on the axons of Kenyon cells. Inhibitory feedback onto Kenyon cells is thus thought to be important for the control of odor-evoked oscillation patterns in the mushroom body. In the visual system, the strong RDL–IR was in layer 1 of the medulla and in the vicinity of the dendrites of the lobula plate tangential cells (Brotz et al., 2001), implicating GABA receptors in visual motion responses. With an antibody raised against the C-terminal peptide of *Drosophila* RDL, strong reactivity was observed in the calyces of the mushroom bodies, glomeruli of the antennal lobes, and several cell layers in the medulla and lobula of the optic lobe of *P. americana*, indicating important roles in the processing of visual, olfactory, and mechanosensory information (Sattelle et al., 2000). The most intense reactivity was detected in the subesophageal ganglion. In *A. domesticus,* the upper part of the peduncle of the mushroom body, which plays a critical role in learning and memory, was strongly positive for IR with the *Drosophila* RDL C-terminal antibody, suggesting that GABA controls these processes (Strambi et al., 1998). The glomeruli of the antennal lobe and the ellipsoid body of the central complex were stained, indicating a role of GABA in olfactory information processing.

5. NEUROPHYSIOLOGY OF RDL GABA RECEPTORS

Studies using genetic tools for *Drosophila* have recently enhanced our knowledge of the physiological roles of RDL GABA receptors. GABA and RDL GABA receptors regulate sleep in *Drosophila* by affecting sleep latency and the duration of sleep episodes (Agosto et al., 2008). When GABA release was suppressed by expressing hyperpolarizing potassium channels in

GABAergic neurons of *Drosophila*, total sleep was reduced. Flies with the homozygous A2′S mutation in the RDL subunit showed an increase in total daily sleep because the mutant RDL channels evoke slow desensitized currents due to longer single channel open durations relative to wild-type channels. The RDL-containing cells were identified as ventral lateral neurons (LNvs) of the circadian clock, which contained the neuropeptide pigment-dispersing factor (PDF) (Chung et al., 2009; Parisky et al., 2008). The RDL knockdown using tissue-specific RNA interference (RNAi) led to a significant reduction in total sleep time. Simultaneous dual-cell patch–clamp recordings demonstrated that pairs of large LNvs exhibit synchronous rhythmic membrane activity (McCarthy et al., 2011). GABA modulated action potential firing, but not synchrony, in large LNvs, regulating excitability in LNvs and PDF release.

Olfactory receptor neurons, which are activated by odor molecules, send axons to the antennal lobe, and form synapses on projection neurons (PNs) and local interneurons (LNs) within the glomeruli. PNs convey odor information to higher order neuropils. Patch–clamp recordings indicated that LNs express picrotoxinin-sensitive GABA receptors and that PNs express both picrotoxinin-sensitive and -insensitive GABA receptors (Wilson and Laurent, 2005). Knockdown of RDL receptors in PNs blocked habituation to carbon dioxide or ethyl butyrate in *Drosophila* (Das et al., 2011; Sudhakaran et al., 2012). LNs provide feedback inhibition on PNs, which underlies olfactory habituation.

RDL receptors are expressed at high levels in the mushroom bodies, which are important structures associated with olfactory associative learning in the insect brain. Tissue-specific knockdown of GABA synthesis in an anterior paired lateral (APL) neuron sending a projection to the mushroom body enhanced olfactory learning, suggesting that GABA release from the neuron plays a role in learning (Liu and Davis, 2009). Reduction of GABA synthesis in APL neurons eliminated the suppression of initial memory and impaired olfactory reversal learning (Wu et al., 2012). APL neurons expressing RDL receptors also mediate behavioral flexibility during visual reversal learning (Ren et al., 2012). RDL knockdown in APL neurons did not affect short-term memory and initial learning, but impaired behavioral flexibility. Studies using the GAL4–UAS system with expression of transgenes for *rdl* or its RNAi constructs demonstrated that overexpression of *rdl* in the mushroom body impairs memory acquisition but not memory stability, while knockdown of *rdl* enhanced memory acquisition via the conditioned stimulus pathway (Liu et al., 2007, 2009).

6. RECEPTOR LIGANDS OF GABACLs AND GLuCLs

6.1. Noncompetitive antagonists

NCAs allosterically block GABA-induced chloride currents by binding to a site different from the GABA binding site. NCAs are also termed channel blockers or negative allosteric modulators based on their function. NCAs include structurally diverse classes of compounds ranging from naturally occurring substances to synthetic compounds. NCAs are practically used as insecticides and ectoparasiticides but are also useful molecular tools in physiological and biochemical studies of GABA receptors.

6.1.1 Terpenoids and other naturally occurring NCAs

A number of terpenoid NCAs of GABA receptors were isolated from plants. The sesquiterpene picrotoxin isolated from Menispermaceae plants is a representative of NCAs and has been used as a standard molecular tool to detect GABA systems (Florey, 1954; Smart and Constanti, 1986; Takeuchi and Takeuchi, 1969; Usherwood and Grundfest, 1965). Picrotoxin contains two constituents, picrotoxinin (Fig. 4.2) and picrotin, in a 1:1 molar ratio (Jarboe and Porter, 1965). Picrotoxinin is a more potent antagonist of GABA receptors than picrotin. Picrotoxinin and its derivatives produce convulsions in animals and have been tested against houseflies (Miller et al., 1979). Picrotoxane terpernoids also include compounds like coryamyrtin, tutin, mellitoxin, and hyenanchin (Porter, 1967).

Although picrotoxin was first highlighted as an antagonist for GABA receptors, the antagonistic action of picrotoxin proved not to be highly specific for GABA receptors. Picrotoxinin blocks other ligand-gated chloride channels such as glycine-gated chloride channels (Pribilla et al., 1992; Yoon et al., 1998) and GluCls (Cully et al., 1996; Eguchi et al., 2006). Histamine-gated chloride channels are also subject to antagonism by picrotoxin (Gisselmann et al., 2002; Hatton and Yang, 2001). Picrotoxin antagonizes ligand-gated cation channels such as $\alpha3\beta4$ and $\alpha7$ nAChRs and type-3 serotonin receptors with IC_{50} values in the micromolar range (Das et al., 2003; Erkkila et al., 2008).

Picrodendrins are a series of picrotoxane terpenoids isolated from the bark and stems of the Euphorbiaceae plant, *Picrodendron baccatum*. Leaves of this plant are powdered and used to kill bedbugs and lice in the Dominican Republic (Hayden et al., 1984). Picrodendrin Q (Fig. 4.2), one of the picrodendrin terpenoids, was more potent than picrotoxinin when tested against *Drosophila* RDL homo-oligomers expressed in *Xenopus* oocytes

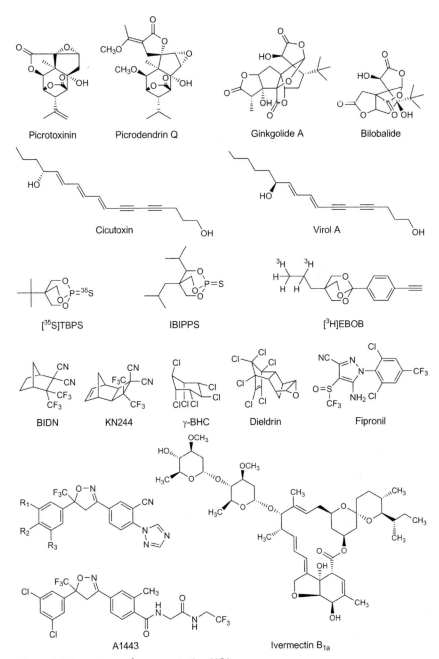

Figure 4.2 Structures of representative NCAs.

and native housefly (*M. domestica*) GABA receptors (Hosie et al., 1996; Ozoe et al., 1998a), with IC_{50} values in the nanomolar range against these GABA receptors. Picrodendrin Q exhibited insecticidal activity against *B. germanica* with an LD_{50} value of 71 ng roach^{-1} (Ozoe et al., 1998a).

Ginkgolides and bilobalide (Fig. 4.2) are diterpene trilactones with distinct structures that are found in Ginkgo (*Ginkgo biloba*). Ginkgo leaf extracts are used as an herbal medicine (Ernst, 2005). These terpenoids showed antagonistic activity against rat neuronal or human recombinant $GABA_A$ receptors (Huang et al., 2003, 2004; Ivic et al., 2003), recombinant ρ1 GABA receptors (Huang et al., 2006), and rat neuronal glycine receptors (Ivic et al., 2003; Kondratskaya et al., 2002). Ginkgolides are selective for glycine receptors, whereas bilobalide is selective for GABA receptors. Ginkgolides and bilobalide blocked human recombinant type-3 serotonin receptors less potently than glycine and GABA receptors (Thompson et al., 2011a,b).

The diterpene bilobalide is highly insecticidal against both susceptible ($LD_{50} = 0.26$ ng female^{-1}) and organophosphate/carbamate-resistant ($0.17–0.28$ ng female^{-1}) strains of brown rice planthoppers (*Nilaparvata lugens*) but it is not effective against tobacco cutworms, houseflies, mosquitoes, German cockroaches, and two-spotted spider mites (Ahn et al., 1997). Female planthoppers topically applied with bilobalide (10 ng individual^{-1}) showed poisoning symptoms typical of those due to GABA receptor antagonism; they quickly died after tremors and paralysis. Bilobalide is nontoxic to mice (LD_{50} of >1000 mg kg^{-1}).

Recently, ginkgolides and bilobalide were tested against *Drosophila* RDL GABA receptors expressed in *Xenopus* oocytes (Thompson et al., 2012). Ginkgolide A and B potently inhibited GABA-induced currents, with IC_{50} values of 3.1 and 19 nM, respectively. The IC_{50} value of ginkgolide A was smaller than the reported IC_{50} value of picrodendrin Q (17 nM) (Hosie et al., 1996). Bilobalide had an IC_{50} value (320 nM) similar to that of picrotoxinin (220 nM). These lactones exhibited IC_{50} values in the micromolar range against human α1β2 and α1β2γ2 GABA receptors. These lactones exhibited insecticidal effects against wild-type *Drosophila*, but their insecticidal activity was not parallel to the inhibitory activity of GABA receptors, most likely due to their difference in penetrability. These compounds were not effective against a *Drosophila* strain with the A2′S mutation. SAR studies indicated that the rigidity of the cage structure and the free hydroxyl groups are important for activity against glycine receptors (Jaracz et al., 2004; Jensen et al., 2007). This may also be the case for GABA receptors and other LGICs.

Other terpenoids derived from plants have been identified or suggested as NCAs, including anisatin and *seco*-prezizaanes (Ikeda et al., 1999; Kuriyama et al., 2002), α-thujone (component of absinth) (Höld et al., 2000), *cis*- and *trans*-3-pinanones (constituents of hyssop oil) (Höld et al., 2002), horminone and taxodione (diterpenes from *Salvia pachystachys*) (Rutherford et al., 1994), and silphinenes (Bloomquist et al., 2008). Anisatin is a poisonous sesquiterpene isolated from the Japanese star anise (*Illcium anisatum*).

Cicutoxin and virol A (Fig. 4.2), which were isolated from the water hemlock *Cicuta virosa*, are straight-chain alcohols that contain uniquely conjugated double and triple bonds. Virol A inhibited GABA-induced chloride currents in rat neurons (Uwai et al., 2001). The potency of cicutoxin and virol A against insect GABA receptors remains to be investigated. Two alkaloids, songorine and alantrypinone, are NCAs of GABA receptors. The diterpene alkaloid songorine isolated from the *Aconitum* plant noncompetitively inhibits [^{3}H]muscimol binding to rat brain membranes and GABA-induced currents in rat neurons (Zhao et al., 2003b). Alantrypinone, which belongs to the spiroquinazoline family of alkaloids, was identified by screening metabolites from the broths of the mold *Aspergillus terreus* using [^{3}H]EBOB binding assays (Kuriyama et al., 2004). Alantrypinone is a selective NCA against insect GABA receptors and has insecticidal activity against cockroaches and aphids. In an attempt to obtain compounds with greater activity, a series of racemic alantrypinone derivatives were systematically synthesized via hetero Diels–Alder reactions, and a total of 34 compounds were examined for their ability to inhibit the specific binding of [^{3}H]EBOB to housefly head membranes (Watanabe et al., 2009). Unusual ergostanoid lactones, petuniolides, which contain an orthoester and spirolactone moieties, induced toxic symptoms similar to those induced by picrotoxinin in desert locust nymphs (Isman et al., 1997). These compounds inhibited [^{35}S] *tert*-butylbicyclophosphorothionate (TBPS) binding to rat brain membranes and were ineffective against cyclodiene-resistant German cockroaches. These findings suggest that these compounds are active as GABA receptor NCAs. Dehydroepiandrosterone sulfate, a neurosteroid found in the brain, has also been suggested to act as an NCA (Sousa and Ticku, 1997).

6.1.2 Bicyclophosphates and bicyloorthocarboxylates

Bicyclophosphates and their thiono (P=S) analogs are a second class of compounds identified as NCAs of GABA receptors. An analog of bicyclophosphates was discovered as a toxic substance that was produced when fire-retardant polyurethane foam was burned. Among the analogs synthesized to study their mode of action and SARs, *tert*-butylbicyclophosphate was found to possess the highest toxicity against mice (Eto et al., 1976; Ozoe and Eto, 1986). The action of ethyl, isopropyl and pentylbicyclophosphates on the depressant action of GABA on single neurons in the rat brain and the depolarizing action of GABA on the isolated rat superior cervical ganglion were investigated. The results suggested that the convulsant action of the bicyclophosphates may be related to their antagonism of the action of the inhibitory neurotransmitter GABA, which is released at synapses in the brain (Bowery et al., 1976). The effects of bicyclophosphates on the miniature inhibitory junction potentials (mijps) of the longitudinal muscle of the earthworm *Pheretima communissima* were investigated. Methyl- and isopropylbicyclophosphates reduced the frequency and mean amplitude of mijps, which was due to the release of GABA from the peripheral nerves, at a concentration of $10~\mu g~mL^{-1}$; at this concentration, picrotoxin exhibited a moderate effect on the mijp (Korenaga et al., 1977). These phosphates at higher concentrations (20–$50~\mu g~mL^{-1}$) completely blocked the generation of mijps. Radioligand binding experiments with rat brain membranes clearly demonstrated that the ^{35}S-labled thiono analog (TBPS; Fig. 4.2) specifically labels the NCA binding site of GABA receptors (Squires et al., 1983).

Earlier ligand-binding experiments were performed with [^{3}H] dihydropicrotoxinin, which has low affinity, but with the advent of [^{35}S] TBPS, which exhibits high affinity and high levels of specific binding, great progress was made in biochemical pharmacology of mammalian GABA receptors. [^{35}S]TBPS was also successfully used for molecular biochemical studies of cloned GABA receptor subtypes (Im et al., 1994). Electrophysiological evidence for GABA receptor antagonism by this class of compounds was also obtained (Van Renterghem et al., 1987). TBPS inhibited GABA-evoked currents in a dose-dependent manner in *Xenopus* oocytes injected with mRNA isolated from chick embryo brains. TBPS blocked spontaneous chloride activities recorded from neuroendocrine cells from the porcine intermediate lobe (Hamann et al., 1990). Recently, the mechanisms of TBPS blockade of GABA receptor channels were examined by patch–clamp analysis and [^{35}S]TBPS binding to HEK293 cells expressing murine $\alpha 1\beta 2\gamma 2$ receptors (Othman et al., 2012). TBPS was shown to bind to the spontaneous–gating and agonist–activation states of the receptors. By contrast, desensitization of the receptors was suggested to reduce the accessibility of TBPS to the channel.

Attempts were made to reduce the mammalian toxicity of TBPS analogs and increase their insecticidal activity by modifying the structure of TBPS, although various insect species were found to be insensitive to bicyclophosphates (Casida et al., 1976). The introduction of an appropriate 3-substituent renders bicyclophosphorothionates more active toward housefly GABA receptors than rat GABA receptors. Bicyclophosphorothionates with a 3-isopropyl group and a 4-alkyl group of five carbon atoms exhibited greater than 10-fold selectivity for housefly versus rat GABA receptors. The selectivity value of an analog that showed the highest selectivity was greater than 97 (Ju and Ozoe, 1999, 2000; Ju et al., 2010). The most potent analog was 4-isobutyl-3-isopropylbicyclophosphorothionate (IBIPPS; Fig. 4.2). Patch-clamp analysis was performed to confirm that IBIPPS blocks GABA-induced currents in neurons isolated from the thoracic ganglia of *P. americana* (Ju et al., 2010). The amplitude of GABA-induced currents gradually decreased during perfusion with IBIPPS before reaching a steady state. The IC_{50} value of IBIPPS was estimated to be 31.7 nM from the dose–response relationship determined at the steady state. SAR analysis has revealed that there are optimum sizes for the 3- and 4-substituents, as exemplified by an isopropyl group at the 3-position and a four- to five-carbon chain or an equivalent cycloalkyl group at the 4-position for housefly GABA receptors. Because the 3-isopropyl group is not tolerated in rat GABA receptors, the introduction of this substituent alone results in a greater than 200-fold net increase in selectivity for housefly versus rat GABA receptors. Naturally occurring quassinoids (e.g., samaderine B), which potentially overlap structurally with bicyclophosphates, exhibited nematocidal activity (Kuriyama et al., 2005).

Bicycloorthocarboxylates that were designed by modifying the structure of bicyclophosphates are high-affinity NCAs for both mammalian and insect GABA receptors, and are therefore toxic to both mammals and insects (Palmer et al., 1991). [^{3}H]EBOB (Fig. 4.2) is a useful molecular tool for labeling the NCA binding sites of insect as well as vertebrate GABA receptors (Cole and Casida, 1992). This radioligand was applied in studies with recombinant receptors expressed in HEK293 cells (Yagle et al., 2003). EBOB blocked both GABA- and glutamate-induced currents in *P. americana* neurons, with approximately 22-fold selectivity for GABA receptors (Ihara et al., 2005). At high concentrations, EBOB labels both GABA and inhibitory glutamate receptors, although at a concentration as low as 0.5 nM used in a [^{3}H]EBOB binding assay it should label GABA receptors selectively. Care must be taken when EBOB is used in studies with insect tissues that contain both receptors. A variety of insecticidal NCAs such as dithianes, phenylthiophosphonates, and acyclic esters and ethers have been derived from bicycloorthocarboxylates (Hamano et al., 2000; Ozoe et al., 1998b; Palmer and Casida, 1992). These NCAs should prove to be useful tools to investigate the molecular basis of ligand selectivity for insect GABA receptors. However, these study efforts have not yet led to the development of novel insecticides.

6.1.3 Organochlorine insecticides and related cycloalkanes

Organochlorine insecticides such as γ-BHC, dieldrin (Fig. 4.2), endosulfan, and other cyclodienes were once used, but have been banned because of their persistent properties in the environment. γ-BHC is an isomer of 1,2,3,4,5,6-hexachlorocyclohexane and is also known as lindane or γ-HCH. The formal nomenclature is $1\alpha,2\alpha,3\beta,4\alpha,5\alpha,6\beta$-hexachlorocyclohexane. The organochlorine insecticides elicit their insecticidal effects by blocking the GABACls as NCAs. Molecular mode–of–action studies of this class of insecticides began with the initial observations that they have a close structural resemblance to the NCA picrotoxinin, and that cyclodiene-resistant insects showed cross–resistance to picrotoxinin (Matsumura and Ghiasuddin, 1983; Tanaka et al., 1984). The chlorine atoms are not essential for activity because dechlorinated cyclic hydrocarbons or lactones also exhibit some insecticidal and NCA activities against GABA receptors (Ozoe and Matsumura, 1986; Ozoe et al., 1993b).

Dieldrin initially enhances and subsequently suppresses GABA-induced currents in rat native and recombinant GABA receptors (Nagata and Narahashi, 1994; Nagata et al., 1994). It was also shown that dieldrin suppresses the amplitude of GABA responses in *D. melanogaster* RDL channels expressed in S2 cells and *Xenopus* oocytes (Buckingham et al., 1996; ffrench-Constant et al., 1993a). Organochlorine insecticides exhibit greater activity in the coxal muscle of *P. americana* than in motor neurons (Schnee et al., 1997). The reason for these differences in potency remains unclear.

γ-BHC and δ-BHC enhanced GABA-induced currents in rat native and recombinant GABA receptors at the onset of action, followed by more critical suppression, whereas α- and β-BHC had little or no effect (Nagata and Narahashi, 1995; Nagata et al., 1996). These enhancements and inhibition occurred in the micromolar range. δ-BHC induced potentiation in *Drosophila* RDL GABA receptors expressed in *Xenopus* oocytes (Hosie and Sattelle, 1996a). The potentiating effect of δ-BHC may be due to the interaction with the barbiturate binding site on the GABA receptors (Aspinwall et al., 1997; Maskell et al., 2001). The potentiating effects of organochlorine insecticides are considered toxicologically minor events. This action of δ-BHC could be similar to that of a strong depressant (Mullins, 1955). γ-BHC inhibited GABA-activated chloride currents in dissociated locust neurons with an IC_{50} value of <10 nM (Lees and Calder, 1996).

γ-BHC also blocked glutamate-gated chloride currents in neurons of *P. americana*, although GABA-induced currents were two orders of magnitude more sensitive to the block, with an IC_{50} value of approximately 1.7 nM (Ihara et al., 2005). Recently, it was reported that γ-BHC also acts at other members of the Cys-loop receptor family, with IC_{50} values in the high nanomolar to low micromolar range. These Cys-loop receptors include glycine receptors containing α subunits (Islam and Lynch, 2012; Vale et al., 2003) and a ligand-gated proton channel from *Gleobacter violaceus* (Alqazzaz et al., 2011). γ-BHC did not block a hetero-oligomeric glycine receptor containing the α1 subunit.

3,3-Bis(trifluoromethyl)bicyclo[2.2.1]heptane-2,2-dicarbonitrile (BIDN) and its analog KN244 (Fig.4.2) are insecticidal compounds (convulsants) with unique structures and modes of action. Although this group of compounds has NCA activities similar to the other NCAs described above, a unique aspect of the activity of this group of compounds is that [^{3}H]BIDN binding to membranes prepared from the whole body of the southern corn rootworm *Diabrotica undecimpunctata howardi* was inhibited by organochlorine insecticides, but not by bicycloorthocarboxylates (Rauh et al., 1997). Application of the phenylpyrazole insecticide fipronil with BIDN yielded additive effects on the reduction of the maximal probability of channel opening in a single channel analysis of *D. melanogaster* RDL homo-oligomers (Grolleau and Sattelle, 2000). KN244 blocked GABA-induced currents in *D. melanogaster* RDL homo-oligomers, with lower activity against the dieldrin-insensitive A2′S mutant (Matsuda et al., 1999). These results suggest that these compounds may bind to a site that is close to the binding site of organochlorine insecticides but different from the binding site of picrotoxinin, trioxabicyclooctanes, and fipronil. These binding sites apparently overlap. GluCls and glycine-gated chloride channels are insensitive to BIDN, indicating that BIDN is highly selective for GABACls of insects and vertebrates (Hamon et al., 1998).

6.1.4 Phenylpyrazoles and other phenylheterocycles

Fipronil (Fig. 4.2) is the first commercially used phenylpyrazole insecticide and has high activity and a broad spectrum against insect pests, with relatively low toxicity in mammals (Colliot et al., 1992). Several other phenylpyrazoles, such as ethiprole and pyriprole, have been developed by changing substituents on the pyrazole ring with fixed 2,6-dichloro-4-trifluoromethylphenyl substitution (Zhao et al., 2010). Some of these phenylpyrazoles are now used not only as insecticides against agricultural pest insects and noxious insects but also as ectoparasiticides. Fipronil differs from classical organochlorine insecticides in terms of structure and actions. Thus, this class of insecticides is the second generation of GABA receptor NCA insecticides that emerged after the classical organochlorine insecticides. In addition to high potency against insect pests, fipronil toxicity in nontarget insects such as honeybees has also been reported. The action of a low dose of fipronil (0.1 ng bee^{-1}) on both GABA receptors and GluCls is responsible for the impairment of honeybee olfactory memory (El Hassani et al., 2009).

In electrophysiological investigations, fipronil blocked GABA-induced currents in homo-oligomeric *D. melanogaster* RDL GABA receptors expressed in S2 cells and native GABA receptors expressed in *P. americana* neurons (Grolleau and Sattelle, 2000; Zhao et al., 2003a). The expression of *rdl* in the DUM neurons of the migratory locust *Locusta migratoria*, which show sensitivity to fipronil and picrotoxin, was confirmed by RT-PCR (Janssen et al., 2010). In biochemical assays, fipronil inhibited [^{3}H]EBOB binding to membranes prepared from housefly heads in a noncompetitive fashion, indicating that fipronil and EBOB may bind to a different or overlapping site (Cole et al., 1993). In electrophysiological studies using *Drosophila* RDL receptors expressed in *Xenopus* oocytes, fipronil produced a rightward shift of the GABA dose–response curve with a decreased maximal amplitude of GABA-induced currents, indicating that fipronil is neither a pure competitive, nor pure noncompetitive antagonist (Hosie et al., 1995). Houseflies are 160- and 315-fold more sensitive to fipronil than mice with respect to receptor affinity and toxicity, respectively (Hainzl and Casida, 1996).

Fipronil blocked GABA-induced chloride currents in both resting and activated GABA receptors in neurons of *P. americana* (Zhao et al., 2003a). This finding indicates that the inhibitory action of fipronil does not require GABA receptor channel opening. The IC$_{50}$ values of fipronil in the resting and activated receptors were estimated to be approximately 28 and 35 nM, respectively. Kinetic analysis showed that fipronil binds to the activated receptor 41-fold faster than to the resting receptor and that it dissociates from the activated receptor 23-fold faster than from the resting receptor. Overall, receptor activation facilitates the block by fipronil. Similar results were obtained with rat dorsal root ganglion neurons (Ikeda et al., 2001). In single-channel patch-clamp studies with rat dorsal root ganglion neurons, fipronil reduced both the frequency and duration of opening of GABACls (Ikeda et al., 2004). The IC_{50} values of fipronil inhibition of rat resting and activated $GABA_A$ receptors were 1.66 and 1.61 μM, respectively (Ikeda et al., 2001). These data indicate that the *P. americana* GABA receptors are 46- to 59-fold more sensitive to fipronil than rat $GABA_A$ receptors.

Earlier electrophysiological studies indicated the existence of two pharmacologically different subtypes of GABA receptors in cockroach cercal-afferent GI synapses (Hue, 1998) and DUM neurons (Le Corronc et al., 2002). The two subtypes in GI synapses differ in their response to GABA, picrotoxin and CACA and their desensitization processes. An initial transient and a subsequent slow phase of hyperpolarization evoked by long pressure applications of muscimol and CACA to DUM neurons revealed different sensitivities to fipronil, picrotoxinin, and dieldrin. Fipronil and picrotoxinin blocked both responses, whereas dieldrin reduced only the transient phase. The *rdl* mutation (amino acid point mutation at the 2'-position of RDL) of GABA receptor subunits is known to be responsible for the insensitivity to NCAs including dieldrin. The insensitivity of the slow phase to dieldrin is probably not caused by the *rdl* mutation because fipronil and picrotoxin inhibited the slow phase of hyperpolarization. Coexpression and localization studies of the subunits suggest that it is unlikely that the heteromultimeric receptors formed by the combination of RDL and other subunits such as GRD and LCCH3 produce the slow phase in the neurons (Aronstein et al., 1996; Zhang et al., 1995).

The receptors are regulated by phosphorylation and dephosphorylation of serine/threonine residues, and it is reported that the sensitivity of the receptors to insecticides is altered by these processes. The existence of two subtypes of GABA receptors in *P. americana* DUM neurons was confirmed by patch-clamp techniques (Murillo et al., 2011). Application of GABA to DUM neurons produced a biphasic current–voltage relationship that arises from GABA receptor-1 and -2. One of the responses due to GABA receptor-2 was regulated by phosphorylation by protein kinase C and CaM kinase II, both of which were in turn controlled by calcium flux through channel proteins of the transient receptor potential family. GABA receptor-2 was more sensitive to fipronil than was GABA receptor-1. Inhibition of protein kinase C but not CaM kinase II rendered GABA receptor-2 10-fold less sensitive to fipronil. GABA receptor modulation is also reported in honeybee Kenyon cells. In this case, calcium-sensitive GABA-induced currents were modulated by the application of a nonspecific kinase inhibitor but not by inhibitors of CaM kinase II or protein kinase A or C (Grünewald and Wersing, 2008).

RNA editing, which occurs at position 122 of the extracellular domain of *Drosophila* RDL subunit, affected the potency of fipronil (Es-Salah et al., 2008). R122G (RNA editing), A301G/T350M (fipronil-resistant mutation), and A301G/T350M/R122G RDL mutants were expressed in *Xenopus* oocytes, and their fipronil sensitivity was evaluated in comparison with that of the wild-type RDL receceptor. The fipronil sensitivity was two- to sixfold decreased by R121G modification compared with that of the wild-type and A301G/T350M receptors.

In addition to phenylpyrazoles, a number of other insecticidal phenylheterocycles have been reported to have the same mode of action as fipronil (Alam et al., 2007; Lyga et al., 2007; Ozoe et al., 2000; Pulman et al., 1996). Most of these reported heterocyclic NCAs exhibit selectivity for insect versus mammalian GABA receptors.

6.1.5 Isoxazolines

A novel class of NCAs, isoxazolines with a unique structure, exhibits antiparasitic activity against cat fleas and dog ticks. Isoxazoline A1443 (Fig. 4.2) potently inhibited specific binding of the GABA receptor NCA [^{3}H]EBOB to housefly head membranes, with an IC_{50} value of 0.455 nM (Ozoe et al., 2010). By contrast, this isoxazoline was a poor inhibitor of [^{3}H]EBOB binding to rat brain membranes, with an IC_{50} value of >10 μM. (*S*)-A1443, which had an IC_{50} value of 83.1 pM, was approximately 1300-fold and 41-fold more potent than (*R*)-A1443 and fipronil, respectively.

Because [^{3}H]EBOB is not highly specific for GABA receptors (Ihara et al., 2005), cDNAs that encode the subunits of the housefly GABACl and GluCl were used to compare the actions of A1443 on GABA and inhibitory glutamate receptors. Two-electrode voltage clamp experiments showed that A1443 blocked both GABA- and glutamate-induced chloride currents in *Xenopus* oocytes expressing *M. domestica* GABACls or GluCls, with IC_{50} values of 5.32 and 79.9 nM, respectively. A1443 was found to be a selective blocker of insect GABA receptors relative to inhibitory glutamate receptors. The 2′ amino acid of ligand-gated chloride channels greatly affects the potency of NCAs such as dieldrin and fipronil. Although A2′S RDL GABA receptors reduce the potency of most NCAs, this mutation did not affect the potency of A1443. Whereas S2′A GluCl (reverse *rdl* mutation) channels were more sensitive to fipronil than the wild-type channels (Hirata et al., 2008), the sensitivity of these channels to A1443 was not different from that of the wild-type channels. Moreover, dieldrin-resistant houseflies expressing A2′S RDL channels were as susceptible to A1443 as standard houseflies. These findings indicate that A1443 is a novel and selective blocker of insect GABACls.

Isoxazolines containing an azole ring in place of the carboxamide side chain of A1443 (Fig. 4.2) have insecticidal activity across a broad range of pest insects and mites. These isoxazolines were also reported to block GABA-induced currents in native cockroach neurons and *Xenopus* oocytes expressing *Drosophila* GABA receptors composed of wild-type and A2′S RDL subunits (Cordova et al., 2010). Glutamate-induced currents were weakly inhibited. A tritiated isoxazoline exhibited specific binding to cockroach head membranes. This binding was not inhibited by γ–BHC, heptachlor epoxide, fipronil, and BIDN, indicating a binding site for isoxazolines that is distinct from that of previously reported NCAs.

6.2. Positive allosteric modulators

6.2.1 Avermectins, milbemycins, and nodulisporic acid

Avermectins and milbemycins are insecticidal, endectocidal 16-membered macrocyclic lactones isolated from *Streptomyces* sp. (Rugg, 2010; Rugg et al., 2010; Shoop et al., 1995). These two groups of lactones differ by the presence or absence of an α-L-oleandrosyl-α-L-oleandrosyl moiety, which is contained in avermectins but not in milbemycins. The insecticide/anthelmintic abamectin is a mixture of avermectin B_{1a} and B_{1b}. Emamectin benzoate is the benzoic salt of an abamectin derivative that differs from abamectin in that it contains a methyamino group in the terminal oleandrose. Ivermectin (Fig. 4.2) is an anthelmintic that comprises 22,23-dihydroavermectin B_{1a} and B_{1b}. The miticide milbemectin is a mixture of milbemycin A_3 and A_4 (Ide et al., 1993). Lepimectin, a recently marketed insecticide, is a milbemectin derivative that has a (methoxyimino)phenylacetyloxy group at the 13-position of the 16-membered ring. Other known derivatives include doramectin, eprinomectin, selamectin, milbemycin oxime, and moxidectin. These lactones exert their anthelmintic and insecticidal effects primarily by potentiating the agonist action of glutamate on GluCls or directly activating GluCls (Cully et al., 1996; Wolstenholme and Rogers, 2005). These effects on GluCls are generally slow and irreversible. Although it was found that avermectin B_{1a} stimulates chloride uptake by the leg muscles of *P. americana* by a mechanism that is not mediated by GABA receptors (Tanaka and Matsumura, 1985), solid evidence for the involvement of GluCls in their mode of action was provided by their cDNA cloning (Cully et al., 1994, 1996).

Studies to identify the binding site of macrocyclic lactones are advancing. Identification of mutations in the target molecules of macrolide-resistant organisms would be helpful to identify the binding site. Point mutations associated with macrolide resistance in pest insects or nematodes were identified in GluCls. The L256F mutation in the β10 loop of the AVR–14 subunit of an ivermectin-resistant strain of the nematode *Cooperia oncophora* rendered the channel 2.4-fold less sensitive to activation by ivermectin in *Xenopus* oocyte functional assays (Njue et al., 2004). The P299S mutation in the TM2–TM3 linker of the GLC–1 subunit of nodulisporic acid (NA)- and ivermectin-resistant *D. melanogaster* caused a 14-fold reduction in sensitivity to ivermectin (Kane et al., 2000). Recently, a G323D mutation was identified in the TM3 segment of GluCls of abamectin-resistant two-spotted spider mites (Kwon et al., 2010).

Studies of homomeric α1 glycine receptors, which are directly activated by high concentrations of ivermectin, demonstrated that the introduction of a glycine residue at the equivalent position in TM3 was required for the high ivermectin sensitivity of glycine receptor channels (Lynagh and Lynch, 2010). A comparison of the sequences of other ivermectin-sensitive and -insensitive channels of the same family revealed that a glycine residue is likely required at this position for high ivermectin sensitivity. Conversely, substitution of other amino acids for a glycine present at the equivalent position of the ivermectin-sensitive AVR–14B channel of the nematode *Haemonchus contortus* rendered the channel less sensitive to ivermectin, suggesting that the glycine is important for high sensitivity to ivermectin. The ability of milbemycin to interact with this transmembrane site was investigated by examining the ability of the G293D mutant of the *H. contortus* AVR–14B GluCl to bind [^{3}H]milbemycin A$_4$ (Yamaguchi et al., 2012). The results suggested that milbemycin A$_4$ interacts not only with the transmembrane site but also with an extracellular site.

Recently, Hibbs and Gouaux (2011) published the three-dimensional structures of ivermectin-bound forms of the homo-oligomeric GluCl of the free-living nematode *Caenorhabditis elegans* at 3.3 Å resolution. In this structure, ivermectin bound to the subunit interfaces in the TM with the oleandrose moiety directed toward the membrane lipid (Fig. 4.3). It was hypothesized that ivermectin stabilizes the open-pore conformation of GluCl by interacting with TM2 and the extracellular linker of TM2 and TM3 (Fig. 4.3). In other studies, ivermectin was proposed to act by binding to the TMs of other LGICs, such as α7 nAChRs and purinergic P2X$_4$ receptors

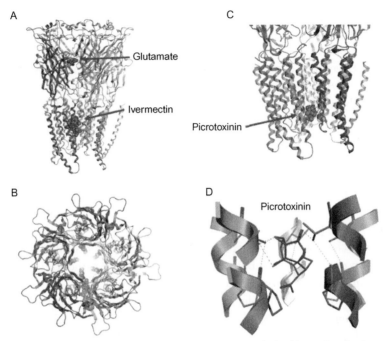

Figure 4.3 Structures of *C. elegans* GluCl. (A) Side view of GluCl bound with glutamate and ivermectin. Illustration is based on X-ray crystal structure (PDB code 3RHW). (B) Top view. (C) Side view of GluCl transmembrane domain bound with picrotoxinin. Illustration is based on X-ray crystal structure (PDB code 3RI5). (D) Close-up of the picrotoxinin-binding site. (See Color Plate 18 at the back of the book.)

(Collins and Millar, 2010; Silberberg et al., 2007). This ivermectin binding site is near the cavity in which general anesthetics and allosteric modulators bind (Bertrand and Gopalakrishnan, 2007; Nury et al., 2011).

Although a number of studies have indicated that the primary mode of action of the macrocyclic lactones is the positive allosteric modulation of GluCls, several studies have shown that avermectins act at vertebrate GABA receptors. Avermectin B_{1a} enhanced GABA–induced chloride currents in *Xenopus* oocytes injected with chick brain mRNA (Sigel and Baur, 1987), and ivermectin activated rat $\alpha1\beta2\gamma2S$ $GABA_A$ receptors in the absence of GABA (Adelsberger et al., 2000). Ivermectin not only induced direct currents in the absence of GABA but also potentiated GABA receptor responses in rat cortical neurons and Ltk cells stably expressing human $\alpha1\beta3\gamma2S$ GABA receptors (Dawson et al., 2000). Ivermectin and moxidectin potentiated the action of GABA in rat $\alpha1\beta2\gamma2$ $GABA_A$ receptors expressed in *Xenopus* oocytes (Ménez et al., 2012). By contrast, 22,23-

dihydroavermectin B_{1a} and ivermectin depressed GABA-induced currents in extrasynaptic GABA receptors in the muscle of the parasitic worm *Ascaris suum* (Martin and Pennington, 1989) and GABA receptors in *L. migratoria* neurons (Bermudez et al., 1991), respectively. Ivermectin had different effects on GABA-induced currents in *Xenopus* oocytes expressing two types of the α/γ subunit from the parasitic nematode *H. contortus* with the β subunit from *C. elegans*; ivermectin enhanced the amplitude of GABA-induced currents in one hetero-oligomer and attenuated the GABA receptor response of another (Feng et al., 2002). Changing only four amino acids in the *H. contortus* subunits drastically changed the responses of the hetero-oligomers to ivermectin. At low concentrations (75 pg–7.5 ng mL^{-1}), 22,23-dihydroavermectin B_{1a} induced irreversible increases in chloride permeability and partially blocked GABA-induced chloride conductance in *S. gregaria* muscle (Duce and Scott, 1985). In contrast, it induced an irreversible increase in chloride permeability and potentiated GABA-induced chloride conductance at high concentrations (10 ng–1 μg mL^{-1}).

NA is an insecticidal indole terpene isolated from the woody plant fungus *Nodulispolium* sp. (Ondeyka et al., 1997). NA induced large steady-state currents similar to those induced by ivermectin in grasshopper neurons (Smith et al., 2000). [^{3}H]*N*-(2-hydroxyethyl)nodulisporamide bound with high affinity to a specific site coupled to a [^{3}H]ivermectin binding site in *Drosophila* GluCls (Smith et al., 2000). Antibodies raised against the *Drosophila* RDL and the GluCl α subunits immunoprecipitated the NA binding protein solubilized from *D. melanogaster* heads, suggesting that NA acts at a receptor that contains both GluCl α and RDL subunits (Ludmerer et al., 2002). The NA receptor was suggested to be one of the two populations of ivermectin receptors on the basis of biochemical and physiological observations.

6.2.2 Benzodiazepines and other neuroactive compounds

Benzodiazepines are anticonvulsant, antianxiety drugs that act at mammalian GABA$_A$ receptors as positive allosteric modulators. Several electrophysiological and biochemical studies have demonstrated that benzodiazepines also act at insect GABA receptors. For example, submicromolar concentrations of diazepam and Ro 5-4864 (4'-chlorodiazepam) enhanced GABA-evoked responses in an identified motor neuron of *P. americana* (Buckingham et al., 2009). This finding from electrophysiological studies may explain a previous biochemical finding that Ro 5-4864 is the most potent in inhibiting [^{3}H]diazepam and [^{3}H]Ro 5-4864 binding to a specific site in the housefly thorax and abdomen. (Ozoe et al., 1987, 1989). These *in vitro* data are also consistent with the *in vivo* data that houseflies were immobilized by the injection of diazepam and Ro 5-4864 (Ozoe et al., 1987). Apart from benzodiazepines, the acaricide bifenazate was found to potentiate GABA-induced currents by increasing the affinity of GABA in *T. urticae* RDL GABA receptors expressed in *Xenopus* oocytes (Hiragaki et al., 2012). Barbiturates such as pentobarbitone, anesthetics such as propofol, the anticonvulstant loreclezole, and δ-BHC enhanced GABA-induced currents in *Drosophila* RDL receptors at high concentrations (Belelli et al., 1996).

6.2.3 Terpenoids

In contrast to convulsant-type terpenoids described in Section 6.1.1, monoterpenes (essential oils) such as thymol, borneol, and menthol showed positive allosteric modulation of GABA receptors. Thymol is a biologically active (e.g., insecticidal) monoterpene that is isolated from the herbaceous plant *Thymus vulgaris*. The structure of thymol (2-isopropyl-5-methylphenol) is analogous to that of the anesthetic propofol (2,6-diisopropylphenol). Although the site of action of these modulators remains elusive, it may be related to the site for menthol (Watt et al., 2008) and valerenic acid (Khom et al., 2007), which likely act via the propofol and loreclezole sites on GABA receptors, respectively. Thymol potentiated GABA-induced currents in human $GABA_A$ receptors and *Drosophila* GABA receptors at $1-100$ μM and elicited small currents in the absence of GABA at 100 μM (Priestley et al., 2003). This study suggests that thymol does not share sites of action with other known allosteric modulators such as benzodiazepines, barbiturates, propofol, and neurosteroids. Thymol interfered with flight motor activity and wing beat frequency in the blowfly, and flies treated with thymol showed patterns of activity most similar to those of flies treated with GABA, suggesting that thymol acts at GABA receptors (Waliwitiya et al., 2010). Twenty-two monoterpenes, including thymol, were tested for inhibition of [³H]TBOB (4-*tert*-butylbicycloorthobenzoate) binding to houseflies (Tong and Coats, 2012). Nine terpenes either increased or inhibited binding. The percentage differences between control binding and the increase/decrease were correlated with insecticidal activity. This finding may suggest that both positive and negative effects of simple monoterpenes lead to insecticidal effects via a common site in housefly GABA receptors.

6.3. Agonists and competitive antagonists

Mammalian GABA receptors are classified into three types: $GABA_A$, $GABA_B$, and $GABA_C$ receptors, as described in Section 3.1. Muscimol and isoguvacine are specific $GABA_A$ receptor agonists, and baclofen and 3-aminopropyl-

(methyl)phosphinic acid (3-APMPA) are selective $GABA_B$ receptor agonists. CACA is a selective $GABA_C$ receptor agonist. Competitive antagonists are also used to characterize GABA receptors. Specific competitive $GABA_A$ receptor antagonists include bicuculline and SR95531, and representative competitive $GABA_B$ receptor antagonists are phaclofen and saclofen.

The $GABA_A/GABA_C$ pharmacological classification is not appropriate for the GABA receptors of insects and other arthropods (Benson, 1989; Jackel et al., 1994; Sattelle, 1990). The pharmacological profiles of GABA receptors differ even within the Arthropoda. For example, the heart of *Lymulus* and the visual interneurons of *C. erythrocephala* (Benson, 1989; Brotz and Borst, 1996) were insensitive to CACA, whereas cultured crustacean thoracic neurons responded to CACA (Jackel et al., 1994). Cultured insect neurons isolated from *P. americana* were activated by both $GABA_A$ and $GABA_C$ receptor agonists but were not inhibited by competitive $GABA_A$ receptor antagonists (Aydar and Beadle, 1999). Both cloned and native GABA receptors from *D. melanogaster*, *A. mellifera*, *L. striatellus*, and *L. migratoria* were sensitive to both $GABA_A$ and $GABA_C$ receptor agonists (Barbara et al., 2005; Grünewald and Wersing, 2008; Hosie and Sattelle, 1996b; Janssen et al., 2010; Millar et al., 1994; Narusuye et al., 2007). Muscimol was the most potent agonist when tested with *Drosophila* RDL receptors expressed in *Xenopus* oocytes, with an EC_{50} value of 9.04 μM, followed by GABA with an EC_{50} value of 19.3 μM. (McGonigle and Lummis, 2010). The rank order of potency of other agonists was TACA > isoguvacine > THIP > β-alanine > 5-aminovaleric acid. A dipole separation of ∼5 Å between the protonated amino group and the dissociated carboxyl group of agonists (or its isostere) appears to be optimal.

A homology model could be used to illustrate the interaction mechanism of these agonists with amino acid residues in the "aromatic box" in the agonist binding site. The agonist binding site is located at the extracellular interface between the principal and the complementary subunits in mammalian GABA receptors (Miller and Smart, 2010; Zhang et al., 2009). The three loops (A–C) of the principal subunit and the three loops (D–F) of the complementary subunit surround the agonist binding pocket. A recent analysis of the interaction between GABA and *Drosophila* RDL receptors incorporating unnatural fluorinated tyrosine in the agonist binding site demonstrated that two aromatic residues in loops B and C contribute to the cation–π interaction with agonists (Lummis et al., 2011). Molecular dynamics simulations based on a homology model of the extracellular domain of the RDL receptor and functional analyses of RDL receptor mutants revealed important cation-π, hydrogen bonding, and electrostatic interactions with amino acid residues in the orthosteric binding site (Ashby et al., 2012). Although the insect GABA receptor shares a common structural feature with its mammalian counterparts, the agonist-interacting amino acids are not necessarily conserved. This may provide unique opportunities to design insect receptor-selective insecticides. Interestingly, the parasitoid wasp *Ampulex compressa* utilizes a mixture of GABA, β-alanine, and taurine for prey capture (Moore et al., 2006). The wasp injects the cocktail into the cockroach prothoracic ganglion to paralyze the front legs.

Competitive antagonists bind to the agonist binding site but inhibit the function of the receptors unlike agonists. Little is known about competitive antagonists of insect GABA receptors. Whereas bicuculline was inactive against most reported invertebrate GABA receptors, SR95531 (gabazine) functioned as a competitive antagonist of cloned GABA receptors of *L. striatellus* and *D. melanogaster* and native GABA receptors of *A. suum*, *S. litura*, and *L. migratoria* (Duittoz and Martin, 1991; Hosie and Sattelle, 1996b; Janssen et al., 2010; Narusuye et al., 2007; Satoh et al., 2005). Several gabazine analogues exhibited greater antagonism of insect pest GABA receptors than gabazine (Rahman et al., 2012). Bicuculline showed a weak antagonistic activity against nicotinic acetylcholine and glutamate receptors in honeybee antennal lobe neurons (Barbara et al., 2005).

As regards metabotropic GABA receptors, 3-APMPA acted as an agonist in cloned *D. melanogaster* $GABA_B$ receptors, whereas the standard mammalian $GABA_B$ receptor agonist baclofen had no effect (Mezler et al., 2001; see Section 3.2.4). Interestingly, 3-APMPA showed insecticidal activity (Fukunaga et al., 1999).

7. MOLECULAR BASIS OF NATIVE INSECT GABA RECEPTORS

The RDL subunit forms a homo-oligomeric ionotropic GABA receptor when heterologously expressed in *Xenopus* oocytes, Sf21 cells, or S2 cells (Eguchi et al., 2006; ffrench-Constant et al., 1993a; Lee et al., 1993, Millar et al., 1994; Narusuye et al., 2007; Shotkoski et al., 1994; Wolff and Wingate, 1998). The pharmacology of the RDL homo-oligomers resembles that of native receptors in terms of agonist pharmacology and insensitivity to bicuculline (Sattelle et al., 1988, 1991, 2003). However, the RDL homo-oligomers differ from native receptors in benzodiazepine sensitivity. For example, flunitrazepam and Ro 5-4865 (4'-chlorodiazepam) had negligible effects on RDL homo-oligomers (Buckingham et al., 1996; Hosie and Sattelle, 1996a), whereas coapplication of flunitrazepam with GABA potentiated the GABA-induced response of identified *P. americana* neurons (Sattelle et al., 1988). This discrepancy between recombinant and native GABA receptors suggests that the RDL subunit may coassemble with other protein(s) to fulfill normal receptor function. Alternatively, this discrepancy may be explained by the splice variants of the RDL subunit (see Section 3.2.3). This remains to be investigated.

Coexpression of RDL with LCCH3 in Sf21 cells formed robust ion channels that have pharmacological properties different from that of the RDL homo-oligomer (Zhang et al., 1995). The RDL/LCCH3 receptor was insensitive to picrotoxin. In addition, RDL and LCCH3 are expressed with different spatial patterns in insects (Hosie et al., 1997), suggesting that these two subunits do not coassemble to form a normal channel. However, recent analysis using *in situ* hybridization and immunolabeling against green fluorescent protein (GFP) expressed with cell-type-specific GAL4 driver strains showed that RDL and LCCH3 were coexpressed in the same cells in the *Drosophila* antennal lobe (Okada et al., 2009). This finding indicates that the coassembly of these subunits cannot be ruled out, at least in some cases.

The GRD subunit formed a heteromeric GABA-gated cation channel when coexpressed with LCCH3 in *Xenopus* oocytes (Gisselmann et al., 2004). GABA acts as an excitatory neurotransmitter in nematodes (Beg and Jorgensen, 2003; Schuske et al., 2004). However, such cationic GABA receptor channels seem to be much less abundant than GABACls because no cationic response has been observed from isolated native neurons. Interestingly, the GRD/LCCH3 receptor was sensitive to picrotoxinin but not γ-BHC, dieldrin, or bicuculline.

8. INHIBITORY GLUTAMATE RECEPTORS (GLuCLs)

Both excitatory and inhibitory glutamate receptors are present in the central neuropil and neuromuscular junction of insects (Featherstone et al., 2005). The presence of inhibitory glutamate (H-response) receptors was first identified on the nonsynaptic (extrajunctional) membrane of locust muscle fibers (Cull-Candy, 1976; Cull-Candy and Usherwood, 1973; Lea and Usherwood, 1973) and later in identified motor neurons and DUM neurons of *P. americana* and *L. migratoria* (Janssen et al., 2010; Sattelle, 1992; Washio, 1994). These receptors produced hyperpolarization in muscle cells and neurons upon activation by glutamate and ibotenate.

The molecular entity of inhibitory responses to glutamate has been unveiled by Cully et al. (1994). Two cDNA clones encoding α and β subunits that constitute an avermectin-sensitive GluCl were first cloned from

C. elegans. This channel may be evolutionarily most closely related to the glycine receptor, which is expressed in vertebrates (Vassilatis et al., 1997b; Xu, 1998). No GluCl has been discovered in vertebrates, and no glycine receptor has been reported in invertebrates. The coexpression of α and β subunits of *C. elegans* GluCl in *Xenopus* oocytes resulted in the formation of a heteromeric chloride channel that was sensitive to glutamate, ibotenate, and avermectin analogs. Following this cloning, a cDNA clone encoding the α subunit was isolated from *D. melanogaster* (Cully et al., 1996). RNA transcripts of *Drosophila glucl* were detected in embryo, larvae, and adult by Northern blot analysis (Cully et al., 1996). Glutamate evoked rapid inward chloride currents in oocytes injected with the α subunit cRNA, whereas ivermectin $4''-O$-phosphate induced slow irreversible currents. The currents were only weakly blocked by picrotoxin at 500 μM.

Orthologues and paralogues of these genes have since been identified by several genome sequencing projects (Delany et al., 1998; Dent et al., 1997; Horoszok et al., 2001; Raymond and Sattelle, 2002; Tandon et al., 2006; Vassilatis et al., 1997a). Although multiple paralogous genes encoding GluCl subunits have been cloned from nematodes (Glendinning et al., 2011), only a single orthologous gene was isolated from most insects. However, the analysis of *Drosophila glucl* α cDNA clones revealed sequence variations by RNA editing and alternative splicing (Semenov and Pak, 1999). RNA editing generates five site-specific A to G changes, three of which alter the coded amino acid residues. Variations by alternative splicing were found at two sites in the 5′ and 3′ regions. A variation generated by the addition of two amino acids was observed at the C-terminus of the GluCl subunit. Combinations of these posttranslational modifications of mRNA would lead to extensive diversity of GluCl with respect to its structure, functional properties, and expression patterns, as in the case of RDL GABA receptors.

Recent genome analyses revealed that a single copy of *glucl* is present in the *A. mellifera, T. cantaneum,* and *N. vitripennis* genomes (Jones and Sattelle, 2006, 2007; Jones et al., 2010). The *A. mellifera* and *N. vitripennis glucl* genes produce two splice GluCl variants of exon 3, whereas the *T. cantaneum glucl* generates three variants. According to recent annotation of the *T. urticae* genome, two-spotted spider mites have six orthologous *glucl* genes (Dermauw et al., 2012). Alternative splicing was not observed in *T. urticae glucl.* Four of the six genes were transcribed at relatively high levels throughout the life cycle. The previously reported mutation G324D (G314D in the new annotation) in TM3 was found in one of the GluCls of a highly (>2000-fold) abamectin-resistant

T. urticae strain, although susceptible and resistant strains had a glutamate at the same position in three of the GluCls. In addition to this mutation, the G326E mutation was also identified at the identical position of one of the GluCls in the resistant strain.

Fipronil also blocks GluCls. In contrast to the action at GABA receptors, however, fipronil acted as an open channel blocker in GluCls of *P. americana*, indicating that channel opening is required for the inhibition of GluCls by fipronil (Zhao et al., 2004b). Interestingly, two types of responses to glutamate, that is, desensitizing and nondesensitizing responses, were recorded in *P. americana* neurons, a phenomenon previously reported in *P. americana* DUM neurons (Raymond et al., 2000), in which two components were discriminated in terms of sensitivity to picrotoxin, BIDN, and ibotenate. The desensitizing responses were 80-fold less sensitive to fipronil than were the nondesensitizing responses, with IC_{50} values of 801 and 10 nM, respectively. This finding can be explained by the action of fipronil as an open channel blocker, that is, channel opening facilitates access to a site within the channel. The IC_{50} value (10 nM) of fipronil against nondesensitizing GluCls is comparable to that of GABA receptors ($\sim$28 nM), indicating that both GABA receptors and GluCls are important targets of fipronil in insects. This finding is inconsistent with the finding that fipronil is approximately 160-fold more potent in cloned housefly RDL receptors than in cloned housefly GluCls (Eguchi et al., 2006). The difference may be due to the difference in insect species or may be caused by the difference between native and cloned receptors.

Trypsin treatment of neurons eliminated the nondesensitizing responses in *P. americana* native GluCls, while the desensitizing responses were unaffected (Narahashi et al., 2010). Intriguingly, treatments with soybean trypsin inhibitor or polyvinylpyrolidone eliminated the desensitizing responses without affecting the nondesensitizing responses. These findings may suggest the existence of protease-modifiable subunits or the proteolytic regulation of receptors. Two subtypes of inhibitory glutamate receptors, which differ in picrotoxin sensitivity, are also present in *A. mellifera* antennal lobe neurons, the primary olfactory neuropil of the brain (Barbara et al., 2005). In this case, fipronil (10 μM) produced not only complete inhibition of GABA- and glutamate-induced currents but also slight inhibition of acetylcholine-induced currents in these neurons.

GluCl subunits are as abundant as RDL subunits in the nervous systems of insects, and both proteins form ligand–gated chloride channels with similar functions and structures. Furthermore, NCAs cross-react with both channels, although most reported NCAs are selective for GABA receptors with the exception of fipronil action on cockroach channels (Eguchi et al., 2006; Ihara et al., 2005; Zhao et al., 2003a, 2004b). It is logical to infer that subunits of both channels might coassemble in the insect body. With immunoprecipitation experiments using antibodies specific for *Drosophila* RDL and GluCl α subunits, one research group demonstrated that the RDL subunit may assemble with the GluCl subunit to form a heteromeric ivermectin receptor that constitutes two-thirds of the population of the solubilized ivermectin receptors in *Drosophila* head membranes (Ludmerer et al., 2002). By contrast, evidence against the existence of the RDL/GluCl receptor was obtained in studies with a *P. americana* thoracic ganglion neuron, in which both GABA and glutamate responses can be independently recorded (Zhao et al., 2004a). The experimental results with this neuron clearly indicate that the GABA receptor subunit and the GluCl subunit do not coassemble, because (1) the inhibition of GABA responses of a single neuron by fipronil did not interfere with the glutamate responses of the neuron in patch-clamp recordings; (2) GABA and glutamate responses were additive in the neurons; and (3) GABA responses were observed in neurons in which GluCls were desensitized by the prolonged application of glutamate. These findings clearly indicate that the receptors responding to GABA and glutamate binding are different entities.

If the RDL and GluCl subunits coassemble to form a channel, the *rdl* mutation, which is responsible for dieldrin resistance, should affect the glutamate response and pharmacology of GluCls. This hypothesis was tested with dieldrin-resistant German cockroaches (*B. germanica*) that have the *rdl* mutation in the GABA receptor subunit (Zhao and Salgado, 2010). The EC_{50} value of the desensitized responses to glutamate in the dieldrin-resistant cockroaches was significantly reduced compared to that of the susceptible counterparts. The Hill coefficient of glutamate activation of the desensitized responses of the resistant cockroaches was reduced to half that of susceptible cockroaches. The time constant of the decay of the desensitizing currents of the resistant cockroaches was greater than that of susceptible cockroaches. These changes imply that the desensitizing type of GluCls may contain the RDL subunit. However, there was no difference in the sensitivity to fipronil sulfone, which is an active metabolite of fipronil, between neurons from the dieldrin-resistant and -susceptible strains of cockroaches.

To test coassembly of the RDL and GluCl subunits, the cRNAs transcribed from housefly *rdl* and *glucl* were injected separately or together into

Xenopus oocytes (Eguchi et al., 2006). The sensitivity of the oocytes coinjected with both cRNAs to GABA and picrotoxinin were not significantly different than that of oocytes injected with *rdl* cRNA. This finding is in contrast to the result that would be expected if the RDL and GluCl subunits coassemble to form a channel.

The physiological role of GluCls in insects is poorly understood, although they are the primary target of macrolide insecticides and parasiticides. Because GluCls are not expressed in vertebrates, the action of fipronil at GluCls may be at least partly responsible for its selective toxicity. However, it is important to examine the physiological relevance of GluCls to understand the extent to which the inhibition of GluCls contributes to insecticidal action because none of the tested NCAs showed specificity against GluCls.

Synthesis of juvenile hormone in the corpus allatum of the Pacific beetle cockroach *Diploptera punctata* is regulated by both excitatory and inhibitory glutamate receptors (Liu et al., 2005). Both glutamate–induced depolarization and hyperpolarization were observed in the cells of the corpus allatum gland using intracellular recordings. The hyperpolarization was also induced by ibotenate and blocked by picrotoxin. Synthesis of juvenile hormone in the gland decreases in the presence of ibotenate and ivermectin, whereas the synthesis was increased by picrotoxin.

Recently, a GluCl α subunit cDNA was cloned from the honeybee *A. mellifera* (El Hassani et al., 2012). The expression of GluCl α subunit was detected in the antennal lobe, thoracic muscle, mushroom body, and optic lobe, using an antibody raised against *H. contortus* GluCl α subunit and *in situ* hybridization. RNAi studies indicated that GluCls are necessary for the retrieval of olfactory memory. Recently, the existence of three isoforms of the honeybee GluCl generated by alternative splicing was reported (Démares et al., 2012). Immunohistochemical analysis showed that they are differentially distributed in the brain.

9. MECHANISM OF NCA ACTIONS

9.1. Structural requirements for NCA actions

SAR studies of NCAs began with picrotoxinin, followed by each class of NCAs. Mouse toxicological studies indicated that the following moieties of picrotoxinin are required for its high toxicity: the unmodified bridgehead hydroxyl group, the lactone group connecting the bridgehead carbons, and the axial isopropenyl group (Jarboe et al., 1968). Consistent with this finding, [^{35}S]TBPS binding studies of simplified picrotoxinin analogs showed that the axial (*trans*) isopropenyl group of this class of NCAs is more suitable than the equatorial (*cis*) isopropenyl group for interaction with rat GABA receptors (Ozoe et al., 1993a). An analogous SAR was observed for the toxicity of bicyclophosphate NCAs, in which a compact hydrophobic substituent combined with the electronegative oxygen atoms was important (Eto et al., 1976). A similar SAR has been consistently observed for other recently reported classes of NCAs, such as picrodendrane and *seco*-prezizaane terpenoids (Kuriyama et al., 2002; Schmidt et al., 2004), and ginkgolides and bilobalide (Huang et al., 2003; Ivic et al., 2003).

The SAR of cyclodienes does not simply match that of the terpenoids and the bicyclophosphates described above because cyclodienes contain chlorine atoms. SAR studies on the insecticidal activity of cyclodienes suggested that active analogs possess two electronegative centers: one composed of chlorines of the hexachloronorbornene nucleus (depicted as P2 and P3 in Fig. 4.4) and the other composed of a double bond (π electron) or oxygen atom (lone pair electron) (depicted as P1 in Fig. 4.4) of the second ring system (Soloway, 1965). This hypothesis permits the structural overlay of three different classes of compounds (cyclodienes, picrotoxinin, and TBPS) to reveal the presence of an additional pharmacophore in NCAs. In this overlay model the above-mentioned compact substituent (e.g., the *tert*-butyl group of TBPS) plays a role as a hydrophobic center that may overlap with the *gem*-chlorines of cyclodienes (Ozoe and Matsumura, 1986). Quantitative SAR studies have demonstrated that the van der Waals radius

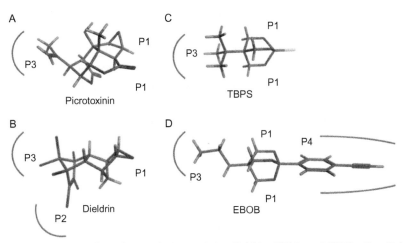

Figure 4.4 Structural similarity of picrotoxinin, dieldrin, TBPS, and EBOB. (See Color Plate 19 at the back of the book.)

and molar volume are important determinants of the insecticidal activity of γ-BHC analogs (Kiso et al., 1978). Given that the chlorine atoms of γ-BHC are able to play roles as hydrogen bond bases (Abraham et al., 2002), the analogs of γ-BHC could fit our model.

The bicycloorthobenzoate NCAs such as EBOB share a cage structure with bicyclophosphates but have an additional phenyl group at the bridgehead. To reconcile the heretofore described model and the SARs of EBOB analogs, many hybrids of cyclodienes and EBOB analogs were synthesized (Ozoe et al., 1990, 1993b). These SAR studies led to a novel pharmacophore model in which the NCA binding site includes an additional pocket (P4) to accommodate the phenyl group of EBOB and related analogs as well as that of 2-phenyl-1,3-dithianes. As a consequence, the generated NCA binding site model consisted of four subsites (P1–P4) (Ozoe et al., 1993b).

It is then necessary to test the hypothesis that structurally diverse NCAs share a common site of action, using methods of quantitative analysis. Computational analysis lends support to the NCA binding-site model. Comparative molecular field analysis (CoMFA) is one of the methods that permit quantitative three-dimensional SAR analysis to verify whether bioactive compounds with different skeletons share the same site of action (Cramer et al., 1988). When the CoMFA method was applied to a set of structurally diverse NCAs, statistically significant results were obtained for the majority of the NCAs used, indicating the validity of the proposed model (Akamatsu et al., 1997). Furthermore, the CoMFA analysis revealed a variety of differences between the NCA binding sites of rat and housefly GABA receptors. One of the marked differences is that the rat binding site is subject to greater steric constraint compared with the housefly binding site (Ozoe and Akamatsu, 2001). In a different method of SAR analysis, this characteristic is represented as shorter distances between pharmacophores in the rat receptor relative to those in the housefly receptor (Ju et al., 2007). This may reflect the difference in amino acids at the $2'$-position of TM2, which will be detailed in Section 10.

The phenylpyrazole insecticides such as fipronil include a structural feature that is different from those of picrotoxinin, bicyclophosphates, bicyloorthobenzoates, and cyclodienes. Phenylpyrazoles contain nitrogen atoms and unique substituent patterns on the pyrazole and benzene rings. Nevertheless, the SAR analysis of phenylpyrazoles suggested that they can be overlaid on the common pharmacophores of bicycloorthobenzoates (Sammelson et al., 2004). By contrast, photoaffinity labeling and ligand competition studies with a photoreactive NCA demonstrated that fipronil binds at a site distinct from that of other NCAs such as picrotoxinin and EBOB in the rat brain (Shimotahira et al., 2011). Little has been documented to date on the SAR of the novel isoxazoline class of NCAs.

9.2. Location of the NCA binding site

9.2.1 Site-directed mutagenesis and chemical labeling

In the late 1900s, important insights into the location of the NCA binding site were obtained by biochemical and electrophysiological experiments. Specific binding of [^{35}S]TBPS to rat brain membranes was observed to be entirely dependent on halide ions and was potently inhibited by GABA (Squires et al., 1983). The inhibition of GABA-induced currents by picrotoxin in native vertebrate neurons was dependent on the presence of GABA and was noncompetitive with GABA; the application of additional GABA reversed this inhibition (Inoue and Akaike, 1988; Yoon et al., 1993). These findings suggest that picrotoxin accesses the binding site through the open chloride channel to stabilize the closed conformation and that it is trapped within the closed channel. In whole-cell patch-clamp experiments with rat dissociated sympathetic ganglion neurons, the rate of onset of the block of picrotoxin-mediated GABA-induced inward current responses was accelerated by GABA (Newland and Cull-Candy, 1992). The maximal picrotoxin-mediated block was delayed in the absence of GABA. These findings support a mechanism that picrotoxin preferentially binds to an agonist-bound open state of the receptor and stabilizes an agonist-bound closed state.

Site-directed mutagenesis and chemical labeling can aid in determining the location of the ligand-binding site. Site-directed mutagenesis approaches to locate the NCA binding site were first performed with human glycine receptors expressed in HEK293 cells (Pribilla et al., 1992). Glycine receptors are members of the Cys-loop LGIC family and share a common feature with GABA receptors. Homomeric glycine receptors containing α subunits are sensitive to picrotoxinin, whereas heteromeric glycine receptors containing α and β subunits are insensitive. Substitution of the amino acids in the TM2 region of the β subunit with those of the α subunit conferred picrotoxinin sensitivity on the heteromeric receptors. These findings indicate that the TM2 segment forming the channel pore is the major determinant of picrotoxinin binding. Similar experiments were performed with a *C. elegans* GluCl α/β chimera subunit (Etter et al., 1999). The substitution of Thr for Ala at the 2′-position in TM2 reduced the picrotoxinin sensitivity at least by 10,000-fold, indicating that picrotoxinin binds to this region deep within the channel pore.

The channel lumen of Cys-loop LGICs is formed by five TM2 segments of five subunits. As the TM2 segment adopts an α-helical structure, amino acids in this region face the channel pore every three or four residues. Therefore, these amino acid residues from the five subunits form rings that are piled up at several levels as exemplified by the X-ray structure of *C. elegans* GluCl (PDB code 3RI5) (Fig. 4.5). These residues in the TM2 segment are designated with an index numbering system to facilitate the comparison of the residues, where the conserved positively charged residue at the cytoplasmic end of TM2 is numbered $0'$ and the number increases toward the extracellular end (the C-terminus). Thus, amino acids at the $-2'$-, $2'$-, $6'$-, $9'$-, $13'$-, and $16'$-positions are thought to be pore-facing amino acids (Fig. 4.5).

With a combination of site-directed mutagenesis, ligand binding, and two-electrode voltage clamp technologies, numerous studies were performed to identify the binding site of a variety of NCAs in GABA receptors. The substitution of a single Thr6$'$ of the β2 subunit with a Phe was found to

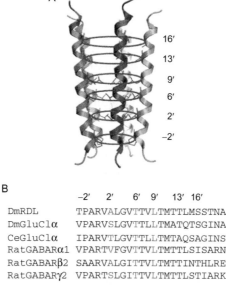

A

16′
13′
9′
6′
2′
−2′

B

	−2′	2′	6′	9′	13′	16′
DmRDL	TPARVALGVTTVLTMTTLMSSTNA
DmGluCl α	VPARVSLGVTTLLTMATQTSGINA
CeGluCl α	IPARVTLGVTTLLTMTAQSAGINS
RatGABARα1	VPARTVFGVTTVLTMTTLSISARN
RatGABARβ2	SAARVALGITTVLTMTTINTHLRE
RatGABARγ2	VPARTSLGITTVLTMTTLSTIARK

Figure 4.5 Channel pore of *C. elegans* GluCl. (A) Channel-lining amino acid residues of the TM2 region of the *C. elegans* GluCl α subunit. Illustration is based on X-ray crystal structure (PDB code 3RI5). (B) Alignment of the TM2 sequences of GABACl (GABAR) and GluCl subunits. Dm, *Drosophila melanogaster*; Ce, *Caenorhabditis elegans*. (See Color Plate 20 at the back of the book.)

confer insensitivity to picrotoxin in the rat $\alpha 1\beta 2\gamma 2$ GABA receptor (Gurley et al., 1995). Because Thr6' is conserved in the three subunits, receptors with one to five mutated 6'-amino acids can be generated. The sensitivity of the receptor to picrotoxin decreased with an increased number of mutated subunits (Erkkila et al., 2008). The lactone and epoxy oxygen atoms of picrotoxinin may form hydrogen bonds with the three Thr6's of the three adjacent subunits. In other studies, picrotoxin protected the $\alpha 1V2'C$ mutant of the rat $\alpha 1\beta 2\gamma 2$ GABA receptor from modification by an extracellularly applied sulfhydryl reagent that irreversibly inhibited GABA-induced currents in the mutant (Xu et al., 1995). Fipronil analogs modified with a chemically reactive functional group irreversibly reacted with $\alpha 1V2'C$ and $\alpha 1S17'C$ mutants of the rat $\alpha 1\beta 1\gamma 2$ GABA receptor, and these reactions were prevented in the presence of picrotoxinin or TBPS (Perret et al., 1999). In contrast to this finding, the inhibition of the rat $\alpha 1\beta 2\gamma 2L$ GABA receptor by fipronil was not affected by the V2'S mutation of the $\alpha 1$ subunit, although fipronil accelerated the current decay compared with the wild type (Li and Akk, 2008). In experiments with hetero-oligomeric GABA receptors formed by chimeric $\beta 3$ (N-terminus)/$\alpha 1$ (C-terminus) and $\alpha 1$ subunits, the introduction of Ala2' and Leu3' of the $\beta 3$ subunit into the chimeric subunit was necessary to form the [^{35}S]TBPS binding site (Jursky et al., 2000). Furthermore, the A1'S, A2'S, and T6'V mutations of the human $\beta 3$ homo-oligomeric GABA receptor were found to affect the binding of [^{3}H]EBOB; the T6'V substitution had the most profound effect (Hisano et al., 2007). In the same receptor, although mutations at eight of 15 positions in the cytoplasmic half of TM2 reduced [^{3}H]EBOB binding, Ala2', Thr6' and Leu9' were identified as proposed interacting sites (Chen et al., 2006a). Recently, similar site-directed mutagenesis experiments were conducted to identify the amino acid residues of GABA receptors that are involved in interactions with *G. biloba* constituents (Thompson et al., 2012). A2'V and T6'V mutations in *Drosophila* RDL receptors dramatically reduced sensitivity to ginkgolides and bilobalide. The replacement of Val2' of the $\alpha 1$ subunit of human $\alpha 1\beta 2$- and $\alpha 1\beta 2\gamma 2$-receptors with Ala increased sensitivity to ginkgolides and bilobalide in most cases. Most of these results indicate that the mutations at the 2'- and 6'-positions in the intracellular half of the TM2 segment significantly affect the sensitivity of GABA receptors to structurally diverse NCAs, and that NCAs bind to the channel-lining region formed by 2' and 6' amino acids.

The importance of 2' and 6' amino acids in NCA activity has been reported for other LGICs, such as glycine and serotonin type 3 receptors

(Das and Dillon, 2005; Hawthorne et al., 2006; Heads et al., 2008; Jensen et al., 2010; Shan et al., 2001; Thompson et al., 2011a; Wang and Slaughter, 2005; Yang et al., 2007). Fipronil and γ-BHC inhibited glycine receptors with potencies similar to those for mammalian GABA receptors (high nanomolar to low micromolar IC_{50} values). Site-directed mutagenesis experiments using $2'$ and $6'$ mutants of $\alpha1$ glycine receptors demonstrated that both compounds interacted with Thr6$'$ through hydrophobic and hydrogen bonding interactions (Islam and Lynch, 2012). However, in contrast to fipronil, γ-BHC did not interact with $2'$ amino acid residues. It was also proposed that fipronil binds to crevices between or behind TM2 in the closed-channel state and then is released to interact with the pore-lining $2'$ and $6'$ amino acids in TM2 in the open-channel state. One possible crevice is the alcohol/anesthetic binding site formed by a critical amino acid residue in TM3 (Islam and Lynch, 2012). In another study that used a spontaneously open homomeric $\beta3$ channels, two pathways of EBOB access to the NCA binding site were suggested. One is direct access through the channel pore, and the other is an indirect pathway through the water cavities between adjacent subunits, as suggested by Islam and Lynch (2012) (Chen et al., 2006b).

In addition to the actions on the $2'$ and $6'$ amino acids, actions of NCAs on other amino acid residues have been reported. Chemically reactive fipronil analogs reacted with a $17'$ amino acid (Perret et al., 1999), and the N19$'$R mutation of the RDL homo-oligomer enhanced the potency of the NCA action of penicillin G (Hosie et al., 2006).

9.2.2 X-ray structure, homology modeling and docking simulation

Homology modeling and docking simulation have become valuable tools for gaining information about molecular interactions between NCAs and their binding site. Using a $\rho1$ GABA receptor homology model constructed by applying the kinked-helices model of the nAChR in the open state as a template, calculation of the Monte Carlo-minimized energy profiles for picrotoxinin pulled through the channel pore led to the conclusions that (1) the ether and carbonyl oxygen atoms of picrotoxinin accept hydrogen atoms from Thr6$'$ and (2) the isopropenyl group enters the rings formed by Ala2$'$ (Zhorov and Bregestovski, 2000).

The X-ray crystallographic structure of acetylcholine binding protein (AChBP), which was determined at 2.7 Å resolution (Brejc et al., 2001), defines the organization of the N-terminal orthosteric binding site. The electron microscopic structure of the TM of the closed-state *Torpedo* nAChR at

4 Å resolution (PDB entry 1OED) was reported (Miyazawa et al., 2003), and a refined structure was later derived from this structure (Unwin, 2005). This refined structure enables a detailed description of the whole receptor in a closed-channel state. The AChBP and the nAChR TM structures have frequently been used as homology modeling templates to study the interactions between agonists and the agonist binding site and the interactions between NCAs and the NCA binding site, respectively.

Recently, more X-ray structures of LGICs have been published. The X-ray structure of a cationic LGIC from the bacterium *Erwinia chrysanthemi* was determined at 3.3 Å resolution (Hilf and Dutzler, 2008). Subsequently, a proton-gated channel from the cyanobacterium *Gloebacter violaceus* was solved in an open channel conformation at 2.9 and 3.1 Å resolution (Bocquet et al., 2008; Hilf and Dutzler, 2009). These high-resolution structures of LGICs may be used as templates for homology modeling. More recently, the X-ray structure of a homomeric *C. elegans* GluCl α-Fab complex was determined in the ivermectin-, glutamate- and picrotoxinin-bound form (Hibbs and Gouaux, 2011). Ivermectin bound between TM3 of the principal subunit and TM1 of the complementary subunit in the membrane domain (Fig. 4.3). Picrotoxinin bound at the cytoplasmic base within the channel pore. The oxygen atoms interacted with the hydroxyl groups of Thr2′ and the isopropenyl group was surrounded by the methylenes of Pro-2′. Picrotoxinin was located closer to the cytoplasm by one turn of the α-helix compared with its possible location in GABA receptors. This may be partly ascribed to the differences between receptors in the amino acid residues in this region. It is also important to note that this binding pose is for the open conformation of the channel.

The docking of NCAs into a β3 GABA receptor homology model based on the protein data 1OED demonstrated that a wide range of NCAs fit the same site (2′–9′) within the channel (Chen et al., 2006a). Consistent with the three-dimensional quantitative SAR model, hydrophobic substituents such as the isopropenyl group of picrotoxinin interacted with the methyl group of Ala2′, and the electronegative portion of NCAs interacted with the hydroxyl group of Thr6′. The 4-ethynylphenyl group of EBOB interacted with the methyl group of Thr6′. The β3 subunit has high sequence identity to the RDL subunit in the TM2 region and can be easily expressed in cell lines. Therefore, the homomeric β3 GABA receptor proved to be a useful insect GABA receptor model to study the mechanisms underlying the interactions of NCAs with insect GABA receptors. It is important to note that the homomeric β3 GABA receptor does not exist as a homomer in humans, and that it thus can be used as a molecular model for insect studies.

The potencies of insecticidal NCAs at the β3 GABA receptor correlated well with the potencies at *M. domestica* native GABA receptors (Alam et al., 2006; Ratra and Casida, 2001). When phenylpyrazole and phenyltriazole NCAs were docked into the β3 GABA receptor homology model, a good correlation between the docking scores (energies) and IC_{50} values (or K_is) obtained from [³H]EBOB binding to housefly head membranes was observed, indicating that the binding mode of certain NCAs can be predicted from this model (Alam et al., 2007; Ci et al., 2007).

9.3. Structural determinants of NCA selectivity

The 2' and 6' amino acids of the TM2 segment of the GABA receptor subunits are critical for the potency of NCAs, and therefore the sensitivity of the receptors to NCAs depends on the characteristics of the side chains of the amino acid residues that constitute the binding site (Ratra et al., 2001). Thr6' is conserved in both mammalian and insect GABA receptors, but amino acids at the 2'-position differ from one subunit to another. The amino acids of the α1, β2, and γ2 subunits at the 2'-position are Val, Ala, and Ser, respectively. The RDL subunits of most insect species have Ala at the equivalent position. Lepidopteran insects, however, have three types of RDL paralogues carrying Ala, Ser, or Gln. The pore size at the 2' ring formed by two valines, two alanines, and one serine of the α1β2γ2 GABA receptor is assumed to be smaller than that formed by five alanines of the homo-oligomeric RDL GABA receptor, although the hydrophobicity and the electronic properties also differ. This might be one of the reasons that the homomeric β3 channel, which has an alanine at the 2'-position, is the most sensitive of the channels to NCA insecticides (Ratra and Casida, 2001). These differences in channels may be taken into consideration when designing NCA insecticides that selectively act on the channels of insect pests.

The first example of target-selective ligands can be seen in the modification of bicyclophosphates. The enlargement of the molecular size of TBPS-type NCAs by introducing a small alkyl group into the ring endocyclic methylene led to NCAs that bound to housefly GABA receptors with higher affinity than to rat GABA receptors (Ju and Ozoe, 1999). IBIPPS exhibited greater than 50-fold selectivity against housefly GABA receptors versus rat GABA receptors, while TBPS was 26-fold more potent against rat receptors than housefly receptors. However, not only the pore-facing amino acid residues in TM2 but also the functional states of the receptor channels, that is, open, closed or desensitized, may affect the affinity and orientation of NCAs for the binding site. The functional states thus may be a determinant of the potency of NCAs.

Another example of the interaction between the 2′ amino acids and NCAs can be seen in ginkgolides and bilobalide. These terpenes contain a *tert*-butyl group as does TBPS. This *tert*-butyl group, thus, may interact with the 2′ amino acid residues in TM2. The A2′V mutation of *Drosophila* RDL receptors caused a drastic decrease in sensitivity to these terpenoids; for example, the IC_{50} value of ginkgolide A was increased from 3.1 nM to 360 μM by this amino acid substitution (Thompson et al., 2012). The change in the side chain from the methyl group to the isopropyl group may produce a steric hindrance for the *tert*-butyl group of ginkgolide A.

10. TARGET-SITE INSENSITIVITY UNDERLYING RESISTANCE TO NCA INSECTICIDES

Target-site insensitivity is one of the principal mechanisms by which insect pests acquire resistance to insecticides. The identity of the amino acid at the 2′-position is a key determinant for conferring resistance to NCA insecticides in insects. A single A2′S or A2′G mutation of the *Drosophila* RDL subunit, which was shown to confer resistance to dieldrin (ffrench-Constant et al., 1991; Thompson et al., 1993a), was widespread all over the world (ffrench-Constant et al., 1993b). This same amino acid substitution is conserved in resistant strains of *P. americana, Bemisia tabaci, M. domestica, T. castaneum, Hypothenemus hampei,* and *B. germanica* (Anthony et al., 1995; ffrench-Constant et al., 1994; Kaku and Matsumura, 1994; Miyazaki et al., 1995; Thompson et al., 1993a). After year 2000, the widespread existence of this mutation associated with dieldrin and fipronil resistance was confirmed in field samples of a variety of insect species such as cockroaches, mosquitoes, and aphids that were collected from a variety of countries (Asih et al., 2012; Gondhalekar and Scharf, 2012; Guillemaud et al., 2003; Hansen et al., 2005; Holbrook et al., 2003; Kolaczinski and Curtis, 2001; Navarro et al., 2010; Tantely et al., 2010).

The resistance of *D. melanogaster* to dieldrin was found to be semidominant (ffrench-Constant et al., 1990). The cross-resistance of the dieldrin-resistant *M. domestica* strain to fipronil is incompletely recessively inherited, in contrast to the inheritance of cyclodiene-resistance, which is intermediately dominantly inherited (Wen and Scott, 1999). The OCR strain of houseflies, which is homozygous for the A2′S allele, showed 3100-fold resistance to dieldrin, whereas this strain was only 31-fold resistant to fipronil (Scott and Wen, 1997). A housefly strain highly resistant to γ-BHC exhibited a high (430-fold) resistance to fipronil by topical application, but a moderate (23-fold) resistance by feeding application (Kristensen et al., 2004). The mechanism underlying the high resistance in this strain was not identified. Dieldrin-resistant strains of *D. melanogaster* and *B. germanica* showed cross-resistance to experimental phenylpyrazole, bicyclophosphorothionate, and bicycloorthobenzoates insecticides (Bloomquist, 1994). Although *B. germanica* strains with high resistance to dieldrin showed low levels of cross-resistance to fipronil, there is a good correlation ($r^2 = 0.93$) between the LD_{50} values of dieldrin and fipronil against these strains, indicating a causal connection of dieldrin and fipronil resistance (Kristensen et al., 2005).

Patch-clamp analysis of the action of NCAs on cultured neurons from homozygous dieldrin-resistant *D. melanogaster* suggested that this mutation leads to two changes in GABACls: (1) structural changes that weaken the affinity of NCAs to the binding site and (2) destabilization of the NCA-favored (desensitized) conformation of GABA receptors (Zhang et al., 1994). This proposed mechanism implies that the mutant receptor may exhibit different levels of insensitivity to NCAs with different structures, due to the structural changes in the binding site (ffrench-Constant and Roush, 1991; Zhang et al., 1994). The dieldrin-resistant OCR strain of houseflies, which were homozygous for the 2′S *rdl* allele, had a EBOB binding site with a decreased affinity compared with wild-type houseflies (Deng et al., 1991; Ozoe et al., 2007), whereas [^{3}H]EBOB binding was not detected in fruitflies with an equivalent mutation (Cole et al., 1995; Lee et al., 1995).

The resistance level of dieldrin-resistant insects to fipronil is lower than the level of dieldrin resistance (Cole et al., 1993, 1995; Colliot et al., 1992; Ozoe et al., 2007; Wen and Scott, 1999). Biochemical studies with the OCR strain of houseflies indicated that a similar level of affinity of fipronil for wild-type and A2′S mutant GABA receptors was responsible for the low resistance to fipronil (Ozoe et al., 2007). The high affinity of fipronil for GABA receptors from dieldrin-resistant houseflies most likely arises from its unique substitution pattern on the pyrazole ring (Ozoe et al., 2007).

Housefly GluCls have a Ser at the 2′-position of the TM2 segment, the same substitution as the *rdl* mutation in dieldrin-resistant *Drosophila*. This may explain why the known NCAs show selectivity toward GABA receptors over GluCls (Eguchi et al., 2006; Ihara et al., 2005). To verify this hypothesis, a GluCl with an S2′A mutation at this position was generated and examined for its sensitivity to NCAs (Hirata et al., 2008). The reverse *rdl* (S2′A) mutant exhibited enhanced sensitivity to γ-BHC, picrotoxinin, and fipronil, supporting the conclusion of the GABA receptor-based studies that the amino acid at the 2′-position is a determinant of NCA action.

Individual clones of the green peach aphid *Myzus persicae* contain up to four different *rdl*-like alleles: A, G, S^TCG, and S′^AGT (Anthony et al., 1998). The *rdl* was found to be duplicated, with one locus carrying A or G alleles and the other carrying two S alleles. It was suggested that cyclodiene resistance is associated with replacements at the first but not the second locus. The A2′S mutation is conserved in laboratory strains of fleas (*Ctenocephalides felis*). The persistence of this mutation without selection with NCAs such as dieldrin in the laboratory suggests that this mutation has low or no fitness costs. A *C. felis* strain homozygous for the mutant 2′ allele was less susceptible to fipronil than other strains, whereas a strain homozygous for the wild-type 2′ allele was the most susceptible to fipronil (Bass et al., 2004). However, all tested cat flea strains that were fully susceptible to fipronil were homozygous for the mutant allele, carrying the A2′S mutation, suggesting that the 2′ mutation may not determine the susceptibility of fleas to fipronil (Brunet et al., 2009). The prevalence of dieldrin resistance was found to be high in natural and laboratory strains of the malaria vector *An. gambiae*. The A2′S and A2′G mutations were detected in resistant strains of *An. gambiae* and *Anopheles arabiensis*, respectively (Du et al., 2005). The A2′S mutation was observed in the GABA receptor subunit of dieldrin-resistant *An. funestus* from West and Central Africa, but not from East and South Africa (Wondji et al., 2011). An additional mutation (V327I) in the TM2–TM3 linker was associated with the A2′S mutation in resistant *An. funestus*. A novel two base pair mutation causing an amino acid change from threonine to leucine at the 6′-position was detected in dieldrin-resistant southern cattle ticks (*Rhipicephalus microplus*), a cattle disease vector (Hope et al., 2010). This mutation has apparently been maintained without selection pressure for more than 20 years.

High levels (505- and 398-fold) of resistance to fipronil compared with a laboratory strain were reported in diamondback moths (*P. xylostella*) collected in the field in India (collected in 2000) and Malaysia (collected in 2003) (Mohan and Gujar, 2003; Sayyed and Wright, 2004). This resistance ratio is one order of magnitude higher than that reported previously. The inheritance of resistance to fipronil is an incompletely recessive trait. The dieldrin-resistant strain generated by dieldrin selection of field-collected populations over 25 generations contained the A2′S mutation in the α1 subunit, whereas no such mutation was detected in the susceptible strain. The strain with homozygous A2′S mutation in the α1 subunit exhibited 7- and 11-fold greater resistance to dieldrin and fipronil, respectively (Yuan et al., 2010). Selection of *P. xylostella* with fipronil over 20 generations led to a strain with 300-, 3.5-, and 6.5-fold decreased susceptibility to fipronil, dieldrin and endosulfan, respectively (Li et al., 2006). The A2′S mutation was identified in fipronil-selected *P. xylostella*, but other mutation(s) or other subunit(s) may contribute to the high level of resistance.

The small brown planthopper (*L. striatellus*) is an economically important insect pest that causes serious damage to rice plants by sucking their sap in the temperate regions of East Asia. This species is the major vector of rice stripe virus, one of the most severe diseases of rice. Populations collected in Japan and Vietnam in 2008 and 2009 showed extremely high resistance to fipronil (Sanada-Morimura et al., 2011). One population in Japan had an approximately 10,000 higher 24-h LD_{50} value compared to the susceptible population. Sequencing of cDNA clones encoding the RDL subunit from a population of field-collected *L. striatellus* that exhibited 11-fold greater resistance to fipronil showed that Ala2′ in TM2 was mutated to Asn (Nakao et al., 2011). Nineteen of twenty-one individuals were genotyped as Ala/Asn and the other two were genotyped as Ala/Ala. Fipronil exerted no antagonistic effect on GABA receptors composed of A2′N mutant RDL subunit expressed alone or together with wild-type RDL subunit in a cell line.

The whitebacked rice planthopper (*S. furcifera*) is also a serious insect pest of rice in Asia. Populations of *S. furcifera* collected from East and Southeast Asia in 2006 were extremely highly resistant to fipronil (Matsumura et al., 2008), with 24-h LD_{50} values of 19.9–239 $\mu g\ g^{-1}$ or greater. A population of *S. furcifera* collected in Japan in 2007, which was 70-fold less susceptible to fipronil than a laboratory strain, was analyzed for mutations in the RDL subunit. A comparison of the deduced amino acid sequences of the cDNA clones of a susceptible strain and a resistant population demonstrated that Ala2′ in TM2 was changed to Asn in the RDL subunit of resistant planthoppers (Nakao et al., 2010). Furthermore, another novel mutation (R340Q) was identified in the RDL subunit of resistant *S. furcifera* (Nakao et al., 2012). This mutation in the cytoplasmic TM3–TM4 loop was always found to be associated with the A2′N mutation in cDNA clones. When expressed alone in a cell line, the A2′N mutant channel was insensitive to fipronil, while the mutant channel coexpressed with the wild type was moderately sensitive to fipronil. The A2′N/R340Q double mutant was insensitive to fipronil when transfected alone or together with wild-type RDL. The R340Q mutant channel was slightly less sensitive to fipronil when expressed alone or with wild-type RDL.

A population of *O. oryzae* collected in Japan in 2008 showed approximately 58-fold resistance to fipronil. The A2'S mutation was found in RDL cDNAs cloned from the resistant beetles (Nakao et al., 2012b). Insensitivity of the mutant receptor to fipronil was observed only when the channels were highly activated.

A strain of *D. simulans* with approximately 20,000-fold resistance to fipronil was generated by laboratory selection (Le Goff et al., 2005). The RDL subunit of this strain carried two mutations, A301G at the 2′-position in TM2 and T350M in TM3. Each mutation caused a decrease in the sensitivity of the *Drosophila* RDL homo-oligomer to fipronil and picrotoxin. The double mutant was less sensitive to fipronil than the single mutants. It is unclear if the amino acid residue in the TM3 segment is involved in the fipronil binding site.

11. SUMMARY AND FUTURE PERSPECTIVES

GABA receptors have long been considered an important target for the development of insecticides (Eto, 1983; Matsumura and Ghiasuddin, 1983; Ozoe et al., 1983). Forty years ago, the receptor entity was unknown, and the naturally occurring convulsant picrotoxinin was the only known NCA. Subsequent discoveries of synthetic chemicals (bicyclophosphates, γ-BHC, and cyclodienes) as NCAs shed light on insect GABA receptors as targets of insecticides (Bowery et al., 1976; Ghiasuddin and Matsumura, 1982; Korenaga et al., 1977). These compounds share some common structural features. However, the discovery of a practical insecticide, fipronil, in 1987 was not made by extension of these compounds, that is, optimization of their structures, but from screening tests. Meanwhile, pharmacological studies have unveiled the presence of multiple ligand-binding sites in GABA receptors and other LGICs, which suggest the potential for the development of novel classes of insecticides. Although no GABAergic or anti-GABAergic practical insecticides other than the phenylpyrazole insecticides have been developed for 25 years, strenuous efforts are leading to the discovery of new chemistries.

Biological milestones in GABA and glutamate receptor studies include the molecular cloning of the cDNAs encoding bovine GABA$_A$ receptor subunits (Schofield et al., 1987), followed by the isolation of *Drosophila* RDL and GluCl subunit cDNA clones (Cully et al., 1994; ffrench-Constant et al., 1991). Cloning of these genes accelerated receptor/channel research from a molecular perspective and led to an enhanced understanding of the structures and functions of insect GABACls and GluCls.

The binding site of NCA insecticides was identified within LGICs, and the interspecies selectivity of NCAs, which is necessary for safety, was found to be determined by the composition of receptor subunits and differences in amino acid residues in the binding site. The function of receptors is diversified by their posttranslational modification. Even though a single gene encodes the RDL or GluCl subunit, the gene undergoes alternative splicing and RNA editing. Lepidopteran insects and Acari (*T. urticae*) possess multiple RDL orthologues. The combination of diversified subunits may generate the physiological and pharmacological diversity of pentameric receptor channels.

While the physiological importance of RDL GABA receptors is a focus of research, GluCls have been less well studied in this respect. Differences in the physiological roles of GluCls and GABACls are of interest because they have similar functions and structures. Although fipronil potently acts not only at GABA receptors but also at GluCls in native cockroach neurons *in vitro*, the toxicological importance of GluCls remains to be studied further.

Insect pests in certain regions have developed resistance to fipronil due to point mutation(s) of amino acid(s) in the NCA binding site or in its proximity. The development of resistance is a serious problem for any chemical control of insect pests. The future of NCAs as commercial insecticides looked bleak for a number of years because of *rdl* resistance. However, advances in molecular and structural biology could permit specific designs of insect pest control agents aimed at coping with crop protection problems. The ability of isoxazoline-type compounds to circumvent this resistance makes the future exploitation of GABA receptors feasible.

ACKNOWLEDGMENTS

I would like to dedicate this chapter to the late Professor Morifusa Eto for his outstanding support and encouragement. I wish to express my gratitude to Professor Ephraim Cohen for excellent editing and invaluable advice. I am grateful to Professor Toshio Narahashi, Professor Fumio Matsumura, Professor Jeffrey G. Scott, and anonymous reviewers for reading the text and making helpful comments. I also thank Mr. Tomo Kita for his assistance in preparing this manuscript.

REFERENCES

Abraham, M.H., Enomoto, K., Clarke, E.D., Sexton, G., 2002. Hydrogen bond basicity of the chlorogroup; hexachlorocyclohexanes as strong hydrogen bond bases. J. Org. Chem. 67, 4782–4786.

Adelsberger, H., Lepier, A., Dudel, J., 2000. Activation of rat recombinant $\alpha_1\beta_2\gamma_{2S}$ GABA$_A$ receptor by the insecticide ivermectin. Eur. J. Pharmacol. 394, 163–170.

Agosto, J., Choi, J.C., Parisky, K.M., Stilwell, G., Rosbash, M., Griffith, L.C., 2008. Modulation of GABA$_A$ receptor desensitization uncouples sleep onset and maintenance in Drosophila. Nat. Neurosci. 11, 354–359.

Ahn, Y.J., Kwon, M., Park, H.M., Han, C.K., 1997. Potent insecticidal activity of Ginkgo biloba derived trilactone terpenes against Nilaparvata lugens. In: Hedin, P.A., Hollingworth, R.M., Masler, E.P., Miyamoto, J., Thompson, D.G. (Eds.), Phytochemicals for Pest Control. ACS Symposium Series 658. American Chemical Society, Washington, DC, pp. 90–105.

Akamatsu, M., Ozoe, Y., Ueno, T., Fujita, T., Mochida, K., Nakamura, T., Matsumura, F., 1997. Sites of action of noncompetitive GABA antagonists in houseflies and rats: three-dimensional QSAR analysis. Pestic. Sci. 49, 319–332.

Akinci, M.K., Schofield, P.R., 1999. Widespread expression of GABA$_A$ receptor subunits in peripheral tissues. Neurosci. Res. 35, 145–153.

Alam, M.S., Kajiki, R., Hanatani, H., Kong, X., Ozoe, F., Matsui, Y., Matsumura, F., Ozoe, Y., 2006. Synthesis and structure-activity relationships of 1-phenyl-1H-1,2,3-triazoles as selective insect GABA receptor antagonists. J. Agric. Food Chem. 54, 1361–1372.

Alam, M.S., Huang, J., Ozoe, F., Matsumura, F., Ozoe, Y., 2007. Synthesis, 3D-QSAR, and docking studies of 1-phenyl-1H-1,2,3-triazoles as selective antagonists for β3 over α1β2γ2 GABA receptors. Bioorg. Med. Chem. 15, 5090–5104.

Alix, P., Grolleau, F., Hue, B., 2002. Ca^{2+}/calmodulin-dependent protein kinase regulates GABA-activated Cl⁻ current in cockroach dorsal unpaired median neurons. J. Neurophysiol. 87, 2972–2982.

Alqazzaz, M., Thompson, A.J., Price, K.L., Breitinger, H.-G., Lummis, S.C.R., 2011. Cys-loop receptor channel blockers also block GLIC. Biophys. J. 101, 2912–2918.

Amat, C., Hue, B., 1997. Activation of picrotoxin-resistant GABA receptors by GABA and related compounds induces modulation of cockroach dorsal paired median (DPM) neuron firing. J. Insect Physiol. 43, 1125–1131.

Anthony, N.M., Brown, J.K., Markham, P.G., ffrench-Constant, R.H., 1995. Molecular analysis of cyclodiene resistance-associated mutations among populations of the sweetpotato whitefly Bemisia tabaci. Pestic. Biochem. Physiol. 51, 220–228.

Anthony, N., Unruh, T., Ganser, D., ffrench-Constant, R., 1998. Duplication of the Rdl GABA receptor subunit gene in an insecticide-resistant aphid, Myzus persicae. Mol. Gen. Genet. 260, 165–175.

Aronstein, K., ffrench-Constant, R., 1995. Immunocytochemistry of a novel GABA receptor subunit Rdl in Drosophila melanogaster. Invert. Neurosci. 1, 25–31.

Aronstein, K., Auld, V., ffrench-Constant, R., 1996. Distribution of two GABA receptor-like subunits in the Drosophila CNS. Invert. Neurosci. 2, 115–120.

Ashby, J.A., McGonigle, I.V., Price, K.L., Cohen, N., Comitani, F., Dougherty, D.A., Molteni, C., Lummis, S.C.R., 2012. GABA binding to an insect GABA receptor: a molecular dynamics and mutagenesis study. Biophys. J. 103, 2071–2081.

Asih, P.B.S., Syahrani, L., Rozi, I.E.P., Pratama, N.R., Marantina, S.S., Arsyad, D.S., Mangunwardoyo, W., Hawley, W., Laihad, F., Shinta, Sukowati, S., Lobo, N.F., Syafruddin, D., 2012. Existence of the rdl mutant alleles among the Anopheles malaria vector in Indonesia. Malar. J. 11, 57.

Aspinwall, L.S., Bermudez, I., King, L.A., Wafford, K.A., 1997. The interactions of hexa-chlorocyclohexane isomers with human γ-aminobutyric acid$_A$ receptors expressed in *Xenopus* oocytes. J. Pharmacol. Exp. Ther. 282, 1557–1564.

Awapara, J., Landua, A.J., Fuerst, R., Seale, B., 1950. Free γ-aminobutyric acid in brain. J. Biol. Chem. 187, 35–39.

Aydar, E., Beadle, D.J., 1999. The pharmacological profile of GABA receptors on cultured insect neurones. J. Insect Physiol. 45, 213–219.

Bai, D., Sattelle, D.B., 1995. A GABA$_B$ receptor on an identified insect motor neurone. J. Exp. Biol. 198, 889–894.

Barbara, G.S., Zube, C., Rybak, J., Gauthier, M., Grünewald, B., 2005. Acetylcholine, GABA and glutamate induce ionic currents in cultured antennal lobe neurons of the honeybee, *Apis mellifera*. J. Comp. Physiol. A 191, 823–836.

Barnard, E.A., Skolnick, P., Olsen, R.W., Möhler, H., Sieghart, W., Biggio, G., Braestrup, C., Bateson, A.N., Langer, S.Z., 1998. International Union of Pharmacology. XV. Subtypes of γ-aminobutyric acid$_A$ receptors: classification on the basis of subunit structure and receptor function. Pharmacol. Rev. 50, 291–313.

Bass, C., Schroeder, I., Turnberg, A., Field, L.M., Williamson, M.S., 2004. Identification of the *Rdl* mutation in laboratory and field strains of the cat flea, *Ctenocephalides felis* (Siphonaptera: Pulicidae). Pest Manag. Sci. 60, 1157–1162.

Bazemore, A.W., Elliott, K.A.C., Florey, E., 1957. Isolation of factor I. J. Neurochem. 1, 334–339.

Beg, A.A., Jorgensen, E.M., 2003. EXP-1 is an excitatory GABA-gated cation channel. Nat. Neurosci. 6, 1145–1152.

Belelli, D., Callachan, H., Hill-Venning, C., Peters, J.A., Lambert, J.J., 1996. Interaction of positive allosteric modulators with human and *Drosophila* recombinant GABA receptors expressed in *Xenopus laevis* oocytes. Br. J. Pharmacol. 118, 563–576.

Benson, J.A., 1989. A novel GABA receptor in the heart of a primitive arthropod, *Limulus polyphemus*. J. Exp. Biol. 147, 421–438.

Bermudez, I., Hawkins, C.A., Taylor, A.M., Beadle, D.J., 1991. Actions of insecticides on the insect GABA receptor complex. J. Recept. Res. 11, 221–232.

Bertrand, D., Gopalakrishnan, M., 2007. Allosteric modulation of nicotinic acetylcholine receptors. Biochem. Pharmacol. 74, 1155–1163.

Bettler, B., Kaupmann, K., Mosbacher, J., Gassmann, M., 2004. Molecular structure and physiological functions of GABA$_B$ receptors. Physiol. Rev. 84, 835–867.

Bloomquist, J.R., 1994. Cyclodiene resistance at the insect GABA receptor/chloride channel complex confers broad cross resistance to convulsants and experimental phenylpyrazole insecticides. Arch. Insect Biochem. Physiol. 26, 69–79.

Bloomquist, J.R., 2001. GABA and glutamate receptors as biochemical sites for insecticide action. In: Ishaaya, I. (Ed.), Biochemical Sites of Insecticide Action and Resistance. Springer-Verlag, Berlin, pp. 17–41.

Bloomquist, J.R., Boina, D.R., Chow, E., Carlier, P.R., Reina, M., Gonzalez-Coloma, A., 2008. Mode of action of the plant-derived silphinenes on insect and mammalian GABA$_A$ receptor/chloride channel complex. Pestic. Biochem. Physiol. 91, 17–23.

Bocquet, N., Nury, H., Baaden, M., Le Poupon, C., Changeux, J.-P., Delarue, M., Corringer, P.-J., 2008. X-ray structure of a pentameric ligand-gated ion channel in an apparently open conformation. Nature 457, 111–114.

Bonanno, G., Fassio, A., Sala, R., Schmid, G., Raiteri, M., 1998. GABA$_B$ receptors as potential targets for drugs able to prevent excessive excitatory amino acid transmission in the spinal cord. Eur. J. Pharmacol. 362, 143–148.

Bowery, N.G., Collins, J.F., Hill, R.G., 1976. Bicyclic phosphorus esters that are potent convulsants and GABA antagonists. Nature 261, 601–603.

Bowery, N.G., Bettler, B., Froestl, W., Gallagher, J.P., Marshall, F., Raiteri, M., Bonner, T.I., Enna, S.J., 2002. International Union of Pharmacology. XXXIII. Mammalian γ-aminobutyric acid$_B$ receptors: structure and function. Pharmacol. Rev. 54, 247–264.

Brejc, K., van Dijk, W.J., Klaassen, R.V., Schuurmans, M., van der Oost, J., Smit, A.B., Sixma, T.K., 2001. Crystal structure of an ACh-binding protein reveals the ligand-binding domain of nicotinic receptors. Nature 411, 269–276.

Brotz, T.M., Borst, A., 1996. Cholinergic and GABAergic receptors on fly tangential cells and their role in visual motion detection. J. Neurophysiol. 76, 1786–1799.

Brotz, T.M., Bochenek, B., Aronstein, K., ffrench-Constant, R.H., Borst, A., 1997. γ-Aminobutyric acid receptor distribution in the mushroom bodies of a fly (*Calliphora erythrocephala*): a functional subdivision of Kenyon cells? J. Comp. Neurol. 383, 42–48.

Brotz, T.M., Gundelfinger, E.D., Borst, A., 2001. Cholinergic and GABAergic pathways in fly motion vision. BMC Neurosci. 2, 1.

Brunet, S., Le Meter, C., Murray, M., Soll, M., Audonnet, J.-C., 2009. *Rdl* gene polymorphism and sequence analysis and relation to in vivo fipronil susceptibility in strains of the cat flea. J. Econ. Entomol. 102, 366–372.

Buckingham, S.D., Sattelle, D.B., 2004. GABA receptors of insects. In: Iatrou, K., Gill, S.S., Gilbert, L.I. (Eds.), Comprehensive Molecular Insect Science. Elsevier Pergamon, Amsterdam, The Netherland, pp. 107–142.

Buckingham, S.D., Sattelle, D.B., 2010. GABA receptors of insects. In: Gilbert, L.I., Gill, S.S. (Eds.), Insect Pharmacology: Channels, Receptors, Toxins and Enzymes. Elsevier Academic Press, London, pp. 29–64.

Buckingham, S.D., Matsuda, K., Hosie, A.M., Baylis, H.A., Squire, M.D., Lansdell, S.J., Millar, N.S., Sattelle, D.B., 1996. Wild-type and insecticide-resistant homo-oligomeric GABA receptors of *Drosophila melanogaster* stably expressed in a *Drosophila* cell line. Neuropharmacology 35, 1393–1401.

Buckingham, S.D., Biggin, P.C., Sattelle, B.M., Brown, L.A., Sattelle, D.B., 2005. Insect GABA receptors: splicing, editing, and targeting by antiparasitics and insecticides. Mol. Pharmacol. 68, 942–951.

Buckingham, S.D., Higashino, Y., Sattelle, D.B., 2009. Allosteric modulation by benzodiazepines of GABA-gated chloride channels of an identified insect motor neurone. Invert. Neurosci. 9, 85–89.

Casida, J.E., Eto, M., Moscioni, A.D., Engel, J.L., Milbrath, D.S., Verkade, J.G., 1976. Structure-toxicity relationships of 2,6,7-trioxabicyclo[2.2.2]octanes and related compounds. Toxicol. Appl. Pharmacol. 36, 261–279.

Cayre, M., Buckingham, S.D., Yagodin, S., Sattelle, D.B., 1999. Cultured insect mushroom body neurons express functional receptors for acetylcholine, GABA, glutamate, octopamine and dopamine. J. Neurophysiol. 81, 1–14.

Chen, R., Belelli, D., Lambert, J.J., Peters, J.A., Reyes, A., Lan, N.C., 1994. Cloning and functional expression of a *Drosophila* γ-aminobutyric acid receptor. Proc. Natl. Acad. Sci. U.S.A. 91, 6069–6073.

Chen, L., Durkin, K.A., Casida, J.E., 2006a. Structural model for γ-aminobutyric acid receptor noncompetitive antagonist binding: widely diverse structures fit the same site. Proc. Natl. Acad. Sci. U.S.A. 103, 5185–5190.

Chen, L., Durkin, K.A., Casida, J.E., 2006b. Spontaneous mobility of GABA$_A$ receptor M2 extracellular half relative to noncompetitive antagonist action. J. Biol. Chem. 281, 38871–38878.

Choudhary, A.F., Laycock, I., Wright, G.A., 2012. γ-Aminobutyric acid receptor A-mediated inhibition in the honeybee's antennal lobe is necessary for the formation of configural olfactory percepts. Eur. J. Neurosci. 35, 1718–1724.

Chung, B.Y., Kilman, V.L., Keath, J.R., Pitman, J.L., Allada, R., 2009. The GABA_A receptor RDL acts in peptidergic PDF neurons to promote sleep in *Drosophila*. Curr. Biol. 19, 386–390.

Ci, S., Ren, T., Su, Z., 2007. Modeling the interaction of fipronil-related non-competitive antagonists with the GABA β3-receptor. J. Mol. Model. 13, 457–464.

Cleland, T.A., 1996. Inhibitory glutamate receptor channels. Mol. Neurobiol. 13, 97–136.

Cole, L.M., Casida, J.E., 1992. GABA-gated chloride channel: binding site for 4'-ethynyl-4-n-[2,3-^{3}H$_2$]propylbicycloorthobenzoate ([^{3}H]EBOB) in vertebrate brain and insect head. Pestic. Biochem. Physiol. 44, 1–8.

Cole, L.M., Nicholson, R.A., Casida, J.E., 1993. Action of phenylpyrazole insecticides at the GABA-gated chloride channel. Pestic. Biochem. Physiol. 46, 47–54.

Cole, L.M., Roush, R.T., Casida, J.E., 1995. *Drosophila* GABA-gated chloride channel: modified [^{3}H]EBOB binding site associated with Ala→Ser or Gly mutants of *Rdl* subunit. Life Sci. 56, 757–765.

Collins, T., Millar, N.S., 2010. Nicotinic acetylcholine receptor transmembrane mutations convert ivermectin from a positive to a negative allosteric modulator. Mol. Pharmacol. 78, 198–204.

Colliot, F., Kukorowski, K.A., Hawkins, D.W., Roberts, D.A., 1992. Fipronil: a new soil and foliar broad spectrum insecticide. In: Brighton Crop Protection Conference—Pest and Disease 2-1, 29–34.

Cordova, D., Rauh, J.J., Benner, E.A., Schroeder, M.E., Sopa, J.S., Lahm, G.P., Pahutski, T.F., Long, J.K., Holyoke, C.W., Smith, R.M., Barry, J.D., Dung, M.H., 2010. Isoxazoline insecticides: a novel class of GABA-gated chloride channel blockers. In: 12th IUPAC International Congress of Pesticide Chemistry. Royal Australian Chemical Institute, North Melbourne No. 265.

Cramer III, R.D., Patterson, D.E., Bunce, J.D., 1988. Comparative molecular field analysis (CoMFA). 1. Effect of shape on binding of steroids to carrier proteins. J. Am. Chem. Soc. 110, 5959–5967.

Cull-Candy, S.G., 1976. Two types of extrajunctional L-glutamate receptors in locust muscle fibres. J. Physiol. 255, 449–464.

Cull-Candy, S.G., Miledi, F.R.S., 1981. Junctional and extrajunctional membrane channels activated by GABA in locust muscle fibres. Proc. R. Soc. Lond. B Biol. Sci. 211, 527–535.

Cull-Candy, S.G., Usherwood, P.N.R., 1973. Two populations of L-glutamate receptors on locust muscle fibres. Nat. New Biol. 246, 62–64.

Cully, D.F., Vassilatis, D.K., Liu, K.K., Paress, P.S., Van der Ploeg, L.H.T., Schaeffer, J.M., Arena, J.P., 1994. Cloning of an avermectin-sensitive glutamate-gated chloride channel from *Caenorhabditis elegans*. Nature 371, 707–711.

Cully, D.F., Paress, P.S., Liu, K.K., Schaeffer, J.M., Arena, J.P., 1996. Identification of a *Drosophila melanogaster* glutamate-gated chloride channel sensitive to the antiparasitic agent avermectin. J. Biol. Chem. 271, 20187–20191.

Das, P., Dillon, G.H., 2005. Molecular determinants of picrotoxin inhibition of 5-hydroxytryptamine type 3 receptors. J. Pharmacol. Exp. Ther. 314, 320–328.

Das, P., Bell-Horner, C.L., Machu, T.K., Dillon, G.H., 2003. The GABA_A receptor antagonist picrotoxin inhibits 5-hydroxytryptamine type 3A receptors. Neuropharmacology 44, 431–438.

Das, S., Sadanandappa, M.K., Dervan, A., Larkin, A., Lee, J.A., Sudhakaran, I.P., Priya, R., Heidari, R., Holohan, E.E., Pimentel, A., Gandhi, A., Ito, K., Sanyal, S., Wang, J.W., Rodrigues, V., Ramaswami, M., 2011. Plasticity of local GABAergic interneurons drives olfactory habituation. Proc. Natl. Acad. Sci. USA 108, E646–E654.

Dawson, G.R., Wafford, K.A., Smith, A., Marshall, G.R., Bayley, P.J., Schaeffer, J.M., Meinke, P.T., McKernan, R.M., 2000. Anticonvulsant and adverse effects of avermectin analogs in mice are mediated through the γ-aminobutyric acid$_A$ receptor. J. Pharmacol. Exp. Ther. 295, 1051–1060.

Delany, N.S., Laughton, D.L., Wolstenholme, A.J., 1998. Cloning and localisation of an avermectin receptor-related subunit from *Haemonchus contortus*. Mol. Biochem. Parasitol. 97, 177–187.

Deng, Y., Palmer, C.J., Casida, J.E., 1991. House fly brain γ-aminobutyric acid-gated chloride channel: target for multiple classes of insecticides. Pestic. Biochem. Physiol. 41, 60–65.

Dent, J.A., Davis, M.W., Avery, L., 1997. *avr-15* encodes a chloride channel subunit that mediates inhibitory glutamatergic neurotransmission and ivermectin sensitivity in *Caenorhabditis elegans*. EMBO J. 16, 5867–5879.

Démares, F., Raymond, V., Armengaud, C., 2012. Expression and localization of glutamate-gated chloride channel variants in honeybee brain (*Apis mellifera*). Insect Biochem. Mol. Biol. http://dx.doi.org/10.1016/j.ibmb.2012.10.003.

Dermauw, W., Ilias, A., Riga, M., Tsagkarakou, A., Grbić, M., Tirry, L., Van Leeuwen, T., Vontas, J., 2012. The cys-loop ligand-gated ion channel gene family of *Tetranychus urticae*: implications for acaricide toxicology and a novel mutation associated with abamectin resistance. Insect Biochem. Mol. Biol. 42, 455–465.

Dimitrijevic, N., Dzitoyeva, S., Satta, R., Imbesi, M., Yildiz, S., Manev, H., 2005. *Drosophila* GABA$_B$ receptors are involved in behavioral effects of γ-hydroxylbutyric acid (GHB). Eur. J. Pharmacol. 519, 246–252.

Du, W., Awolola, T.S., Howell, P., Koekemoer, L.L., Brooke, B.D., Benedict, M.Q., Cowtzee, M., Zheng, L., 2005. Independent mutations in the *Rdl* locus confer dieldrin resistance to *Anopheles gambiae* and *An. arabiensis*. Insect Mol. Biol. 14, 179–183.

Dubreil, V., Sinakevitch, I.G., Hue, B., Geffard, M., 1994. Neuritic GABAergic synapses in insect neurosecretory cells. Neurosci. Res. 19, 235–240.

Duce, I.R., Scott, R.H., 1985. Actions of dihydroavermectin B$_{1a}$ on insect muscle. Br. J. Pharmacol. 85, 395–401.

Duittoz, A.H., Martin, R.J., 1991. Antagonist properties of arylaminopyridazine GABA derivatives at the *Ascaris* muscle GABA receptor. J. Exp. Biol. 159, 149–164.

Dupuis, J.P., Bazelot, M., Barbara, G.S., Paute, S., Gauthier, M., Raymond-Delpech, V., 2010. Homomeric RDL and heteromeric RDL/LCCH3 GABA receptors in the honeybee antennal lobes: two candidates for inhibitory transmission in olfactory processing. J. Neurophysiol. 103, 458–468.

Dzitoyeva, S., Dimitrijevic, N., Manev, H., 2003. γ-Aminobutyric acid B receptor 1 mediates behavior-impairing actions of alcohol in *Drosophila*: adult RNA interference and pharmacological evidence. Proc. Natl. Acad. Sci. U.S.A. 100, 5485–5490.

Dzitoyeva, S., Gutnov, A., Imbesi, M., Dimitrijevic, N., Manev, H., 2005. Developmental role of GABA$_{B(1)}$ receptors in *Drosophila*. Dev. Brain Res. 158, 111–114.

Eguchi, Y., Ihara, M., Ochi, E., Shibata, Y., Matsuda, K., Fushiki, S., Sugama, H., Hamasaki, Y., Niwa, H., Wada, M., Ozoe, F., Ozoe, Y., 2006. Functional characterization of *Musca* glutamate- and GABA-gated chloride channels expressed independently and coexpressed in *Xenopus* oocytes. Insect Mol. Biol. 15, 773–783.

El Hassani, A.K., Dupuis, J.P., Gauthier, M., Armengaud, C., 2009. Glutamatergic and GABAergic effects of fipronil on olfactory learning and memory in the honeybee. Invert. Neurosci. 9, 91–100.

El Hassani, A.K., Schuster, S., Dyck, Y., Demares, F., Leboulle, G., Armengaud, C., 2012. Identification, localization and function of glutamate-gated chloride channel receptors in the honeybee brain. Eur. J. Neurosci. 36, 2409–2420.

Enell, L., Hamasaka, Y., Kolodziejczyk, A., Nässel, D.R., 2007. γ-Aminobutyric acid (GABA) signaling components in *Drosophila*: immunocytochemical localization of GABA$_B$ receptors in relation to the GABA$_A$ receptor subunit RDL and a vesicular GABA transporter. J. Comp. Neurol. 505, 18–31.

Enna, S.J., 2007. The GABA receptors. In: Enna, S.J., Möhler, H. (Eds.), The GABA Receptors. third ed Humana Press, Totowa, NJ, pp. 1–21.

Erkkila, B.E., Sedelnikova, A.V., Weiss, D.S., 2008. Stoichiometric pore mutations of the $GABA_AR$ reveal a pattern of hydrogen bonding with picrotoxin. Biophys. J. 94, 4299–4306.

Ernst, E., 2005. The efficacy of herbal medicine—an overview. Fund. Clin. Pharmacol. 19, 405–409.

Es-Salah, Z., Lapied, B., Le Goff, G., Hamon, A., 2008. RNA editing regulates insect gamma-aminobutyric acid receptor function and insecticide sensitivity. Neuroreport 19, 939–943.

Eto, M., 1983. Development of insecticidal cyclic phosphoryl compounds through chemical and biochemical approaches. J. Environ. Sci. Health B 18, 119–145.

Eto, M., Ozoe, Y., Fujita, T., Casida, J.E., 1976. Significance of branched bridge-head substituent in toxicity of bicyclic phosphate esters. Agric. Biol. Chem. 40, 2113–2115.

Etter, A., Cully, D.F., Liu, K.K., Reiss, B., Vassilatis, D.K., Schaeffer, J.M., Arena, J.P., 1999. Picrotoxin blockade of invertebrate glutamate-gated chloride channels: subunit dependence and evidence for binding within the pore. J. Neurochem. 72, 318–326.

Featherstone, D.E., Rushton, E., Rohrbough, J., Liebl, F., Karr, J., Sheng, Q., Rodesch, C.K., Broadie, K., 2005. An essential *Drosophila* glutamate receptor subunit that functions in both central neuropil and neuromuscular junction. J. Neurosci. 25, 3199–3208.

Feng, X.-P., Hayashi, J., Beech, R.N., Prichard, R.K., 2002. Study of the nematode putative GABA type-A receptor subunits: evidence for modulation by ivermectin. J. Neurochem. 83, 870–878.

Ferraguti, F., Crepaldi, L., Nicoletti, F., 2008. Metabotropic glutamate 1 receptor: current concepts and perspectives. Pharmacol. Rev. 60, 536–581.

ffrench-Constant, R.H., Rocheleau, T.A., 1993. *Drosophila* γ-aminobutyric acid receptor gene *Rdl* shows extensive alternative splicing. J. Neurochem. 60, 2323–2326.

ffrench-Constant, R.H., Roush, R.T., 1991. Gene mapping and cross-resistance in cyclodiene insecticide-resistant *Drosophila melanogaster* (Mg.). Genet. Res. 57, 17–21.

ffrench-Constant, R.H., Roush, R.T., Mortlock, D., Dively, G.P., 1990. Isolation of dieldrin resistance from field populations of *Drosophila melanogaster* (Diptera: Drosophilidae). J. Econ. Entomol. 83, 1733–1737.

ffrench-Constant, R.H., Mortlock, D.P., Shaffer, C.D., MacIntyre, R.J., Roush, R.T., 1991. Molecular cloning and transformation of cyclodiene resistance in *Drosophila*: an invertebrate γ-aminobutyric acid subtype A receptor locus. Proc. Natl. Acad. Sci. U.S.A. 88, 7209–7213.

ffrench-Constant, R.H., Rocheleau, T.A., Steichen, J.C., Chalmers, A.E., 1993a. A point mutation in a *Drosophila* GABA receptor confers insecticide resistance. Nature 363, 449–451.

ffrench-Constant, R.H., Steichen, J.C., Rocheleau, T.A., Aronstein, K., Roush, R.T., 1993b. A single-amino acid substitution in a γ-aminobutyric acid subtype A receptor locus is associated with cyclodiene insecticide resistance in *Drosophila* populations. Proc. Natl. Acad. Sci. U.S.A. 90, 1957–1961.

ffrench-Constant, R.H., Steichen, J.C., Brun, L.O., 1994. A molecular diagnostic for endosulfan insecticide resistance in the coffee berry borer *Hypothenemus hampei* (Coleoptera: Scolytidae). Bull. Entomol. Res. 84, 11–16.

Florey, E., 1954. An inhibitory and an excitatory factor of mammalian central nervous system, and their action of a single sensory neuron. Arch. Int. Physiol. 62, 33–53.

Fukunaga, A., Hasegawa, H., Ogawa, C., Matsuno, A., Imamura, K., Ozoe, Y., 1999. Insecticidal properties of 3-aminopropyl(methyl)phosphinic acid and its effect on K^+-evoked release of acetylcholine from cockroach synaptosomes. Comp. Biochem. Physiol. C 122, 283–286.

Gengs, C., Leung, H.-T., Skingsley, D.R., Iovchev, M.I., Yin, Z., Semenov, E.P., Burg, M.G., Hardie, R.C., Pak, W.L., 2002. The target of *Drosophila* photoreceptor synaptic transmission is a histamine-gated chloride channel encoded by *ort* (*hclA*). J. Biol. Chem. 277, 42113–42120.

Ghiasuddin, S.M., Matsumura, F., 1982. Inhibition of gamma-aminobutyric acid (GABA)-induced chloride uptake by gamma-BHC and heptachlor epoxide. Comp. Biochem. Physiol. C 73, 141–144.

Gill, S., 2010. Addendum: recent progress in insect GABA receptors. In: Gilbert, L.I., Gill, S.S. (Eds.), Insect Pharmacology: Channels, Receptors, Toxins and Enzymes. Elsevier Academic Press, London, pp. 65–67.

Gisselmann, G., Pusch, H., Hovemann, B.T., Hatt, H., 2002. Two cDNAs coding for histamine-gated ion channels in *D. melanogaster*. Nat. Neurosci. 5, 11–12.

Gisselmann, G., Plonka, J., Pusch, H., Hatt, H., 2004. *Drosophila melanogaster* GRD and LCCH3 subunits form heteromultimeric GABA-gated cation channels. Br. J. Pharmacol. 142, 409–413.

Glendinning, S.K., Buckingham, S.D., Sattelle, D.B., Wonnacott, S., Wolstenholme, A.J., 2011. Glutamate-gated chloride channels of *Haemonchus contortus* restore drug sensitivity to ivermectin resistant *Caenorhabditis elegans*. PLoS One 6, e22390.

Gondhalekar, A.D., Scharf, M.E., 2012. Mechanisms underlying fipronil resistance in a multiresistant field strain of the German cockroach (Blattodea: Blattellidae). J. Med. Entomol. 49, 122–131.

Grolleau, F., Sattelle, D.B., 2000. Single channel analysis of the blocking actions of BIDN and fipronil on a *Drosophila melanogaster* GABA receptor (RDL) stably expressed in a *Drosophila* cell line. Br. J. Pharmacol. 130, 1833–1842.

Grünewald, B., Wersing, A., 2008. An ionotropic GABA receptor in cultured mushroom body Kenyon cells of the honeybee and its modulation by intracellular calcium. J. Comp. Physiol. A 194, 329–340.

Guillemaud, T., Brun, A., Anthony, N., Sauge, M.-H., Boll, R., Delorme, R., Fournier, D., Lapchin, L., Vanlerberghe-Masutti, F., 2003. Incidence of insecticide resistance alleles in sexually-reproducing populations of the peach–potato aphid *Myzus persicae* (Hemiptera: Aphididae) from southern France. Bull. Entomol. Res. 93, 289–297.

Gurley, D., Amin, J., Ross, P.C., Weiss, D.S., White, G., 1995. Point mutations in the M2 region of α, β, or γ subunit of the $GABA_A$ channel that abolish block by picrotoxin. Receptors Channels 3, 13–20.

Hainzl, D., Casida, J.E., 1996. Fipronil insecticide: novel photochemical desulfinylation with retention of neurotoxicity. Proc. Natl. Acad. Sci. U.S.A. 93, 12764–12767.

Hamann, M., Desarmenien, M., Vanderheyden, P., Piguet, P., Feltz, P., 1990. Electrophysiological study of *tert*-butylbicyclophosphorothionate-induced block of spontaneous chloride channels. Mol. Pharmacol. 37, 578–582.

Hamano, H., Nagata, K., Fukuda, N., Shimotahira, H., Ju, X.-L., Ozoe, Y., 2000. 5-[4-(3,3-Dimethylbutoxycarbonyl)phenyl]-4-pentynoic acid and its derivatives inhibit ionotropic γ-aminobutyric acid receptors by binding to the 4'-ethynyl-4-*n*-propylbicycloorthobenzoate site. Bioorg. Med. Chem. 8, 665–674.

Hamasaka, Y., Wegener, C., Nässel, D.R., 2005. GABA modulates *Drosophila* circadian clock neurons via $GABA_B$ receptors and decreases in calcium. J. Neurobiol. 65, 225–240.

Hamon, A., Le Corronc, H., Hue, B., Rauh, J.J., Sattelle, D.B., 1998. BIDN, a bicyclic dinitrile convulsant, selectively blocks GABA-gated Cl⁻ channels. Brain Res. 780, 20–26.

Hansen, K.K., Kristensen, M., Jensen, K.-M.V., 2005. Correlation of a resistance-associated *Rdl* mutation in the German cockroach, *Blattella germanica* (L.), with persistent dieldrin resistance in two Danish field populations. Pest Manag. Sci. 61, 749–753.

Harrison, J.B., Chen, H.H., Sattelle, E., Barker, P.J., Huskisson, N.S., Rauh, J.J., Bai, D., Sattelle, D.B., 1996. Immunocytochemical mapping of a C-terminus anti-peptide antibody to the GABA receptor subunit, RDL in the nervous system of *Drosophila melanogaster*. Cell Tissue Res. 284, 269–278.

Harvey, R.J., Schmitt, B., Hermans-Borgmeyer, I., Gundelfinger, E.D., Betz, H., Darlison, M.G., 1994. Sequence of a *Drosophila* ligand-gated ion-channel polypeptide with an unusual amino-terminal extracellular domain. J. Neurochem. 62, 2480–2483.

Hatton, G.I., Yang, Q.Z., 2001. Ionotropic histamine receptors and H_2 receptors modulate supraoptic oxytocin neuronal excitability and dye coupling. J. Neurosci. 21, 2974–2982.

Hawthorne, R., Cromer, B.A., Ng, H.-L., Parker, M.W., Lynch, J.W., 2006. Molecular determinants of ginkgolide binding in the glycine receptor pore. J. Neurochem. 98, 395–407.

Hayden, W.J., Gillis, W.T., Stone, D.E., Broome, C.R., Webster, G.L., 1984. Systematics and palynology of *Picrodendron*: further evidence for relationship with the Oldfieldioideae (Euphorbiaceae). J. Arnold Arbor. 65, 105–127.

Heads, J.A., Hawthorne, R.L., Lynagh, T., Lynch, J.W., 2008. Structure-activity analysis of ginkgolide binding in the glycine receptor pore. J. Neurochem. 105, 1418–1427.

Henderson, J.E., Soderlund, D.M., Knipple, D.C., 1993. Characterization of a putative γ-aminobutyric acid (GABA) receptor β subunit gene from *Drosophila melanogaster*. Biochem. Biophys. Res. Commun. 193, 474–482.

Henderson, J.E., Knipple, D.C., Soderlund, D.M., 1994. PCR-based homology probing reveals a family of GABA receptor-like genes in *Drosophila melanogaster*. Insect Biochem. Mol. Biol. 24, 363–371.

Hevers, W., Lüddens, H., 1998. The diversity of $GABA_A$ receptors. Mol. Neurobiol. 18, 35–86.

Hibbs, R.E., Gouaux, E., 2011. Principles of activation and permeation in an anion-selective Cys-loop receptor. Nature 474, 54–60.

Hilf, R.J.C., Dutzler, R., 2008. X-ray structure of a prokaryotic pentameric ligand-gated ion channel. Nature 452, 375–379.

Hilf, R.J.C., Dutzler, R., 2009. Structure of a potentially open state of a proton-activated pentameric ligand-gated ion channel. Nature 457, 115–118.

Hille, B., 2001. Ionic Channels of Excitable Membranes, third ed. Sinauer Associates, Sunderland.

Hiragaki, S., Kobayashi, T., Ochiai, N., Toshima, K., Dekeyser, M.A., Matsuda, K., Takeda, M., 2012. A novel action of highly specific acaricide; bifenazate as a synergist for a GABA-gated chloride channel of *Tetranychus urticae* [Acari: Tetranychidae]. Neurotoxicology 33, 307–313.

Hirata, K., Ishida, C., Eguchi, Y., Sakai, K., Ozoe, F., Ozoe, Y., Matsuda, K., 2008. Role of a serine residue (S278) in the pore-facing region of the housefly L-glutamate-gated chloride channel in determining sensitivity to noncompetitive antagonists. Insect Mol. Biol. 17, 341–350.

Hisano, K., Ozoe, F., Huang, J., Kong, X., Ozoe, Y., 2007. The channel-lining 6' amino acid in the second membrane-spanning region of ionotropic GABA receptors has more profound effects on 4'-ethynyl-4-*n*-propylbicycloorthobenzoate binding than the 2' amino acid. Invert. Neurosci. 7, 39–46.

Holbrook, G.L., Roebuck, J., Moore, C.B., Waldvogel, M.G., Schal, C., 2003. Origin and extent of resistance to fipronil in the German cockroach, *Blattella germanica* (L.) (Dictyoptera: Blattellidae). J. Econ. Entomol. 96, 1548–1558.

Höld, K.M., Sirisoma, N.S., Sparks, S.E., Casida, J.E., 2000. Metabolism and mode of action of *cis*- and *trans*-3-pinanones (the active ingredients of hyssop oil). Xenobiotica 32, 251–265.

Höld, K.M., Sirisoma, N.S., Ikeda, T., Narahashi, T., Casida, J.E., 2002. α-Thujone (the active component of absinthe): γ-aminobutyric acid type A receptor modulation and metabolic detoxification. Proc. Natl. Acad. Sci. U.S.A. 97, 3826–3831.

Hope, M., Menzies, M., Kemp, D., 2010. Identification of a dieldrin resistance-associated mutation in *Rhipicephalus* (*Boophilus*) *microplus* (Acari: Ixodidae). J. Econ. Entomol. 103, 1355–1359.

Horoszok, L., Raymond, V., Sattelle, D.B., Wolstenholme, A.J., 2001. GLC-3: a novel fipronil and BIDN-sensitive, but picrotoxinin-insensitive, L-glutamate-gated chloride channel subunit from *Caenorhabditis elegans*. Br. J. Pharmacol. 132, 1247–1254.

Hosie, A.M., Baylis, H.A., Buckingham, S.D., Sattelle, D.B., 1995. Actions of the insecticide fipronil, on dieldrin-sensitive and -resistant GABA receptors of *Drosophila melanogaster*. Br. J. Pharmacol. 115, 909–912.

Hosie, A.M., Sattelle, D.B., 1996a. Allosteric modulation of an expressed homo-oligomeric GABA-gated chloride channel of *Drosophila melanogaster*. Br. J. Pharmacol. 117, 1229–1237.

Hosie, A.M., Sattelle, D.B., 1996b. Agonist pharmacology of two *Drosophila* GABA receptor splice variants. Br. J. Pharmacol. 119, 1577–1585.

Hosie, A.M., Ozoe, Y., Koike, K., Ohmoto, T., Nikaido, T., Sattelle, D.B., 1996. Actions of picrodendrin antagonists on dieldrin-sensitive and -resistant *Drosophila* GABA receptors. Br. J. Pharmacol. 119, 1569–1576.

Hosie, A.M., Aronstein, K., Sattelle, D.B., ffrench-Constant, R.H., 1997. Molecular biology of insect neuronal GABA receptors. Trends Neurosci. 20, 578–583.

Hosie, A.M., Buckingham, S.D., Presnail, J.K., Sattelle, D.B., 2001. Alternative splicing of a *Drosophila* GABA receptor subunit gene identifies determinants of agonist potency. Neuroscience 102, 709–714.

Hosie, A.M., Buckingham, S.D., Hamon, A., Sattelle, D.B., 2006. Replacement of asparagine with arginine at the extracellular end of the second transmembrane (M2) region of insect GABA receptors increases sensitivity to penicillin G. Invert. Neurosci. 6, 75–79.

Huang, S.H., Duke, R.K., Chebib, M., Sasaki, K., Wada, K., Johnston, G.A.R., 2003. Bilobalide, a sesquiterpene trilactone from *Ginkgo biloba*, is an antagonist at recombinant $\alpha_1\beta_2\gamma_{2L}$ GABA$_A$ receptors. Eur. J. Pharmacol. 464, 1–8.

Huang, S.H., Duke, R.K., Chebib, M., Sasaki, K., Wada, K., Johnston, G.A.R., 2004. Ginkgolides, diterpene trilactones of *Ginkgo biloba*, as antagonists at recombinant $\alpha_1\beta_2\gamma_{2L}$ GABA$_A$ receptors. Eur. J. Pharmacol. 494, 131–138.

Huang, S.H., Duke, R.K., Chebib, M., Sasaki, K., Wada, K., Johnston, G.A.R., 2006. Mixed antagonistic effects of bilobalide at ρ_1 GABA$_C$ receptor. Neuroscience 137, 607–617.

Hue, B., 1991. Functional assay for GABA receptor subtypes of a cockroach giant interneuron. Arch. Insect Biochem. Physiol. 18, 147–157.

Hue, B., 1998. A picrotoxin-resistant GABA-gated chloride channel receptor subtype in the cockroach central nervous system. Arch. Insect Biochem. Physiol. 37, 231–238.

Hue, B., Pelhate, M., Chanelet, J., 1979. Pre- and postsynaptic effects of taurine and GABA in the cockroach central nervous system. Can. J. Neurol. Sci. 6, 243–250.

Ide, J., Okazaki, T., Ono, M., Saito, A., Nakagawa, K., Naito, S., Sato, K., Tanaka, K., Yoshikawa, H., Ando, M., Katsumi, S., Matsumoto, K., Toyama, T., Shibano, M., Abe, M., 1993. Milbemycin: discovery and development. Annu. Rep. Sankyo Res. Lab. 45, 1–98.

Ihara, M., Ishida, C., Okuda, H., Ozoe, Y., Matsuda, K., 2005. Differential blocking actions of 4'-ethynyl-4-*n*-propylbicycloorthobenzoate (EBOB) and γ-hexachlorocyclohexane (γ-HCH) on γ-aminobutyric acid- and glutamate-induced responses of American cockroach neurons. Invert. Neurosci. 5, 157–164.

Ikeda, T., Ozoe, Y., Okuyama, E., Nagata, K., Honda, H., Shono, T., Narahashi, T., 1999. Anisatin modulation of the γ-aminobutyric acid receptor-channel in rat dorsal root ganglion neurons. Br. J. Pharmacol. 127, 1567–1576.

Ikeda, T., Zhao, X., Nagata, K., Kono, Y., Shono, T., Yeh, J.Z., Narahashi, T., 2001. Fipronil modulation of γ-aminobutyric acid$_A$ receptors in rat dorsal root ganglion neurons. J. Pharmacol. Exp. Ther. 296, 914–921.

Ikeda, T., Nagata, K., Kono, Y., Yeh, J.Z., Narahashi, T., 2004. Fipronil modulation of GABA$_A$ receptor single-channel currents. Pest Manag. Sci. 60, 487–492.

Im, W.B., Pregenzer, J.F., Thomsen, D.R., 1994. Effects of GABA and various allosteric ligands on TBPS binding to cloned rat GABA$_A$ receptor subtypes. Br. J. Pharmacol. 112, 1025–1030.

Inoue, M., Akaike, N., 1988. Blockade of γ-aminobutyric acid-gated chloride current in frog sensory neurons by picrotoxin. Neurosci. Res. 5, 380–394.

Islam, R., Lynch, J.W., 2012. Mechanism of action of the insecticides, lindane and fipronil, on glycine receptor chloride channels. Br. J. Pharmacol. 165, 2707–2720.

Isman, M.B., Jeffs, L.B., Ellinger, C.A., Miyake, T., Matsumura, F., 1997. Petuniolides, natural insecticides from *Petunia parodii*, are antagonists of GABA$_a$ receptors. Pestic. Biochem. Physiol. 58, 103–107.

Ivic, L., Sands, T.T.J., Fishkin, N., Nakanishi, K., Kriegstein, A.R., Strømgaard, K., 2003. Terpene trilactones from *Ginkgo biloba* are antagonists of cortical glycine and GABA$_A$ receptors. J. Biol. Chem. 278, 49279–49285.

Jackel, C., Krenz, W.-D., Nagy, F., 1994. Bicuculline/baclofen-insensitive GABA response in crustacean neurones in culture. J. Exp. Biol. 191, 167–193.

Janssen, D., Derst, C., Rigo, J.-M., Van Kerkhove, E., 2010. Cys-loop ligand-gated chloride channels in dorsal unpaired median neurons of *Locusta migratoria*. J. Neurophysiol. 103, 2587–2598.

Jaracz, S., Nakanishi, K., Jensen, A.A., Strømgaard, K., 2004. Ginkgolides and glycine receptors: a structure-activity relationship study. Chem. Eur. J. 10, 1507–1518.

Jarboe, C.H., Porter, L.A., 1965. The preparative column chromatographic separation of picrotoxin. J. Chromatogr. 19, 427–428.

Jarboe, C.H., Porter, L.A., Buckler, R.T., 1968. Structural aspects of picrotoxinin action. J. Med. Chem. 11, 729–731.

Jensen, A.A., Begum, N., Vogensen, S.B., Knapp, K.M., Gundertofte, K., Dzyuba, S.V., Ishii, H., Nakanishi, K., Kristiansen, U., Strømgaard, K., 2007. Probing the pharmacophore of ginkgolides as glycine receptor antagonists. J. Med. Chem. 50, 1610–1617.

Jensen, A.A., Bergmann, M.L., Sander, T., Balle, T., 2010. Ginkgolide X is a potent antagonist of anionic Cys-loop receptors with a unique selectivity profile at glycine receptors. J. Biol. Chem. 285, 10141–10153.

Jones, A.K., Sattelle, D.B., 2006. The cys-loop ligand-gated ion channel superfamily of the honeybee, *Apis mellifera*. Invert. Neurosci. 6, 123–132.

Jones, A.K., Sattelle, D.B., 2007. The cys-loop ligand-gated ion channel gene superfamily of the red flour beetle, *Tribolium castaneum*. BMC Genomics 8, 327.

Jones, A.K., Buckingham, S.D., Papadaki, M., Yokota, M., Sattelle, B.M., Matsuda, K., Sattelle, D.B., 2009. Splice-variant- and stage-specific RNA editing of the *Drosophila* GABA receptor modulates agonist potency. J. Neurosci. 29, 4287–4292.

Jones, A.K., Bera, A.N., Lees, K., Sattelle, D.B., 2010. The cys-loop ligand-gated ion channel gene superfamily of the parasitoid wasp *Nasonia vitripennis*. Heredity 104, 247–259.

Ju, X.-L., Ozoe, Y., 1999. Bicyclophosphorothionate antagonists exhibiting selectivity for housefly GABA receptors. Pestic. Sci. 55, 971–982.

Ju, X.-L., Ozoe, Y., 2000. Noncompetitive antagonist-binding sites of rat and housefly γ-aminobutyric acid receptors display different enantiospecificities for *tert*-butyl(isopropyl)bicyclophosphorothionate. Bioorg. Med. Chem. 8, 2337–2341.

Ju, X.-L., Hao, Y.-L., Pei, J.-F., Ozoe, Y., 2007. Investigation of structural requirements for inhibitory activity at the rat and housefly picrotoxinin binding sites in ionotropic GABA receptors using DISCOtech and CoMFA. Chemosphere 69, 864–871.

Ju, X.-L., Fusazaki, S., Hishinuma, H., Qiao, X., Ikeda, I., Ozoe, Y., 2010. Synthesis and structure–activity relationship analysis of bicyclophosphorothionate blockers with selectivity for housefly γ-aminobutyric acid receptor channels. Pest Manag. Sci. 66, 1002–1010.

Jursky, F., Fuchs, K., Buhr, A., Tretter, V., Sigel, E., Sieghart, W., 2000. Identification of amino acid residues of GABA$_A$ receptor subunits contributing to the formation and affinity of the *tert*-butylbicyclophosphorothionate binding site. J. Neurochem. 74, 1310–1316.

Kaku, K., Matsumura, F., 1994. Identification of the site of mutation within the M2 region of the GABA receptor of the cyclodiene-resistant German cockroach. Comp. Biochem. Physiol. C 108, 367–376.

Kane, N.S., Hirschberg, B., Qian, S., Hunt, D., Thomas, B., Brochu, R., Ludmerer, S.W., Zheng, Y., Smith, M., Arena, J.P., Cohen, C.J., Schmatz, D., Warmke, J., Cully, D.F., 2000. Drug-resistant *Drosophila* indicate glutamate-gated chloride channels are targets for the antiparasitics nodulisporic acid and ivermectin. Proc. Natl. Acad. Sci. U.S.A. 97, 13949–13954.

Kaupmann, K., Huggel, K., Heid, J., Flor, P.J., Bischoff, S., Mickel, S.J., McMaster, G., Angst, C., Bittiger, H., Froestl, W., Bettler, B., 1997. Expression cloning of GABA$_B$ receptors uncovers similarity to metabotropic glutamate receptors. Nature 386, 239–246.

Kerkut, G.A., Leake, L.D., Shapira, A., Cowan, S., Walker, R.J., 1965. The presence of glutamate in nerve-muscle perfusates of *Helix*, *Carcinus* and *Periplaneta*. Comp. Biochem. Physiol. 15, 485–502.

Khom, S., Baburin, I., Timin, E., Hohaus, A., Trauner, G., Kopp, B., Hering, S., 2007. Valerenic acid potentiates and inhibits GABA$_A$ receptors: molecular mechanism and subunit specificity. Neuropharmacology 53, 178–187.

Kiso, M., Fujita, T., Kurihara, N., Uchida, M., Tanaka, K., Nakajima, M., 1978. Structure-activity relationship of lindane analogs. Pestic. Biochem. Physiol. 8, 33–43.

Kolaczinski, J., Curtis, C., 2001. Laboratory evaluation of fipronil, a phenylpyrazole insecticide, against adult *Anopheles* (Diptera: Culicidae) and investigation of its possible cross-resistance with dieldrin in *Anopheles stephensi*. Pest Manag. Sci. 57, 41–45.

Kolodziejczyk, A., Sun, X., Meinertzhagen, I.A., Nässel, D.R., 2008. Glutamate, GABA and acetylcholine signaling components in the lamina of the *Drosophila* visual system. PLoS One 3, e2110.

Kondratskaya, E.L., Lishko, P.V., Chatterjee, S.S., Krishtal, O.A., 2002. BN52021, a platelet activating factor antagonist, is a selective blocker of glycine-gated chloride channel. Neurochem. Int. 40, 647–653.

Korenaga, S., Ito, Y., Ozoe, Y., Eto, M., 1977. The effects of bicyclic phosphate esters on the invertebrate and vertebrate neuro-muscular junctions. Comp. Biochem. Physiol. C 57, 95–100.

Kristensen, M., Jespersen, J.B., Knorr, M., 2004. Cross-resistance potential of fipronil in *Musca domestica*. Pest Manag. Sci. 60, 894–900.

Kristensen, M., Hansen, K.K., Jensen, K.-M.V., 2005. Cross-resistance between dieldrin and fipronil in German cockroach (Dictyoptera: Blattellidae). J. Econ. Entomol. 98, 1305–1310.

Krnjević, K., 2010. When and why amino acids? J. Physiol. 588, 33–44.

Kuriyama, T., Schmidt, T.J., Okuyama, E., Ozoe, Y., 2002. Structure-activity relationships of *seco*-prezizaane terpenoids in γ-aminobutyric acid receptors of houseflies and rats. Bioorg. Med. Chem. 10, 1873–1881.

Kuriyama, T., Kakemoto, E., Takahashi, N., Imamura, K., Oyama, K., Suzuki, E., Harimaya, K., Yaguchi, T., Ozoe, Y., 2004. Receptor assay-guided isolation of anti-GABAergic insecticidal alkaloids from a fungal culture. J. Agric. Food Chem. 52, 3884–3887.

Kuriyama, T., Ju, X.-L., Fusazaki, S., Hishinuma, H., Satou, T., Koike, K., Nikaido, T., Ozoe, Y., 2005. Nematocidal quassinoids and bicyclophosphorothionates: a possible common mode of action on the GABA receptor. Pestic. Biochem. Physiol. 81, 176–187.

Kwon, D.H., Yoon, K.S., Clark, J.M., Lee, S.H., 2010. A point mutation in a glutamate-gated chloride channel confers abamectin resistance in the two-spotted spider mite, *Tetranychus urticae* Koch. Insect Mol. Biol. 19, 583–591.

Lea, T.J., Usherwood, P.N.R., 1973. The site of action of ibotenic acid and the identification of two populations of glutamate receptors on insect muscle-fibres. Comp. Gen. Pharmacol. 4, 333–350.

Le Corronc, H., Alix, P., Hue, B., 2002. Differential sensitivity of two insect GABA-gated chloride channels to dieldrin, fipronil and picrotoxinin. J. Insect Physiol. 48, 419–431.

Lee, H.-J., Rocheleau, T., Zhang, H.-G., Jackson, M.B., ffrench-Constant, R.H., 1993. Expression of a *Drosophila* GABA receptor in a baculovirus insect cell system. FEBS Lett. 335, 315–318.

Lee, H.-J., Zhang, H.-G, Jackson, M.B., ffrench-Constant, R.H., 1995. Binding and physiology of 4'-ethynyl-4-*n*-propylbicycloorthobenzoate (EBOB) in cyclodiene-resistant *Drosophila*. Pestic. Biochem. Physiol. 51, 30–37.

Lee, D., Su, H., O'Dowd, D.K., 2003. GABA receptors containing Rdl subunits mediate fast inhibitory synaptic transmission in *Drosophila* neurons. J. Neurosci. 23, 4625–4634.

Lees, G., Calder, J., 1996. Interaction of lindane isomers with chloride currents in insect membranes: steric requirements for channel modulation and block. Pestic. Biochem. Physiol. 55, 40–48.

Le Goff, G., Hamon, A., Bergé, J.-B., Amichot, M., 2005. Resistance to fipronil in *Drosophila simulans*: influence of two point mutations in the RDL GABA receptor subunit. J. Neurochem. 92, 1295–1305.

Li, P., Akk, G., 2008. The insecticide fipronil and its metabolite fipronil sulfone inhibit the rat α1β2γ2L GABA receptor. Br. J. Pharmacol. 155, 783–794.

Li, A., Yang, Y., Wu, S., Li, C., Wu, Y., 2006. Investigation of resistance mechanisms to fipronil in diamondback moth (Lepidoptera: Plutellidae). J. Econ. Entomol. 99, 914–919.

Liu, X., Davis, R.L., 2009. The GABAergic anterior paired lateral neuron suppresses and is suppressed by olfactory learning. Nat. Neurosci. 12, 53–59.

Liu, H.-P., Lin, S.-C., Lin, C.-Y., Yeh, S.-R., Chiang, A.-S., 2005. Glutamate-gated chloride channels inhibit juvenile hormone biosynthesis in the cockroach, *Diploptera punctata*. Insect Biochem. Mol. Biol. 35, 1260–1268.

Liu, X., Krause, W.C., Davis, R.L., 2007. GABA$_A$ receptor RDL inhibits *Drosophila* olfactory associative learning. Neuron 56, 1090–1102.

Liu, X., Buchanan, M.E., Han, K.-A., Davis, R.L., 2009. The GABA$_A$ receptor RDL suppresses the conditioned stimulus pathway for olfactory learning. J. Neurosci. 29, 1573–1579.

Ludmerer, S.W., Warren, V.A., Williams, B.S., Zheng, Y., Hunt, D.C., Ayer, M.B., Wallace, M.A., Chaudhary, A.G., Egan, M.A., Meinke, P.T., Dean, D.C., Garcia, M.L., Cully, D.F., Smith, M.M., 2002. Ivermectin and nodulisporic acid receptors in *Drosophila melanogaster* contain both γ-aminobutyric acid-gated Rdl and glutamate-gated GluClα chloride channel subunits. Biochemistry 41, 6548–6560.

Lummis, S.C.R., McGonigle, I., Ashby, J.A., Dougherty, D.A., 2011. Two amino acid residues contribute to a cation-π binding interaction in the binding site of an insect GABA receptor. J. Neurosci. 31, 12371–12376.

Lyga, J.W., Ali, S.F., Kinne, L.P., Marek, F.L., Wusaty, M.A., Staetz, C.A., Willut, J., 2007. Discovery of 3-arylpyrimidin-2,4-diones as GABA-gated chloride channel insecticides: translation from target site to field. In: Lyga, J.W., Theodoridis, G. (Eds.), Synthesis and Chemistry of Agrochemicals VII. ACS Symposium Series 948. American Chemical Society, Washington, DC, pp. 153–166.

Lynagh, T., Lynch, J.W., 2010. A glycine residue essential for high ivermectin sensitivity in Cys-loop ion channel receptors. Int. J. Parasitol. 40, 1477–1481.

MacLeod, K., Laurent, G., 1996. Distinct mechanisms for synchronization and temporal patterning of odor-encoding neural assemblies. Science 274, 976–979.

Martin, R.J., Pennington, A.J., 1989. A patch-clamp study of effects of dihydroavermectin on *Ascaris* muscle. Br. J. Pharmacol. 98, 747–756.

Maskell, P.D., Wafford, K.A., Bermudez, I., 2001. Effects of γ-HCH and δ-HCH on human recombinant $GABA_A$ receptors: dependence on $GABA_A$ receptor subunit combination. Br. J. Pharmacol. 132, 205–212.

Matsuda, K., Hosie, A.M., Holyoke Jr., C.W., Rauh, J.J., Sattelle, D.B., 1999. Cross-resistance with dieldrin of a novel tricyclic dinitrile GABA receptor antagonist. Br. J. Pharmacol. 127, 1305–1307.

Matsumura, F., Ghiasuddin, S.M., 1983. Evidence for similarities between cyclodiene type insecticides and picrotoxinin in their action mechanisms. J. Environ. Sci. Health B 18, 1–14.

Matsumura, M., Takeuchi, H., Satoh, M., Sanada-Morimura, S., Otuka, A., Watanabe, T., Thanh, D.V., 2008. Species-specific insecticide resistance to imidacloprid and fipronil in the rice planthoppers *Nilaparvata lugens* and *Sogatella furcifera* in Eat and South-east Asia. Pest Manag. Sci. 64, 1115–1121.

McCarthy, E.v., Wu, Y., deCarvalho, T., Brandt, C., Gao, G., Nitabach, M.N., 2011. Synchronized bilateral synaptic inputs to *Drosophila melanogaster* neuropeptidergic rest/arousal neurons. J. Neurosci. 31, 8181–8193.

McGonigle, I., Lummis, S.C.R., 2010. Molecular characterization of agonists that bind to an insect GABA receptor. Biochemistry 49, 2897–2902.

Ménez, C., Sutra, J.-F., Prichard, R., Lespine, A., 2012. Relative neurotoxicity of ivermectin and moxidectin in Mdr1ab (-/-) mice and effects on mammalian GABA(A) channel activity. PLoS Negl. Trop. Dis. 6, e1883.

Mezler, M., Müller, T., Raming, K., 2001. Cloning and functional expression of $GABA_B$ receptors from *Drosophila*. Eur. J. Neurosci. 13, 477–486.

Millar, N.S., Buckingham, S.D., Sattelle, D.B., 1994. Stable expression of a functional homo-oligomeric *Drosophila* GABA receptor in a *Drosophila* cell line. Proc. R. Soc. Lond. B 258, 307–314.

Miller, P.S., Smart, T.G., 2010. Binding, activation and modulation of Cys-loop receptors. Trends Pharmacol. Sci. 31, 161–174.

Miller, T.A., Maynard, M., Kennedy, J.M., 1979. Structure and insecticidal activity of picrotoxinin analogs. Pestic. Biochem. Physiol. 10, 128–136.

Miyazaki, M., Matsumura, F., Beeman, R.W., 1995. DNA sequence and site of mutation of the GABA receptor of cyclodiene-resistant red flour beetle, *Tribolium castaneum*. Comp. Biochem. Physiol. B Biochem. Mol. Biol. 111, 399–406.

Miyazawa, A., Fujiyoshi, Y., Unwin, N., 2003. Structure and gating mechanism of the acetylcholine receptor pore. Nature 423, 949–955.

Mohan, M., Gujar, G.T., 2003. Local variation in susceptibility of the diamondback moth, *Plutella xylostella* (Linnaeus) to insecticides and role of detoxification enzymes. Crop Prot. 22, 495–504.

Moore, E.L., Haspel, G., Libersat, F., Adams, M.E., 2006. Parasitoid wasp sting: a cocktail of GABA, taurine and β-alanine opens chloride channels for central synaptic block and transient paralysis of a cockroach host. J. Neurobiol. 66, 811–820.

Mullins, L.J., 1955. Structure-toxicity in hexachlorocyclohexane isomers. Science 122, 118–119.

Murillo, L., Hamon, A., Es-Salah-Lamoureux, Z., Itier, V., Quinchard, S., Lapied, B., 2011. Inhibition of protein kinase C decreases sensitivity of GABA receptor subtype to fipronil insecticide in insect neurosecretory cells. Neurotoxicology 32, 828–835.

Murphy, V.F., Wann, K.T., 1988. The action of GABA receptor agonists and antagonists on muscle membrane conductance in *Schistocerca gregaria*. Br. J. Pharmacol. 95, 713–722.

Nagata, K., Narahashi, T., 1994. Dual action of the cyclodiene insecticide dieldrin on the γ-aminobutyric acid receptor-chloride channel complex of rat dorsal root ganglion neurons. J. Pharmacol. Exp. Ther. 269, 164–171.

Nagata, K., Narahashi, T., 1995. Differential effects of hexachlorocyclohexane isomers on the GABA receptor-chloride channel complex in rat dorsal root ganglion neurons. Brain Res. 704, 85–91.

Nagata, K., Hamilton, B.J., Carter, D.B., Narahashi, T., 1994. Selective effects of dieldrin on the GABA$_A$ receptor-channel subunits expressed in human embryonic kidney cells. Brain Res. 645, 19–26.

Nagata, K., Huang, C.-S., Hamilton, B.J., Carter, D.B., Narahashi, T., 1996. Differential effects of hexachlorocyclohexane isomers on the GABA receptor subunits expressed in human embryonic kidney cell line. Brain Res. 738, 131–137.

Nakao, T., Naoi, A., Kawahara, N., Hirase, K., 2010. Mutation of the GABA receptor associated with fipronil resistance in the whitebacked planthopper, *Sogatella furcifera*. Pestic. Biochem. Physiol. 97, 262–266.

Nakao, T., Kawase, A., Kinoshita, A., Abe, R., Hama, M., Kawahara, N., Hirase, K., 2011. The A2'N mutation of the RDL γ-aminobutyric acid receptor conferring fipronil resistance in *Laodelphax striatellus* (Hemiptera: Delphacidae). J. Econ. Entomol. 104, 646–652.

Nakao, T., Hama, M., Kawahara, N., Hirase, K., 2012a. Fipronil resistance in *Sogatella furcifera*: molecular cloning and functional expression of wild-type and mutant RDL GABA receptor subunits. J. Pestic. Sci. 37, 37–44.

Nakao, T., Naoi, A., Hama, M., Kawahara, N., Hirase, K., 2012b. Concentration-dependent effects of GABA on insensitivity to fipronil in the A2'S mutant RDL GABA receptor from fipronil-resistant *Oulema oryzae* (Coleoptera: Chrysomelidae). J. Econ. Entomol. 105, 1781–1788.

Narahashi, T., Zhao, X., Ikeda, T., Salgado, V.L., Yeh, J.Z., 2010. Glutamate-activated chloride channels: unique fipronil targets present in insects but not in mammals. Pestic. Biochem. Physiol. 97, 149–152.

Narusuye, K., Nakao, T., Abe, R., Nagatomi, Y., Hirase, K., Ozoe, Y., 2007. Molecular cloning of a GABA receptor subunit from *Laodelphax striatella* (Fallén) and patch clamp analysis of the homo-oligomeric receptors expressed in a *Drosophila* cell line. Insect Mol. Biol. 16, 723–733.

Navarro, L., Gongora, C., Benavides, P., 2010. Single nucleotide polymorphism detection at the *Hypothenemus hampei Rdl* gene by allele-specific PCR amplification with T_m-shift primers. Pestic. Biochem. Physiol. 97, 204–208.

Newland, C.F., Cull-Candy, S.G., 1992. On the mechanism of action of picrotoxin on GABA receptor channels in dissociated sympathetic neurones of the rat. J. Physiol. 447, 191–213.

Niswener, C.M., Conn, P.J., 2010. Metabotropic glutamate receptors: physiology, pharmacology, and disease. Annu. Rev. Pharmacol. Toxicol. 50, 295–322.

Njue, A.I., Hayashi, J., Kinne, L., Feng, X.-P., Prichard, R.K., 2004. Mutations in the extracellular domains of glutamate-gated chloride channel α3 and β subunits from ivermectin-resistant *Cooperia oncophora* affect agonist sensitivity. J. Neurochem. 89, 1137–1147.

Nury, H., Van Renterghem, C., Weng, Y., Tran, A., Baaden, M., Dufresne, V., Changeux, J.-P., Sonner, J.M., Delarue, M., Corringer, P.-J., 2011. X-ray structures of general anaesthetics bound to a pentameric ligand-gated ion channel. Nature 469, 428–431.

Okada, R., Awasaki, T., Ito, K., 2009. Gamma-aminobutyric acid (GABA)-mediated neural connections in the *Drosophila* antennal lobe. J. Comp. Neurol. 514, 74–91.

Olsen, R.W., Sieghart, W., 2008. International Union of Pharmacology. LXX. Subtypes of γ-aminobutyric acid$_A$ receptors: classification on the basis of subunit composition, pharmacology, and function. Update. Pharmacol. Rev. 60, 243–260.

Ondeyka, J.G., Helms, G.L., Hensens, O.D., Goetz, M.A., Zink, D.L., Tsipouras, A., Shoop, W.L., Slayton, L., Dombrowski, A.W., Polishook, J.D., Ostlind, D.A., Tsou, N.N., Ball, R.G., Singh, S.B., 1997. Nodulisporic acid A, a novel and potent insecticide from a *Nodulisporium* Sp. Isolation, structure determination, and chemical transformations. J. Am. Chem. Soc. 119, 8809–8816.

Othman, N.A., Gallacher, M., Deeb, T.Z., Baptista-Hon, D.T., Perry, D.C., Hales, T.G., 2012. Influences on blockade by *t*-butylbicyclo-phosphoro-thionate of GABA$_A$ receptor spontaneous gating, agonist activation and desensitization. J. Physiol. 590, 163–178.

Otsuka, M., Iversen, L.L., Hall, Z.W., Kravitz, E.A., 1966. Release of gamma-aminobutyric acid from inhibitory nerves of lobster. Proc. Natl. Acad. Sci. U.S.A. 56, 1110–1115.

Otsuka, M., Kravitz, E.A., Potter, D.D., 1967. Physiological and chemical architecture of a lobster ganglion with particular reference to gamma-aminobutyrate and glutamate. J. Neurophysiol. 30, 725–752.

Ozoe, Y., Akamatsu, M., 2001. Non-competitive GABA antagonists: probing the mechanisms of their selectivity for insect *versus* mammalian receptors. Pest Manag. Sci. 57, 923–931.

Ozoe, Y., Eto, M., 1986. Bridged bicyclic organophosphorus compounds as a probe for toxicological study on GABA synapse. In: Clark, J.M., Matsumura, F. (Eds.), Membrane Receptors and Enzymes as Targets of Insecticidal Action. Plenum Press, New York, pp. 75–105.

Ozoe, Y., Matsumura, F., 1986. Structural requirements for bridged bicyclic compounds acting on picrotoxinin receptor. J. Agric. Food Chem. 34, 126–134.

Ozoe, Y., Mochida, K., Nakamura, T., Shimizu, A., Eto, M., 1983. Toxicity of bicyclic phosphate GABA antagonists to the housefly, *Musca domestica* L. J. Pestic. Sci. 8, 601–605.

Ozoe, Y., Mochida, K., Nakamura, T., Yoyama, A., Matsumura, F., 1987. Actions of benzodiazepines on the housefly: binding to thorax/abdomen extracts and biological effects. Comp. Biochem. Physiol. C 87, 187–191.

Ozoe, Y., Fukuda, K., Mochida, K., Nakamura, T., 1989. Actions of benzodiazepines on the housefly. 3. *In vitro* binding of [^{3}H]Ro 5-4864 responding to GABA receptor ligands. Comp. Biochem. Physiol. C 93, 193–199.

Ozoe, Y., Sawada, Y., Mochida, K., Nakamura, T., Matsumura, F., 1990. Structure-activity relationships in a new series of insecticidally active dioxatricycloalkenes derived by structural comparison of the GABA antagonists bicycloorthocarboxylates and endosulfan. J. Agric. Food Chem. 38, 1264–1268.

Ozoe, Y., Kuwano, E., Eto, M., 1993a. Potency of isomers of 8-isopropyl-6-oxabicyclo [3.2.1]octan-7-one at the picrotoxinin binding site in the GABA-gated chloride channel in rat brain. Biosci. Biotechol. Biochem. 57, 504–505.

Ozoe, Y., Takayama, T., Sawada, Y., Mochida, K., Nakamura, T., Matsumura, F., 1993b. Synthesis and structure-activity relationships of a series of insecticidal dioxatricyclododecenes acting as the noncompetitive antagonist of GABA$_A$ receptors. J. Agric. Food Chem. 41, 2135–2141.

Ozoe, Y., Akamatsu, M., Higata, T., Ikeda, I., Mochida, K., Koike, K., Ohmoto, T., Nikaido, T., 1998a. Picrodendrin and related terpenoid antagonists reveal structural differences between ionotropic GABA receptors of mammals and insects. Bioorg. Med. Chem. 6, 481–492.

Ozoe, Y., Niina, K., Matsumoto, K., Ikeda, I., Mochida, K., Ogawa, C., Matsuno, A., Miki, M., Yanagi, K., 1998b. Actions of cyclic esters, S-esters, and amides of phenyl- and phenylthiophosphonic acids on mammalian and insect GABA-gated chloride channels. Bioorg. Med. Chem. 6, 73–83.

Ozoe, Y., Yagi, K., Nakamura, M., Akamatsu, M., Miyake, T., Matsumura, F., 2000. Fipronil-related heterocyclic compounds: structure-activity relationships for interaction with γ-aminobutyric acid- and voltage-gated ion channels and insecticidal action. Pestic. Biochem. Physiol. 66, 92–104.

Ozoe, Y., Ishikawa, S., Tomiyama, S., Ozoe, F., Kozaki, T., Scott, J.G., 2007. Antagonism of the GABA receptor of dieldrin-resistant houseflies by fipronil and its analogues. In: Lyga, J.W., Theodoridis, G. (Eds.), Synthesis and Chemistry of Agrochemicals VII. ACS Symposium Series 948. American Chemical Society, Washington, DC, pp. 39–50.

Ozoe, Y., Takeda, M., Matsuda, K., 2009. γ-Aminobutyric acid receptors: a rationale for developing selective insect pest control chemicals. In: Ishaaya, I., Horowitz, A.R. (Eds.), Biorational Control of Arthropod Pests. Springer, Heidelberg, pp. 131–162.

Ozoe, Y., Asahi, M., Ozoe, F., Nakahira, K., Mita, T., 2010. The antiparasitic isoxazoline A1443 is a potent blocker of insect ligand-gated chloride channels. Biochem. Biophys. Res. Commun. 391, 744–749.

Palmer, C.J., Casida, J.E., 1992. Insecticidal 1,3-dithianes and 1,3-dithiane 1,1-dioxides. J. Agric. Food Chem. 40, 492–496.

Palmer, C.J., Cole, L.M., Larkin, J.P., Smith, I.H., Casida, J.E., 1991. 1-(4-Ethynylphenyl)-4-substituted-2,6,7-trioxabicyclo[2.2.2]octanes; effect of 4-substituent on toxicity to houseflies and mice and potency at the GABA-gated chloride channel. J. Agric. Food Chem. 39, 1329–1334.

Parisky, K.M., Agosto, J., Pulver, S.R., Shang, Y., Kuklin, E., Hodge, J.J.L., Kang, K., Liu, X., Garrity, P.A., Rosbash, M., Griffith, L.C., 2008. PDF cells are a GABA-responsive wake-promoting component of the Drosophila sleep circuit. Neuron 60, 672–682.

Perret, P., Sarda, X., Wolff, M., Wu, T.-T., Bushey, D., Goeldner, M., 1999. Interaction of non-competitive blockers within the γ-aminobutyric acid type A chloride channel using chemically reactive probes as chemical sensors for cysteine mutants. J. Biol. Chem. 274, 25350–25354.

Pirri, J.K., McPherson, A.D., Donnelly, J.L., Francis, M.M., Alkema, M.J., 2009. A tyramine-gated chloride channel coordinates distinct motor programs of a Caenorhabditis elegans escape response. Neuron 62, 526–538.

Porter, L.A., 1967. Picrotoxinin and related substances. Chem. Rev. 67, 441–464.

Pribilla, I., Takagi, T., Langosch, D., Bormann, J., Betz, H., 1992. The atypical M2 segment of the beta subunit confers picrotoxinin resistance to inhibitory glycine receptor channels. EMBO J. 11, 4305–4311.

Priestley, C.M., Williamson, E.M., Wafford, K.A., Sattelle, D.B., 2003. Thymol, a constituent of thyme essential oil, is a positive allosteric modulator of human GABA$_A$ receptors and a homo-oligomeric GABA receptor from Drosophila melanogaster. Br. J. Pharmacol. 140, 1363–1372.

Pulman, D.A., Smith, I.H., Larkin, J.P., Casida, J.E., 1996. Heterocyclic insecticides acting at the GABA-gated chloride channel: 5-aryl-2-arylpyrimidines and -1,3-thiazines. Pestic. Sci. 46, 237–245.

Putrenko, I., Zakikhani, M., Dent, J.A., 2005. A family of acetylcholine-gated chloride channel subunits in *Caenorhabditis elegans*. J. Biol. Chem. 280, 6392–6398.

Raghu, S.V., Joesch, M., Borst, A., Reiff, D.F., 2007. Synaptic organization of lobula plate tangential cells in *Drosophila*: γ-aminobutyric acid receptors and chemical release sites. J. Comp. Neurol. 502, 598–610.

Rahman, M.M., Akiyoshi, Y., Furutani, S., Matsuda, K., Furuta, K., Ikeda, I., Ozoe, Y., 2012. Competitive antagonism of insect GABA receptors by iminopyridazine derivatives of GABA. Bioorg. Med. Chem. 20, 5957–5964.

Ranganathan, R., Cannon, S.C., Horvitz, H.R., 2000. Mod-1 is a serotonin-gated chloride channel that modulates locomotory behaviour in *C. elegans*. Nature 408, 470–475.

Rao, V.T.S., Siddiqui, S.Z., Prichard, R.K., Forrester, S.G., 2009. A dopamine-gated ion channel (HcGGR3*) from *Haemonchus contortus* is expressed in the cervical papillae and is associated with macrocyclic lactone resistance. Mol. Biochem. Parasitol. 166, 54–61.

Rao, V.T.S., Accardi, M.V., Siddiqui, S.Z., Beech, R.N., Prichard, R.K., Forrester, S.G., 2010. Characterization of a tyramine-gated chloride channel from *Haemonchus contortus*. Mol. Biochem. Parasitol. 173, 64–68.

Ratra, G.S., Casida, J.E., 2001. GABA receptor subunit composition relative to insecticide potency and selectivity. Toxicol. Lett. 122, 215–222.

Ratra, G.S., Kamita, S.G., Casida, J.E., 2001. Role of human $GABA_A$ receptor β3 subunit in insecticide toxicity. Toxicol. Appl. Pharmacol. 172, 233–240.

Rauh, J.J., Benner, E., Schnee, M.E., Cordova, D., Holyoke, C.W., Howard, M.H., Bai, D., Buckingham, S.D., Hutton, M.L., Hamon, A., Roush, R.T., Sattelle, D.B., 1997. Effects of [³H]-BIDN, a novel bicyclic dinitrile radioligand for GABA-gated chloride channels of insects and vertebrates. Br. J. Pharmacol. 121, 1496–1505.

Raymond, V., Sattelle, D.B., 2002. Novel animal-health drug targets from ligand-gated chloride channels. Nat. Rev. Drug Discov. 1, 427–436.

Raymond, V., Sattelle, D.B., Lapied, B., 2000. Co-existence in DUM neurones of two GluCl channels that differ in their picrotoxin sensitivity. Neuroreport 11, 2695–2701.

Raymond-Delpech, V., Matsuda, K., Sattelle, B.M., Rauh, J.J., Sattelle, D.B., 2005. Ion channels: molecular targets of neuroactive insecticides. Invert. Neurosci. 5, 119–133.

Ren, Q., Li, H., Wu, Y., Ren, J., Guo, A., 2012. A GABAergic inhibitory neural circuit regulates visual reversal learning in *Drosophila*. J. Neurosci. 32, 11524–11528.

Ringstad, N., Abe, N., Horvitz, R., 2009. Ligand-gated chloride channels are receptors for biogenic amines in *C. elegans*. Science 325, 96–100.

Robbins, J., 1959. The excitation and inhibition of crustacean muscle by amino acids. J. Physiol. 148, 39–50.

Roberts, E., Frankel, S., 1950. γ-Aminobutyric acid in brain: its formation from glutamic acid. J. Biol. Chem. 187, 55–63.

Rugg, D., 2010. Addendum: the insecticidal macrocyclic lactones. In: Gilbert, L.I., Gill, S.S. (Eds.), Insect Pharmacology: Channels, Receptors, Toxins and Enzymes. Elsevier Academic Press, London, pp. 97–99.

Rugg, D., Buckingham, S.D., Sattelle, D.B., Jansson, R.K., 2010. The insecticidal macrocyclic lactones. In: Gilbert, L.I., Gill, S.S. (Eds.), Insect Pharmacology: Channels, Receptors, Toxins and Enzymes. Elsevier Academic Press, London, pp. 69–96.

Rutherford, D., Nielsen, M., Tokutomi, N., Akaike, N., 1994. Effects of plant diterpenes on the neuronal $GABA_A$ receptor-operated chloride current. Neuroreport 5, 2569–2572.

Sammelson, R.E., Caboni, P., Durkin, K.A., Casida, J.E., 2004. GABA receptor antagonists and insecticides: common structural features of 4-alkyl-1-phenylpyrazoles and 4-alkyl-1-phenyltrioxabicyclooctanes. Bioorg. Med. Chem. 12, 3345–3355.

Sanada-Morimura, S., Sakumoto, S., Ohtsu, R., Otuka, A., Huang, S.-H., Thanh, D.V., Matsumura, M., 2011. Current status of insecticide resistance in the small brown planthopper, *Laodelphax striatellus*, in Japan, Taiwan, and Vietnam. Appl. Entmol. Zool. 46, 65–73.

Satoh, H., Daido, H., Nakamura, T., 2005. Preliminary analysis of the GABA-induced current in cultured CNS neurons of the cutworm moth, *Spodoptera litura*. Neurosci. Lett. 381, 125–130.

Sattelle, D.B., 1990. GABA receptors of insects. Adv. Insect Physiol. 22, 1–113.

Sattelle, D.B., 1992. Receptors for L-glutamate and GABA in the nervous system of an insect (*Periplaneta americana*). Comp. Biochem. Physiol. C 103, 429–438.

Sattelle, D.B., Pinnock, R.D., Wafford, K.A., David, J.A., 1988. GABA receptors on the cell-body membrane of an identified insect motor neuron. Proc. R. Soc. Lond. B Biol. Sci. 232, 443–456.

Sattelle, D.B., Lummis, S.C.R., Wong, J.F.H., Rauh, J.J., 1991. Pharmacology of insect GABA receptors. Neurochem. Res. 16, 363–374.

Sattelle, D.B., Harrison, J.B., Chen, H.H., Bai, D., Takeda, M., 2000. Immunocytochemical localization of putative γ-aminobutyric acid receptor subunits in the head ganglia of *Periplaneta americana* using an anti-RDL C-terminal antibody. Neurosci. Lett. 289, 197–200.

Sattelle, D.B., Bai, D., Chen, H.H., Skeer, J.M., Buckingham, S.D., Rauh, J.J., 2003. Bicuculline-insensitive GABA-gated Cl⁻ channels in the larval nervous system of the moth *Manduca sexta*. Invert. Neurosci. 5, 37–43.

Sayyed, A.H., Wright, D.J., 2004. Fipronil resistance in the diamondback moth (Lepidoptera: Plutellidae): inheritance and number of genes involved. J. Econ. Entomol. 97, 2043–2050.

Schmidt, T.J., Gurrath, M., Ozoe, Y., 2004. Structure-activity relationships of *seco*-prezizaane and picrotoxane/picrodendrane terpenoids by *Quasar* receptor-surface modeling. Bioorg. Med. Chem. 12, 4159–4167.

Schnee, M.E., Rauh, J.J., Buckingham, S.D., Sattelle, D.B., 1997. Pharmacology of skeletal muscle GABA-gated chloride channels in the cockroach *Periplaneta americana*. J. Exp. Biol. 200, 2947–2955.

Schofield, P.R., Darlison, M.G., Fujita, N., Burt, D.R., Stephenson, F.A., Rodriguez, H., Rhee, L.M., Ramachandran, J., Reale, V., Glencorse, T.A., Seeburg, P.H., Barnard, E.A., 1987. Sequence and functional expression of the GABA_A receptor shows a ligand-gated receptor super-family. Nature 328, 221–227.

Schuske, K., Beg, A.A., Jorgensen, E.M., 2004. The GABA nervous system in *C. elegans*. Trends Neurosci. 27, 407–414.

Scott, R.H., Duce, I.R., 1987. Pharmacology of GABA receptors on skeletal muscle fibres of the locust (*Schistocerca gregaria*). Comp. Biochem. Physiol. C 86, 305–311.

Scott, J.G., Wen, Z., 1997. Toxicity of fipronil to susceptible and resistant strains of German cockroaches (Dictyoptera: Blattellidae) and house flies (Diptera: Muscidae). J. Econ. Entomol. 90, 1152–1156.

Semenov, E.P., Pak, W.L., 1999. Diversification of *Drosophila* chloride channel gene by multiple posttranslational mRNA modifications. J. Neurochem. 72, 66–72.

Shan, Q., Haddrill, J.L., Lynch, J.W., 2001. A single β subunit M2 domain residue controls the picrotoxin sensitivity of $\alpha\beta$ heteromeric glycine receptor chloride channels. J. Neurochem. 76, 1109–1120.

Shang, Q.-L., Liang, P., Gao, X.-W., 2009. Cloning, developmental and tissue-specific expression of γ-aminobutyric acid (GABA) receptor alpha2 subunit gene in *Spodoptera exigua* (Hübner). Pestic. Biochem. Physiol. 93, 1–7.

Shi, X., Liang, P., Song, D., Yang, W., Gao, X., 2012. Quantification of γ-aminobutyric acid in the heads of houseflies (*Musca domestica*) and diamondback moths (*Plutella xylostella* (L.)), using capillary electrophoresis with laser-induced fluorescence detection. J. Sep. Sci. 35, 548–555.

Shimotahira, H., Fusazaki, S., Ikeda, I., Ozoe, Y., 2011. A photoreactive probe that differentiates the binding sites of noncompetitive GABA receptor antagonists. Bioorg. Med. Chem. Lett. 21, 1598–1600.

Shoop, W.L., Mrozik, H., Fisher, M.H., 1995. Structure and activity of avermectins and milbemycins in animal health. Vet. Parasitol. 59, 139–156.

Shotkoski, F., Lee, H.-J., Zhang, H.-G., Jackson, M.B., ffrench-Constant, R.H., 1994. Functional expression of insecticide-resistant GABA receptors from the mosquito *Aedes aegypti*. Insect Mol. Biol. 3, 283–287.

Sieghart, W., 1995. Structure and pharmacology of γ-aminobutyric acid$_A$ receptor subtypes. Pharmacol. Rev. 47, 181–234.

Sigel, E., Baur, R., 1987. Effect of avermectin B$_{1a}$ on chick neuronal γ-aminobutyrate receptor channels expressed in *Xenopus* oocytes. Mol. Pharmacol. 32, 749–752.

Silberberg, S.D., Li, M., Swartz, K.J., 2007. Ivermectin interaction with transmembrane helices reveals widespread rearrangements during opening of P2X receptor channels. Neuron 54, 263–274.

Smart, T.G., Constanti, A., 1986. Studies on the mechanism of action of picrotoxinin and other convulsants at the crustacean muscle GABA receptor. Proc. R. Soc. Lond. B Biol. Sci. 227, 191–216.

Smith, M.M., Warren, V.A., Thomas, B.S., Brochu, R.M., Ertel, E.A., Rohrer, S., Schaeffer, J., Schmatz, D., Petuch, B.R., Tang, Y.S., Meinke, P.T., Kaczorowski, G.J., Cohen, C.J., 2000. Nodulisporic acid opens insect glutamate-gated chloride channels: identification of a new high affinity modulator. Biochemistry 39, 5543–5554.

Soloway, S.B., 1965. Correlation between biological activity and molecular structure of the cyclodiene insecticides. In: Metcalf, R.L. (Ed.), Advances in Pest Control Research, vol. VI. Interscience Publishers, New York, pp. 85–126.

Sousa, A., Ticku, M.K., 1997. Interactions of the neurosteroid dehydroepiandrosterone sulfate with the GABA$_A$ receptor complex reveals that it may act *via* the picrotoxin site. J. Pharmacol. Exp. Ther. 282, 827–833.

Squires, R.F., Casida, J.E., Richardson, M., Saederup, E., 1983. [^{35}S]*t*-Butylbicyclophosphorothionate binds with high affinity to brain-specific sites coupled to γ-aminobutyric acid-A and ion recognition sites. Mol. Pharmacol. 23, 326–336.

Stilwell, G.E., ffrench-Constant, R.H., 1998. Transcriptional analysis of the *Drosophila* GABA receptor gene *Resistance to dieldrin*. J. Neurobiol. 36, 468–484.

Strambi, C., Cayre, M., Sattelle, D.B., Augier, R., Charpin, P., Strambi, A., 1998. Immunocytochemical mapping of an RDL-like GABA receptor subunit and of GABA in brain structures related to learning and memory in the cricket *Acheta domesticus*. Learn. Mem. 5, 78–89.

Su, H., O'Dowd, D.K., 2003. Fast synaptic currents in *Drosophila* mushroom body Kenyon cells are mediated by α-bungarotoxin-sensitive nicotinic acetylcholine receptors and picrotoxin-sensitive GABA receptors. J. Neurosci. 23, 9246–9253.

Sudhakaran, I.P., Holohan, E.E., Osman, S., Rodrigues, V., VijayRaghavan, K., Ramaswami, M., 2012. Plasticity of recurrent inhibition in the *Drosophila* antennal lobe. J. Neurosci. 32, 7225–7231.

Takeuchi, A., Takeuchi, N., 1969. A study of the action of picrotoxin on the inhibitory neuromuscular junction of the crayfish. J. Physiol. 205, 377–391.

Tanaka, K., Matsumura, F., 1985. Action of avermectin B$_{1a}$ on the leg muscles and the nervous system of the American cockroach. Pestic. Biochem. Physiol. 24, 124–135.

Tanaka, K., Scott, J.G., Matsumura, F., 1984. Picrotoxinin receptor in the central nervous system of the American cockroach: its role in the action of cyclodiene-type insecticides. Pestic. Biochem. Physiol. 22, 117–127.

Tandon, R., LePage, K.T., Kaplan, R.M., 2006. Cloning and characterization of genes encoding α and β subunits of glutamate-gated chloride channel protein in *Cylicocyclus nassatus*. Mol. Biochem. Parasitol. 150, 46–55.

Tantely, M.L., Tortosa, P., Alout, H., Berticat, C., Berthomieu, A., Rutee, A., Dehecq, J.-S., Makoundou, P., Labbé, P., Pasteur, N., Weill, M., 2010. Insecticide resistance in *Culex pipiens quinquefasciatus* and *Aedes albopictus* mosquitoes from La Réunion Island. Insect Biochem. Mol. Biol. 40, 317–324.

Thompson, S.M., Gähwiler, B.H., 1992. Comparison of the actions of baclofen at pre- and postsynaptic receptors in the rat hippocampus *in vitro*. J. Physiol. 451, 329–345.

Thompson, M., Steichen, J.C., ffrench-Constant, R.H., 1993a. Conservation of cyclodiene insecticide resistance-associated mutations in insects. Insect Mol. Biol. 2, 149–154.

Thompson, M., Shotkoski, F., ffrench-Constant, R., 1993b. Cloning and sequencing of the cyclodiene insecticide resistance gene from the yellow fever mosquito *Aedes aegypti*. FEBS Lett. 325, 187–190.

Thompson, A.J., Duke, R.K., Lummis, S.C.R., 2011a. Binding sites for bilobalide, diltiazem, ginkgolide, and picrotoxinin at the 5-HT$_3$ receptor. Mol. Pharmacol. 80, 183–190.

Thompson, A.J., Jarvis, G.E., Duke, R.K., Johnston, G.A.R., Lummis, S.C.R., 2011b. Ginkgolide B and bilobalide block the pore of the 5-HT$_3$ receptor at a location that overlaps the picrotoxin binding site. Neuropharmacology 60, 488–495.

Thompson, A.J., McGonigle, I., Duke, R., Johnston, G.A.R., Lummis, S.C.R., 2012. A single amino acid determines the toxicity of *Ginkgo biloba* extracts. FASEB J. 26, 1884–1891.

Tong, F., Coats, J.R., 2012. Quantitative structure–activity relationships of monoterpenoid binding activities to the housefly GABA receptor. Pest Manag. Sci. 68, 1122–1129.

Traynelis, S.F., Wollmuth, L.P., McBain, C.J., Menniti, F.S., Vance, K.M., Ogden, K.K., Hansen, K.B., Yuan, H., Myers, S.J., Dingledine, R., 2010. Glutamate receptor ion channels: structure, regulation, and function. Pharmacol. Rev. 62, 405–496.

Udenfriend, S., 1950. Identification of γ-aminobutyric acid in brain by the isotope derivative method. J. Biol. Chem. 187, 65–69.

Unwin, N., 2005. Refined structure of the nicotinic acetylcholine receptor at 4 Å resolution. J. Mol. Biol. 346, 967–989.

Usherwood, P.N.R., 1994. Insect glutamate receptors. Adv. Insect Physiol. 24, 309–341.

Usherwood, P.N.R., Grundfest, H., 1965. Peripheral inhibition in skeletal muscle of insects. J. Neurophysiol. 28, 497–518.

Uwai, K., Ohashi, K., Takaya, Y., Oshima, Y., Furukawa, K., Yamagata, K., Omura, T., Okuyama, S., 2001. Virol A, a toxic *trans*-polyacetylenic alcohol of *Cicuta virosa*, selectively inhibits the GABA-induced Cl⁻ current in acutely dissociated rat hippocampal CA1 neurons. Brain Res. 889, 174–180.

Vale, C., Fonfría, E., Bujons, J., Messeguer, A., Rodríguez-Farré, E., Suñol, C., 2003. The organochlorine pesticides γ-hexachlorocyclohexane (lindane), α-endosulfan and dieldrin differentially interact with GABA$_A$ and glycine-gated chloride channels in primary cultures of cerebellar granule cells. Neuroscience 117, 397–403.

Van Renterghem, C., Bilbe, G., Moss, S., Smart, T.G., Constanti, A., Brown, D.A., Barnard, E.A., 1987. GABA receptors induced in *Xenopus* oocytes by chick brain mRNA: evaluation of TBPS as a use-dependent channel-blocker. Brain Res. 388, 21–31.

Vassilatis, D.K., Arena, J.P., Plasterk, R.H.A., Wilkinson, H.A., Schaeffer, J.M., Cully, D.F., Van der Ploeg, L.H.T., 1997a. Genetic and biochemical evidence for a novel avermectin-sensitive chloride channel in *Caenorhabditis elegans*. Isolation and characterization. J. Biol. Chem. 272, 33167–33174.

Vassilatis, D.K., Elliston, K.O., Paress, P.S., Hamelin, M., Arena, J.P., Schaeffer, J.M., Van der Ploeg, L.H.T., Cully, D.F., 1997b. Evolutionary relationship of the ligand-gated ion channels and the avermectin-sensitive, glutamate-gated chloride channels. J. Mol. Evol. 44, 501–508.

Waldvogel, H.J., Billinton, A., White, J.H., Emson, P.C., Faull, R.L.M., 2004. Comparative cellular distribution of GABA$_A$ and GABA$_B$ receptors in the human basal ganglia: immunohistochemical colocalization of the α_1 subunit of the GABA$_A$ receptor, and the GABA$_B$R1 and GABA$_B$R2 receptor subunits. J. Comp. Neurol. 470, 339–356.

Waliwitiya, R., Belton, P., Nicholson, R.A., Lowenberger, C.A., 2010. Effects of the essential oil constituent thymol and other neuroactive chemicals on flight motor activity and wing beat frequency in the blowfly *Phaenicia sericata*. Pest Manag. Sci. 66, 277–289.

Wang, P., Slaughter, M.M., 2005. Effects of GABA receptor antagonists on retinal glycine receptors and on homomeric glycine receptor alpha subunits. J. Neurophysiol. 93, 3120–3126.

Washio, H., 1994. Effects of putative neurotransmitters on dorsal unpaired median neurons of cockroach (*Periplaneta americana*) thoracic ganglia. J. Insect Physiol. 40, 841–847.

Watanabe, T., Arisawa, M., Narusuye, K., Alam, M.S., Yamamoto, K., Mitomi, M., Ozoe, Y., Nishida, A., 2009. Alantrypinone and its derivatives: synthesis and antagonist activity toward insect GABA receptors. Bioorg. Med. Chem. 17, 94–110.

Watt, E.E., Betts, B.A., Kotey, F.O., Humbert, D.J., Griffith, T.N., Kelly, E.W., Veneskey, K.C., Gill, N., Rowan, K.C., Jenkins, A., Hall, A.C., 2008. Menthol shares general anesthetic activity and sites of action on the GABA$_A$ receptor with the intravenous agent, propofol. Eur. J. Pharmacol. 590, 120–126.

Wen, Z., Scott, J.G., 1999. Genetic and biochemical mechanisms limiting fipronil toxicity in the LPR strain of house fly, *Musca domestica*. Pestic. Sci. 55, 988–992.

Wilson, R.I., Laurent, G., 2005. Role of GABAergic inhibition in shaping odor-evoked spatiotemporal patterns in the *Drosophila* antennal lobe. J. Neurosci. 25, 9069–9079.

Witte, I., Kreienkamp, H.-J., Gewecke, M., Roeder, R., 2002. Putative histamine-gated chloride channel subunits of the insect visual system and thoracic ganglion. J. Neurochem. 83, 504–514.

Wolff, M.A., Wingate, V.P.M., 1998. Characterization and comparative pharmacological studies of a functional γ-aminobutyric acid (GABA) receptor cloned from the tobacco budworm, *Heliothis virescens* (Noctuidae:Lepidoptera). Invert. Neurosci. 3, 305–315.

Wolstenholme, A.J., Rogers, A.T., 2005. Glutamate-gated chloride channels and the mode of action of the avermectin/milbemycin anthelmintics. Parasitology 131, S85–S95.

Wondji, C.S., Davire, R.K., Tukur, Z., Irving, H., Djouaka, R., Morgan, J.C., 2011. Identification and distribution of a GABA receptor mutation conferring dieldrin resistance in the malaria vector *Anopheles funestus* in Africa. Insect Biochem. Mol. Biol. 41, 484–491.

Wu, Y., Ren, Q., Li, H., Guo, A., 2012. The GABAergic anterior paired lateral neurons facilitate olfactory reversal learning in *Drosophila*. Learn. Mem. 19, 478–486.

Xu, H., 1998. Identification of major phylogenetic branches of inhibitory ligand-gated channel receptors. J. Mol. Evol. 47, 323–333.

Xu, M., Covey, D.F., Akabas, M.H., 1995. Interaction of picrotoxin with GABA$_A$ receptor channel-lining residues probed in cysteine mutants. Biophys. J. 69, 1858–1867.

Yagle, M.A., Martin, M.W., de Fiebre, C.M., de Fiebre, N.C., Drewe, J.A., Dillon, G.H., 2003. [^{3}H]Ethynylbicycloorthobenzoate ([^{3}H]EBOB) binding in recombinant GABA$_A$ receptors. Neurotoxicology 24, 817–824.

Yamaguchi, M., Sawa, Y., Matsuda, K., Ozoe, F., Ozoe, Y., 2012. Amino acid residues of both the extracellular and transmembrane domains influence binding of the antiparasitic agent milbemycin to *Haemonchus contortus* AVR-14B glutamate-gated chloride channels. Biochem. Biophys. Res. Commun. 419, 562–566.

Yang, Z., Cromer, B.A., Harvey, R.J., Parker, M.W., Lynch, J.W., 2007. A proposed structural basis for picrotoxinin and picrotin binding in the glycine receptor pore. J. Neurochem. 103, 580–589.

Yoon, K.-W., Covey, D.F., Rothman, S.M., 1993. Multiple mechanisms of picrotoxin block of GABA-induced currents in rat hippocampal neurons. J. Physiol. 464, 423–439.

Yoon, K.-W., Wotring, V.E., Fuse, T., 1998. Multiple picrotoxinin effect on glycine channels in rat hippocampal neurons. Neuroscience 87, 807–815.

Yu, L.-L., Cui, Y.-J., Lang, G.-J., Zhang, M.-Y., Zhang, C.-X., 2010. The ionotropic γ-aminobutyric acid receptor gene family of the silkworm, Bombyx mori. Genome 53, 688–697.

Yuan, G., Gao, W., Yang, Y., Wu, Y., 2010. Molecular cloning, genomic structure, and genetic mapping of two Rdl-orthologous genes of GABA receptors in the diamondback moth, Plutella xylostella. Arch. Insect Biochem. Physiol. 74, 81–90.

Zhang, H.-G., ffrench-Constant, R.H., Jackson, M.B., 1994. A unique amino acid of the Drosophila GABA receptor with influence on drug sensitivity by two mechanisms. J. Physiol. 479, 65–75.

Zhang, H.-G., Lee, H.-J., Rocheleau, T., ffrench-Constant, R.H., Jackson, M.B., 1995. Subunit composition determines picrotoxin and bicuculline sensitivity of Drosophila γ-aminobutyric acid receptors. Mol. Pharmacol. 48, 835–840.

Zhang, D., Pan, Z.-H., Awobuluyi, M., Lipton, S.A., 2001. Structure and function of GABA$_C$ receptors: a comparison of native versus recombinant receptors. Trends Pharmacol. Sci. 22, 121–132.

Zhang, J., Xue, F., Chang, Y., 2009. Agonist- and antagonist-induced conformational changes of loop F and their contributions to the ρ1 GABA receptor function. J. Physiol. 587, 139–153.

Zhao, X., Salgado, V.L., 2010. The role of GABA and glutamate receptors in susceptibility and resistance to chloride channel blocker insecticides. Pestic. Biochem. Physiol. 97, 153–160.

Zhao, X., Salgado, V.L., Yeh, J.Z., Narahashi, T., 2003a. Differential actions of fipronil and dieldrin insecticides on GABA-gated chloride channels in cockroach neurons. J. Pharmacol. Exp. Ther. 306, 914–924.

Zhao, X.-Y., Wang, Y., Li, Y., Chen, X.-Q., Yang, H.-H., Yue, J.-M., Hu, G.-Y., 2003b. Songorine, a diterpenoid alkaloid of the genus Aconitum, is a novel GABA$_A$ receptor antagonist in rat brain. Neurosci. Lett. 337, 33–36.

Zhao, X., Salgado, V.L., Yeh, J.Z., Narahashi, T., 2004a. Kinetic and pharmacological characterization of desensitizing and non-desensitizing glutamate-gated chloride channels in cockroach neurons. Neurotoxicology 25, 967–980.

Zhao, X., Yeh, J.Z., Salgado, V.L., Narahashi, T., 2004b. Fipronil is a potent open channel blocker of glutamate-activated chloride channels in cockroach neurons. J. Pharmacol. Exp. Ther. 310, 192–201.

Zhao, Q., Li, Y., Xiong, L., Wang, Q., 2010. Design, synthesis and insecticidal activity of novel phenylpyrazoles containing a 2,2,2-trichloro-1-alkoxyethyl moiety. J. Agric. Food Chem. 58, 4992–4998.

Zheng, Y., Hirschberg, B., Yuan, J., Wang, A.P., Hunt, D.C., Ludmerer, S.W., Schmatz, D.M., Cully, D.F., 2002. Identification of two novel Drosophila melanogaster histamine-gated chloride channel subunits expressed in the eye. J. Biol. Chem. 277, 2000–2005.

Zhorov, B.S., Bregestovski, P.D., 2000. Chloride channels of glycine and GABA receptors with blockers: Monte Carlo minimization and structure-activity relationships. Biophys. J. 78, 1786–1803.

Zhou, X.-M., Wu, Q.-J., Zhang, Y.-J., Bai, L.-Y., Huang, X.-Y., 2008. Cloning and characterization of a GABA receptor from Plutella xylostella (Lepidoptera: Plutellidae). J. Econ. Entomol. 101, 1888–1896.

CHAPTER FIVE

Calcium Channels as Molecular Target Sites of Novel Insecticides

Peter Lümmen
Bayer CropScience AG, Monheim, Germany

Contents

Abstract

As a key intracellular messenger, calcium is involved in the regulation of diverse cellular processes in vertebrate and invertebrates. In excitable cells, voltage-gated calcium channels transform electrical signals into intracellular calcium transients which trigger physiological processes such as muscle contraction or neurotransmitter release. ATP-driven pumps and ion transporters terminate calcium signals by reducing the intracellular concentrations to resting levels. It is, therefore, conceivable that calcium homeostasis and signaling are attractive targets for future insecticide discovery. The first examples proving the validity of this concept are specific peptide toxins blocking insect voltage-gated calcium channels and the new diamide insecticides targeting ryanodine-sensitive release channels.

Advances in Insect Physiology, Volume 44
ISBN 978-0-12-394389-7
http://dx.doi.org/10.1016/B978-0-12-394389-7.00005-3

ABBREVIATIONS

AChBP acetylcholine-binding protein
ADP adenosine diphosphate
AID α1-interaction domain
AMP-PCP adenylylmethylenediphosphonate
ATP adenosine triphosphate
cADPR cyclic adenosine diphosphate ribose
CaM calmodulin
CaMKII calmodulin-dependent kinase II
Ca$_V$ voltage-gated calcium channels
CBD calcium-binding domain
CCD central core disease
CICR calcium-induced calcium release
DHP dihydropyridine
dsRNAi double-stranded RNA interference
ER endoplasmic reticulum
HEK human embryonic kidney
ICK inhibitor cystine knot
IMP inositol monophosphate
IP$_3$ inositol triphosphate
IP$_3$R inositol triphosphate receptor
kDa kilodalton
MCU mitochondrial calcium uniporter
MH malignant hyperthermia
MTX maitotoxin
nAChR nicotinic acetylcholine receptor
NCS neuronal calcium sensor
NCX sodium–calcium exchanger
PA phenylalkylamine
PKA protein kinase A
PLC phospholipase C
PMCA plasma membrane calcium ATPase
PTP permeability transition pore
RyR ryanodine receptor
SERCA sarcoplasmic reticulum calcium ATPase
SOC store-operated calcium channel
SOCE store-operated calcium entry
SR sarcoplasmic reticulum
TRP transient receptor potential

1. INTRODUCTION

1.1. Insecticides: In search of new modes of action

Food production for the ever-growing world population is the major challenge that agriculture is facing today and will have to cope with in the future. Consequently, agricultural productivity is to be increased by an estimated

70% until 2050 (FAO, 2011). Eventually, environmental resources such as groundwater and natural habitats need to be protected. Insect pests are estimated to cause between 10% and 18% of preharvest losses (Oerke, 2006; Oerke and Dehne, 2004). By using chemical insecticides, acaricides, nematicides, and insect-resistant transgenic crops, farmers can effectively avert crop losses due to insects, mites, and plant-pathogenic nematodes.

The great majority of insecticides currently available on the market are acting through only four major biochemical modes of action, affecting signal transmission in the central nervous system (Casida, 2009). Acetylcholinesterase, the voltage-gated sodium channels, nicotinic acetylcholine receptors (nAChRs), and GABA- or glutamate-gated chloride channels account for roughly 75% of the targets of commercial insecticides. In all these targets, point mutations conferring reduced sensitivity to the corresponding insecticides have been detected (Du and Dong, 2008; ffrench-Constant, 1999; Hemingway et al., 2002; Kono and Tomita, 2006; Perry et al., 2011; Van Leeuwen et al., 2010). Target-site resistance is often combined with increased metabolic detoxification by cytochrome P450 monooxygenases (Bass and Field, 2011; Feyereisen, 2006, 2011), esterases (Cao et al., 2008; Yan et al., 2009), or glutathione-S-transferases (Enayati et al., 2005; Ketterman et al., 2011; Ranson and Hemingway, 2005).

The target-site mutations account for highly resistant field populations of insects and mites that continuously compromise the efficacy of chemical pest control. Therefore, considerable efforts are undertaken in agrochemical research to discover insecticides, acaricides, and nematicides with new modes of action.

Calcium is involved in the regulation of numerous cellular processes in eukaryotic cells, including gene regulation, membrane transport processes, hormone biosynthesis, muscle contraction, neurotransmitter release, and programmed cell-death. Therefore, it can be rationalized that compounds disrupting calcium signaling and homeostasis should be effective insecticides due to the central role that calcium plays as an intracellular messenger, particularly in excitable cells. Such compounds should be highly effective by agonistically activating calcium-signaling cascades or by blocking negative (feedback) regulatory loops.

In the following sections, research focusing on calcium-signaling proteins in excitable cells as new target sites for insect control will be summarized. Recently, two highly promising classes of synthetic insecticides disrupting calcium homeostasis have been introduced into the market place. Their mode of action will be discussed in detail.

1.2. The cellular biochemistry of calcium

The role of calcium as a second messenger functioning in neurotransmission and muscle contraction has been thoroughly studied in vertebrates (Cartwright et al., 2011; Catterall, 1991; Catterall and Few, 2008). Much less is known in invertebrate organisms, but in the meantime, evidence has accumulated that the calcium-regulatory circuits of excitable cells and their central protein elements have been conserved during evolution between invertebrates and vertebrates.

Calcium ions have some peculiar properties in comparison with other divalent cations available in the biosphere. Calcium binds faster to both inorganic and organic anions due to its variable degree of hydration, typically between 6 and 8 rapidly exchanging water molecules per calcium ion. In addition, calcium ions readily form almost insoluble salts with phosphates. It is conceivable that this could have been detrimental during the evolution of a cellular energy conservation chemistry which relies on phosphorylation (Wang et al., 2011; Williams, 2006). Therefore, cells need to adjust the intracellular calcium concentrations, $[Ca^{2+}]_i$, to values of about four orders of magnitude lower than the millimolar concentrations in the extracellular medium. Consequently, proteins that sequester the ion, and membrane pumps that pump calcium out of the cell or into intracellular storage compartments, have evolved. Their diversity reflects the great variety of calcium-regulated biological processes.

Starting from their primordial function in the sequestration of calcium, calcium-handling proteins have taken over new roles to regulate the intracellular calcium concentrations in a temporal and spatial fashion. Functioning as parts of well-orchestrated regulatory circuits, calcium-conducting transmembrane channels mediate tightly controlled calcium entry. The fast switching between ion-conducting and closed conformational states, upon intra-, or extracellular stimuli, is the characteristic feature of these channels (Clapham, 1995). In addition, in the resting state, low intracellular calcium concentrations need to be reestablished by both outward transport and sequestration into intracellular stores. These transport processes are conducted by specialized pumps and transporters. Finally, calcium-binding proteins in the cytoplasm and in the intracellular Ca^{2+} storage compartments, such as calcineurin, calsequestrin, and calreticulin help to maintain low intracellular concentrations of free calcium by functioning as calcium buffers (Yáñez et al., 2012).

In summary, calcium-buffering proteins, calcium channels, and calcium pumps operate as central players in the regulatory networks of excitable cells (as schematically shown in Fig. 5.1).

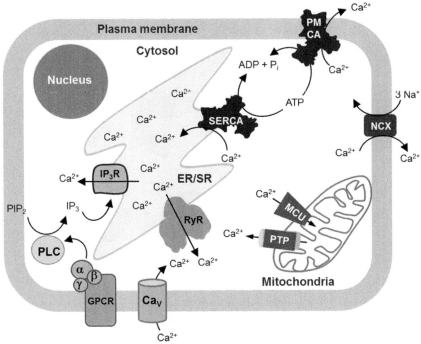

Figure 5.1 Schematic representation of calcium regulation and the roles of calcium-handling proteins. The internal calcium stores are represented by the sarco-/endoplasmatic reticulum and the mitochondria. Voltage-gated calcium channels (Ca_V) mediate a limited calcium entry after depolarization of the cell membrane. This is followed by a massive calcium release from the internal stores mediated by the ryanodine receptor (RyR). The second intracellular calcium-release channel is the IP_3 receptor (IP_3R). IP_3 is produced by phospholipase C (PLC) through hydrolysis of phosphatidylinositol 4,5-bisphosphate. PLC activity is stimulated through certain G-protein-coupled receptors. The sarcoplasmic Ca^{2+}-ATPase (SERCA) refills the internal calcium stores, whereas the plasma membrane Ca^{2+}-ATPase (PMCA) and the Na^+/Ca^{2+} exchanger (NCX) transport calcium to the extracellular space. In mitochondria, the calcium uniporter (MCU) and the permeability transition pore (PTP) carry out calcium uptake and release in/from the mitochondrial stores. (See Color Plate 21 at the back of the book.)

Specific calcium-binding structural motifs have been identified in calcium-binding proteins from bacteria to mammals. The canonical EF-hand motif, for example, is present in many intracellular calcium-binding proteins calcium-sensors such as the eukaryotic calmodulins (CaMs) (Ikura, 1996; Mikhaylova et al., 2011), and several bacterial CaM–like proteins (Michiels et al., 2002; Zhou et al., 2006). It consists of two almost perpendicular α-helices connected by a short loop. Carbonyl groups and side-chain carboxyls of the 11/12-residue loop chelate the calcium ion.

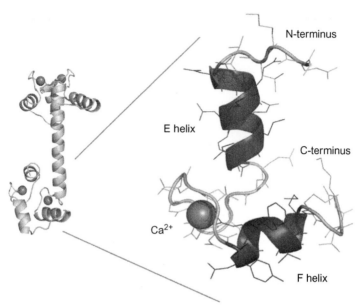

Figure 5.2 Single EF-hand calcium-binding motif of the *Drosophila melanogaster* cal-modulin (CaM). Similar to the mammalian CaMs, the insect calmodulin structure contains four EF-hand motifs (left structure). One of these calcium-binding motifs with bound calcium is depicted in higher magnification (right structure). Coordinates: PDB file number 4CLN. Graphics generated using PyMOL v0.99 (DeLano Scientific). (For color version of this figure, the reader is referred to the online version of this chapter.)

EF-hand motifs are typically found as pairs, or multiple pairs, in calcium-binding proteins (Ikura, 1996). Several EF-hand proteins have been structurally characterized. One of the four EF-hand motifs of the fruit fly *Drosophila melanogaster* CaM structure solved at a resolution of 2.2 Å is depicted in Fig. 5.2 (Taylor et al., 1991). A distinct group of EF-hand Ca^{2+}-binding proteins, termed neuronal calcium sensors (NCSs), has been described in the nervous system. Their sequence differs from the CaMs, and their calcium affinity is typically higher. The specific expression patterns of NCS make them potential candidates for the spatial or temporal regulation of Ca^{2+}-signaling pathways (Mikhaylova et al., 2011).

In insects and mammals, intracellular calcium storage compartments comprise primarily the endoplasmic reticulum or sarcoplasmic reticulum (ER/SR) (Arnon et al., 1997; Davies and Terhzaz, 2009; Messutat et al., 2001; Oberwinkler, 2002; Verkhratsky, 2002). It should be noted for the sake of completeness that mitochondria have been recognized as a second type of intracellular calcium stores. They are likely involved in the fine-tuning and modulation of intracellular calcium signals, probably by direct physical inter-action with the SR. As such, they contain calcium-uptake systems, for exam-ple, the mitochondrial calcium uniporters (MCUs), as well as calcium-release mechanisms, represented mainly by the permeability transition pores (PTPs) (Ruiz-Meana et al., 2010).

2. CALCIUM PUMPS AND TRANSPORTERS

2.1. Plasma membrane Ca^{2+}-ATPase

Excitable cells maintain a steep calcium gradient across the cytoplasmic membrane by the activity of a high-affinity calcium pump, the plasma mem-brane Ca^{2+}-ATPase (PMCA), and by the low-affinity, but high capacity Na^+/Ca^{2+} exchanger (NCX) (Carafoli et al., 2004; Lytton, 2007). PMCAs are proteins of molecular masses of about 134 kDa that are present in essen-tially all eukaryotic cells. Characteristic structural features are the 10 membrane-spanning helices and the extended C-terminal tail (Strehler, 2004). The high calcium affinity of PMCA facilitates catalytic operation even at the submicromolar steady-state intracellular calcium concentrations. During the catalytic cycle, formation of a phosphorylaspartate intermediate is a hallmark of all members of the P-type ATPase family (Yatime et al., 2009). The high-affinity calcium-binding site termed E1 is exposed to the cytoplasmic side. Phosphorylation at the expense of adenosine triphos-phate (ATP) induces a conformational change to the E2 state that exposes the calcium to the extracellular medium. Calcium ions are released due to the reduced calcium affinity of the E2 state, and the enzyme is subse-quently dephosphorylated to the E1 state. Similar to other P-type pumps, PMCAs are sensitive to inhibition by orthovanadate and La^{3+}. CaM is the key regulator of PMCA catalytic activity, reducing the apparent K_m values for Ca^{2+} from $>10\ \mu M$ to submicromolar concentrations. The CaM-binding site is located on the long C-terminal tail of the enzyme. In the unbound state, this site interacts with an internal domain next to the catalytic core resulting in a greatly reduced catalytic activity. Upon CaM binding, a conformational change releases the CaM-binding domain from its internal acceptor site and activates the enzyme (Di Leva et al., 2008).

In mammals, four isoforms of PMCAs are coded by separate genes. Two of them, isoforms PCMA1 and PCMA4, are considered to perform house-keeping roles as they are ubiquitously expressed in all tissues. The other two isoforms (PMCA2 and PMCA3) are expressed predominantly in the ner-vous tissue (Ortega et al., 2007). The genome of *D. melanogaster* contains a single PMCA gene (*CG42314*) located on chromosome 4, and further isoforms generated by alternative splicing are differentially expressed (Desai and Lnenicka, 2011). A number of studies implicate PMCA in the regulation of calcium homeostasis in insect tissues. For instance, muscle-specific expression of double-stranded RNA interference (dsRNAi) targeting the *Drosophila* PMCA revealed a lethal phenotype: although most of the larvae hatched, most of them died during the first instar with characteristic symptoms of contraction paralysis (Bai et al., 2008). In those animals, intracellular calcium levels increased about 30-fold over control. Evidence has been provided that PMCA is probably the molecular target site of maitotoxin (MTX), a highly potent 3500-Da cyclic polyether toxin isolated from dinoflagellates that are known to cause seafood poisoning. MTX preferentially binds to calcium, triggering calcium overload and cell death. Biochemical studies indicate that MTX converts the PMCA pump to a calcium channel by keeping both E1 and E2 ion gates open at the same time (Sinkins et al., 2009). Although the tobacco budworm, *Heliothis virescens* PMCA enzyme has been claimed as a candidate target for insecticide discovery (Federspiel et al., 2006a), no insecticidal PCMA inhibitors have been reported to date.

2.2. Sodium–calcium exchanger

The NCX functionally cooperates with PMCA in the extrusion of calcium ions from excitable cells. Three NCX isoforms, NCX1, NCX2, and NCX3, exist in mammals (Iwamoto, 2007). NCX restores the intracellular Ca^{2+} concentration to resting levels following excitation and the subsequent increase in intracellular Ca^{2+} due to the activity of transmembrane calcium channels (Brini and Carafoli, 2011). The stoichiometry of ion transport reveals that three Na^+ ions are taken up per one Ca^{2+} transported outward, meaning that the NXC activity is electrogenic. Under specific physiological conditions, for example, during the heart systole with massive Na^+ entry through voltage-gated sodium channels, NCX can operate in the reverse mode accounting for a limited Ca^{2+} entry into the cell (Noble and Herchuelz, 2007).

The current topological model of mammalian NCX predicts nine transmembrane α-helices arranged in two modules of five N-terminal helices and four C-terminal helices. The transmembrane domains are connected by an extended cytosolic loop which comprises the regulatory region including two calcium-binding domains (CBDs). NCX transport activity increases with increasing intracellular calcium concentrations. CBD1 is the primary site of calcium regulation due to its high affinity. The recently published structure of a bacterial NCX homolog reveals three putative Na^+-, and one Ca^{2+}-binding sites. In addition, separate intramolecular transport pathways are suggested for the two ions (Liao et al., 2012). A *Drosophila* NCX ortholog, CALX, has been cloned from photoreceptor cells of the visual system (Schwarz and Benzer, 1997). Interestingly, the regulation of CALX activity by calcium is opposite to other members of the NCX family: it is inhibited rather than stimulated by increasing intracellular calcium concentrations. The structures of both calcium sensor domains, CBD1 and CBD2, of the *Drosophila* CALX have been solved recently (Wu et al., 2009, 2010). In the Ca^{2+}-bound CBD1 structure, up to four calcium ions occupy the site and the observed conformational differences between the apostructure and the calcium-bound state can explain its primary role in calcium regulation. Despite the different regulation of insect and mammalian NCXs, the structural basis for calcium binding is similar in both proteins. Calcium binding by CBD2, however, has been excluded due to structural reasons, and a modulatory function has been discussed for this site. Mammalian NCX isoforms have been suggested as potential drug targets, as for example, in the prevention of brain ischemia. However, there is still controversy concerning the relative importance of inhibiting NCX in the forward or reverse mode of operation (Iwamoto, 2007). Specific inhibitors have been identified which have been suggested as potential tools to investigate the physiology and pathophysiology of NCX (Iwamoto et al., 2007). It can be expected that inhibition of insect NCX in the forward mode should lead to calcium overload with detrimental physiological consequences, such as cramping and contraction paralysis, impairment of neurotransmission, or apoptotic neuronal degeneration. However, no insecticides targeting NCX have been reported so far.

2.3. Sarcoplasmic Ca^{2+}-ATPase

The sarcoplasmic Ca^{2+}-ATPase (SERCA) is a highly efficient mechanism to maintain low-level intracellular calcium concentrations by transporting calcium ions into the lumen of the ER/SR. In mammalian heart muscles, it removes more than 70% of the cytosolic calcium, and in skeletal muscles, SERCA-mediated reloading of the ER/SR stores accounts for the clearance of most of intracellular Ca^{2+} (McLennan and Kranias, 2004). Like the PMCAs, SERCA belongs to the P-type family of ATP-driven ion pumps.

Two calcium ions are stoichiometrically transported per hydrolyzed ATP molecule. Phosphorylation of an invariant acidic amino acid residue by ATP yields a high-energy intermediate to provide the free energy driving catalysis. The overall structure of the 110-kDa protein exhibits 10 transmembrane helices and three distinct cytoplasmic domains, A, N, and P (Toyshima et al., 2000). The conserved Asp residue, which is phosphorylated during catalysis, is localized in the P domain, while domain N is the nucleotide-binding region. Structural comparison of the Ca^{2+}-bound and Ca^{2+}-free structures have provided insight into the conformational changes underlying catalysis (Toyoshima and Nomura, 2002). Two calcium ions from the cytosol bind cooperatively to the transmembrane region, followed by phosphorylation of the conserved aspartate residue to generate the phosphoprotein $Ca_2E1{\sim}P$ intermediate. Conformational change to the lower-energy E2P state then releases the two Ca^{2+} ions to the ER/SR lumen. This is accompanied by exchanging two to three luminal protons to the cytosolic side. Finally, hydrolysis of the phosphorylasparate then resets the enzyme E2 state which is ready to start the next catalytic cycle (Møller et al., 2005). The role of SERCA in calcium regulation has been investigated with specific inhibitors such as thapsigargin and cyclopiazonic acid (CPA). Thapsigargin is a sesquiterpene lactone of plant origin that depletes intracellular calcium stores by inhibiting SERCA at subnanomolar concentrations. Therefore, it has become a valuable pharmacological tool to discern intracellular calcium-release processes (Thastrup et al., 1990). The inhibitor locks the enzyme in a calcium-free conformation that prevents Ca^{2+} binding and subsequent phosphorylation, and thus leaves the enzyme in a catalytically inactive dead-end state (Sagara et al., 1992). Mutational and structural data imply that thapsigargin binds in a cavity formed by three transmembrane helices (M3, M5, and M7) located close to the cytosolic interface of the transmembrane domain (Xu et al., 2004).

In insects, thapsigargin-sensitive calcium signaling has been documented in a number of studies. For example, in the Malpighian tubules of *Drosophila*, thapsigargin modifies fluid transport via increasing the intracellular calcium in stellate cells (Dube et al., 2000; Rosay et al., 1997). Additionally, thapsigargin-sensitive SERCA activity has been directly measured in membrane fractions of adult *Drosophila* (Vazquez-Martinez et al., 2003). Cyclopiazonic acid is an indole-tetramic acid secondary metabolite produced by certain *Aspergillus* and *Penicillium* strains. Identification of both thapsigargin- and CPA-sensitive calcium stores in the visual system of *Drosophila* has further indicated that SERCA plays similar roles in calcium signaling both in mammals and insects (Arnon et al., 1997). In *Drosophila*, the dorsal vessel drives the circulation of the hemolymph. The contraction of the dorsal vessel muscles is regulated by calcium similar to the heart contraction in mammals. Studying a SERCA temperature-sensitive (ts) mutant of *Drosophila* both at permissive and restrictive temperatures, it was found that SERCA controls the contraction rate in a concerted fashion with the PMCA and NCX (Desai-Shah et al., 2010). Mutants of the *Drosophila* SERCA-ortholog Ca-P60A (CG3725) displayed characteristic larval phenotypes at restrictive temperatures marked by prolonged contraction of the body-wall muscles and paralysis. This was interpreted as a consequence of intracellular calcium overload in the myocytes and confirmed the central role of SERCA in muscle relaxation (Sanyal et al., 2005). A cDNA coding for an insect SERCA has been cloned from the noctuid moth *H. virescens* which shows 72% sequence identity with avian and mammalian orthologs. However, heterologous expression in COS-7 cells did not result in a catalytically competent enzyme, despite correct localization in the ER (Lockyer et al., 1998). No commercial insecticides inhibiting SERCA are known to date. However, enzymes cloned from the nematode *Caenorhabditis elegans* (Zwaal et al., 2002) and the moth *H. virescens* (Federspiel et al., 2006b) have been claimed in patent applications as target sites to screen for novel pesticides. In a recent report, structurally simplified acyltetramic acid analogs of CPA were synthesized, some of which inhibited the *H. virescens* SERCA with submicromolar affinities. It was not disclosed, however, whether the compounds had insecticidal activity (Yao et al., 2011).

3. CALCIUM CHANNELS

3.1. The heterogeneity of calcium channels

Calcium signals of excitable cells are characterized by fast and reversible increases in intracellular calcium concentrations that often are referred to as calcium transients. Rapid changes in the calcium conductivity of transmembrane channels define the dynamics and shape, and the temporal and spatial patterns of calcium transients in response to external or internal stimuli. Depending on their roles in calcium signaling, either calcium channels are located in the plasma membrane functioning as entry channels or they release calcium-like "gatekeepers" of the intracellular storage compartments (Bootman et al., 2001). The roles of calcium channels in the generation, modulation, and termination of calcium signals appear to be similar in insects and mammals. Examples for intensively studied physiological systems in insects that are regulated by calcium are the *Drosophila* neuromuscular synapses, the central nervous system, and the motoneurons (Desai-Shah and Cooper, 2009; Peng and Wu, 2007; Ryglewski et al., 2012).

Annotation of the *D. melanogaster* genome has revealed members of all calcium channel families (Littleton and Ganetzky, 2000). In addition, voltage-gated calcium channels (Ca_V) and the ryanodine-sensitive calcium-release channels have received some attention in the literature as factual or prospective targets for pest control agents (Bloomquist, 1996; Hall et al., 1995; Usherwood and Vais, 1995).

Ca_V increase their calcium conductivity in response to action potentials, that is, the depolarization of the plasma membrane. In the *Drosophila* genome, the family of voltage-gated channels is represented by four α-subunit genes, one β-subunit gene, three $\alpha 2\delta$ genes, and one γ-subunit gene. Interestingly, Ca_V are targeted by insect-specific spider venom peptides (King, 2007). This topic will be dealt with more detail below.

Ryanodine-sensitive calcium-release channels (RyR) located on the ER/SR membranes respond to calcium entry through voltage-gated channels with a massive release of calcium from the internal stores. Recently, the RyRs have been described as target sites for novel families of synthetic insecticides (Nauen, 2006). Therefore, a more thorough discussion in the last paragraphs of this review (*vide infra*) is justified. Other important families of calcium channels are represented by the transient receptor potential (TRP) channels, the store-operated calcium channels (SOC), and the inositol 1,4,5-triphosphate receptors.

3.2. TRP calcium channels

A specific type of calcium entry channels in the plasma membrane was originally identified in a *Drosophila* mutant defective in the visual system. It was named *transient receptor potential* (*trp*) to earmark the fast decay of photoreceptor membrane potentials following sustained and intense illumination of this mutant (Minke, 1982). Later, sequence analysis of the *trp* gene disclosed a member of a new superfamily of calcium-permeable channels subsequently named TRP (Montell and Rubin, 1989). TRP channels have evolved to a considerable diversity in insects as indicated by the fact that the *Drosophila* genome contains members of seven TRP subfamilies. Interestingly, TRP channels are involved in the translation of a variety of visual, mechanical, and chemical stimuli into neuronal signals, thereby serving key functions in sensory processes (Damann et al., 2008).

TRP channels possess six predicted transmembrane helices and a putative pore loop connecting helices 5 and 6. Diverse experimental evidence suggests a tetrameric quaternary structure. TRP channels lack the positively charged voltage-sensor residues in helix 4, which is a characteristic structural feature of voltage-gated Na^+-, K^+-, and Ca^{2+}-channels (Minke, 2010). Some TRP channel subtypes like the mammalian TRPV1 are activated by capsaicin, vanilloids, and a great variety of potentially noxious or irritating compounds (Kissin and Szallasi, 2011). The insecticidal activity of horseradish or yellow mustard extracts was attributed to the presence of allyl- and benzyl-isothiocyanates. The finding that those compounds activated TRPA1 channels *in vitro* led to the suggestion that TRP channels may be possible target sites for new insecticides (Nagata, 2007). However, no synthetic insecticidal compounds interfering with TRP channels have been developed so far.

3.3. Store-operated calcium channels

The depletion of intracellular calcium stores through the activity of calcium-release channels triggers calcium entry from the extracellular environment. The process is termed store-operated calcium entry (SOCE), and it is mediated by specialized calcium-conducting channels in the plasma membrane. Inhibition of the SERCA is the most commonly used experimental method to deplete the calcium stores and induce SOCE. The molecular nature of SOC has been a controversial issue for some time (Spassova et al., 2004). Originally, members of the TRP channel superfamily had been implicated in SOCE. Later, a genome-wide RNAi screen in *Drosophila* for genes affecting SOCE revealed several genes influencing SOCE. One gene encoding the ER calcium sensor and SOCE regulator, STIM (stromal interaction molecule), was identified. Additionally, a new channel protein candidate, termed Orai, was suggested to mediate the calcium release-activated Ca^{2+} currents (CRAC) (Feske et al., 2006). Secondary structure prediction reveals four transmembrane domains and cytoplasmic N- and C-termini which is in agreement with plasma membrane localization. Three homologs of Orai exist in mammals (Gwack et al., 2007). After store depletion, the regulator protein STIM is relocated close to the plasma membrane where it activates the channel protein Orai/CRAC by a still poorly understood mechanism (Fahrner et al., 2009). Knockdown of either STIM or Orai expression by dsRNAi in *Drosophila* neurons caused flight-defective phenotypes suggesting the contribution of SOCE to the regulation of intracellular calcium levels in the control of flight (Venkiteswaran and Hasan, 2009).

SOC have been suggested as novel targets for therapeutic intervention, for example, in cardiovascular diseases and epilepsy (Leung et al., 2008; Steinbeck et al., 2011). In insects, however, little is known about the pharmacology of SOC and potential lead structures for insecticide discovery are not known.

3.4. Inositol 1,4,5-triphosphate receptors

The second messenger inositol 1,4,5-triphosphate (IP_3) stimulates calcium release by a specific class of intracellular release channels, the IP_3 receptors (IP_3Rs). IP_3 is produced by phospholipase C (PLC) that hydrolyzes the phosphorylated plasma membrane glycolipid phosphatidylinositol 4,5-bisphosphate (PIP_2) to IP_3 and diacylglycerol. PLC activity is stimulated, for example, by agonists of muscarinic acetylcholine receptors via specific G proteins (Lanzafame et al., 2003; Nash et al., 2001). IP_3 thus generates calcium transients by triggering calcium release from the ER mediated by IP_3R. The two families of intracellular calcium–release channels, the IP_3Rs and the ryanodine-sensitive (RyR) release channels, not only exhibit a high degree of structural similarity but also seem to share common regulatory mechanisms (Seo et al., 2012; Yuchi and Van Petegem, 2011). In insects, regulatory roles of IP_3R have been described in the central nervous system (Qazi and Trimmer, 1999) as well as its involvement in certain developmental processes (Venkatesh and Hasan, 1997). The *D. melanogaster* genome contains a single IP_3R gene. Both biophysical and structural studies have emphasized the evolutionary conservation between insect and mammalian IP_3R functions (Srikanth et al., 2004a,b). *Drosophila* IP_3R mutants display complex phenotypes. The severity of these phenotypes largely depends on the degree of reduced functionality: null mutants exhibit growth defects and lethality during the second instar larval stage, while weaker alleles disrupt more specific IP_3R functions, for example, in neurohormone secretion by aminergic cells (Joshi et al., 2004) or in serotonergic and dopaminergic neurons involved in the neuronal control of flight (Banerjee et al., 2004). Evidently, IP_3Rs are essential for viability, but so far no insecticidal compounds targeting IP_3Rs have been reported so far in the literature.

3.5. Voltage-gated calcium channels

3.5.1 Mammalian Ca_V

Ca_V are oligomeric transmembrane protein complexes that conduct calcium into the cells in response to depolarization of the membrane potential. They constitute a key element in the transformation of electrical excitation into intracellular signals that regulate cell-physiological processes, such as the re-lease of neurotransmitters at synapses, and the excitation–contraction (EC) coupling in muscle cells. Mammalian Ca_V exist as diverse isoforms, classified according to their biophysical characteristics, pharmacology, and expression patterns (Catterall, 2011). The high-voltage-activated (HVA) L-type chan-nels occur as four isoforms designated $Ca_V1.1$–$Ca_V1.4$. They are expressed in skeletal ($Ca_V1.1$), cardiac, and smooth muscles ($Ca_V1.2$), and in the ner-vous system ($Ca_V1.3$, $Ca_V1.4$). L-type channels are sensitive to the dihydropyridine (DHP) and phenylalkylamine (PA) channel blockers. In contrast, T-type Ca_V ($Ca_V3.1$–$Ca_V3.3$) are characterized by low–voltage ac-tivation. The voltage-sensitivity of N-type channels ($Ca_V2.1$) is intermedi-ate between L-type and T-type Ca_V. N- and T-type channels are insensitive to DHPs, but can be pharmacologically distinguished by their sensitivity to specific peptide toxins isolated from the venoms of cone snails or spiders (Catterall, 1993; Wu et al., 1999). Three additional Ca^{2+} currents have been found in specific neuronal cell types of the brain: P-, Q-, and R-type are distinguished by their different sensitivities toward the spider toxins ω-agatoxin IVA, ω-agatoxin IIIA, and the ω-conotoxin MVIIC (Randall and Tsien, 1995, 1997).

Recent studies have provided considerable progress in understanding the structure and function of Ca_V. Cryo-electron microscopy combined with single-particle analysis revealed for the first time the global architecture of the Ca_V complex at ~30 Å resolution (Serysheva et al., 2002; Wang et al., 2002). Although secondary structure elements and atomic details are not resolved at this low resolution, they nevertheless provide a useful structural framework for the interpretation of physiological, biochemical, and mutagenesis data that have been accumulated over the years. Mamma-lian Ca_V complexes are assembled from five different types of subunits: α1, α2, δ, β, γ. The large α1 subunit of Ca_V shares some general structural ho-mology (Fig. 5.3) with voltage-gated sodium and potassium channels (Yu and Catterall, 2004). The α-subunits constitute the ion pore and are primar-ily targeted by physiological and pharmacological regulators of channel ac-tivity. The transmembrane part of the 170-kDa α-subunit consists of four repeats (I–IV) with six transmembrane segments (S1–S6) each. Intracellular loops, which connect these repeats, contain protein–protein interaction sites for other Ca_V subunits as well as proteins that regulate channel function (Findeisen and Minor, 2010). The transmembrane helix S4 of each repeat is characterized by positively charged amino acid residues. Together with helices S1–S3, the voltage-sensor S4 helices regulate the potential-sensitive

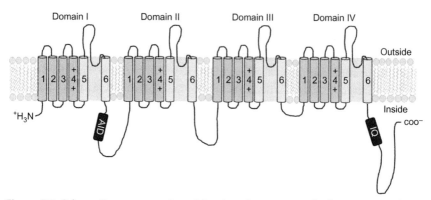

Figure 5.3 Schematic representation of the domain structure of voltage-gated calcium channels. Highlighted are helices 5 and 6 of each domain. The inner pore is formed by helices 6 of domain I–IV. The pore loops connect helices 5 and 6 of each domain. The positive charges depict the voltage-sensor helix 4. The β-subunit interacts with the AID site located in the intracellular loop between domain I and II. The carboxy-terminal IQ-motif binds calmodulin. (See Color Plate 22 at the back of the book.)

gating of the channel. Although the voltage-sensing mechanism is not understood in every detail, current models propose outward movements of the S4 helices in response to depolarization which are conformationally coupled to pore opening (Catterall and Yarov-Yarovoy, 2010; Lee et al., 2005; Yarov-Yarovoy et al., 2006, 2012).

Helices S5 and S6 together with the connecting pore loop constitute the pore domain of the channel architecture. The pore loops of each of the four transmembrane segments contain conserved glutamate residues (EEEE motif) forming a negatively charged ring structure that conveys calcium selectivity to the channel. Interestingly, it is experimentally possible to change the ion selectivity by mutating the charge amino acid residues of the pore loop (Heinemann et al., 1992; Marom et al., 2007). It should be noted at this point that the pharmacologically important lipophilic Ca_V blockers bind within the pore domain.

An extended C-terminal intracellular tail is connected to the S6 segment of repeat IV. Its importance for channel regulation is highlighted by the fact that binding sites for CaM, CaM-dependent kinases, protein kinase C, the β and γ subunits of G proteins (Ca_V2 channels), and SNARE proteins have been identified in the C-terminal region of Ca_V α1 subunits (Catterall, 2010; Zuccotti et al., 2011). As a prerequisite for detailed molecular analysis, recombinant expression of Ca_V α1 subunits alone is sufficient to constitute calcium-conducting channels. However, their expression levels are low and their biophysical properties clearly deviate from those of native channels, immediately posing the question of the physiological relevance of these expression studies (Wei et al., 1991). Coexpression of Ca_V β-subunits generally improves the expression levels and restores the physiological properties of voltage gating and regulation. Cytoplasmic β-subunits consist of five core domains, two of them showing homology to the canonical SH3 and the others to yeast guanylate kinase. The protein–protein interaction site on the Ca_V α1-subunit has been identified as an 18-residue sequence motif termed α1-interaction domain (AID) in the cytoplasmic loop linking S6 of repeat I to S1 of repeat II. X-ray crystallography revealed that AID binds in a hydrophobic cleft of the guanylate kinase domain of the β-subunit (Chen et al., 2004; Richards et al., 2004).

The accessory subunits α2 and δ are encoded by a single gene and proteolytically cleaved into the corresponding subunits following translation. The topological model of the channel complex depicts that δ and γ subunits are transmembrane proteins, while the extracellular α2 subunits are connected via the δ subunit. Coexpression of α2δ increases expression of functional Ca_V channels with physiological gating properties in different expression systems. The effects of γ-subunits in heterologous expression are less understood (Catterall, 2011).

The recently published 2.7-Å X-ray structure of a voltage-gated sodium channel from the bacterium *Arcobacter butzleri* may serve as a topological model for vertebrate and invertebrate voltage-gated sodium and calcium channels. It allows new insights into conserved structural features underlying membrane potential sensing, channel gating, and activation/inactivation mechanisms in the voltage-gated ion channel superfamily (Payandeh et al., 2011). The first bacterial Na_V was characterized from *Bacillus halodurans*, and subsequently, homologous channel proteins have been identified in other bacterial species. The overall sequence identity is only 17–23% with eukaryotic Na_V, and 18–23% with Ca_V. Unlike eukaryotic voltage-gated channels with four transmembrane repeats, the bacterial Na_V proteins contain only one transmembrane region consisting of the pore-forming domain (helices S5, S6, and the pore loop) and the voltage-sensor domain (helices S1–S4). Correspondingly, functional prokaryotic Na_V are homotetramers (Charalambous and Wallace, 2011; Powl et al., 2010). Like in eukaryotic voltage-gated channels, cation selectivity can be altered by mutating specific residues in the selectivity filter. Increasing the negative charge of the filter region by substituting the uncharged Ser^{192} and Ser^{195} residues by aspartate transforms the bacterial Na_V into a calcium channel (Yue et al., 2002), supporting the view of a common structural pattern that has evolved within the voltage-gated cation channel superfamily. Interest-ingly, the pharmacology of bacterial Na_V partly shares some characteristics with mammalian Ca_V, which do not correlate with the different cation se-lectivity. In particular, L-type Ca_V blockers of the DHP type (nifedipine, Fig 5.4D) block the bacterial channel with low-micromolar IC_{50} values. Residues of S5 and S6 of one monomer together with residues from S6 of an adjacent monomer may form a hydrophobic binding site that seems to involve homologous residues of the DHP site of eukaryotic Ca_V (Ren et al., 2001; Shafrir et al., 2008).

It has been suggested to employ bacterial voltage-gated sodium channels as structural models for drug and insecticide discovery, since high-resolution structural information on eukaryotic voltage-gated ion channels is lacking. A similar approach has been followed in the field of the nAChRs. X-ray struc-tures of surrogate receptors, the molluscan acetylcholine-binding proteins (AChBPs), have been successfully employed to describe the binding modes of nicotinic ligands including the neonicotinoid insecticides (Hansen et al., 2005b; Sander et al., 2010; Tomizawa and Casida, 2009, 2011).

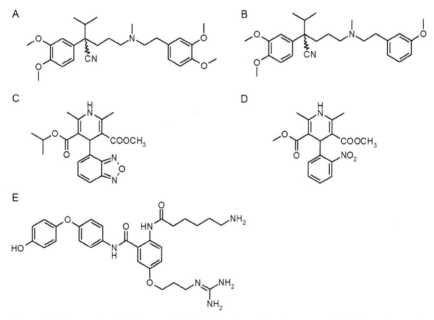

Figure 5.4 Examples of pharmacological blockers of voltage-gated calcium channels. (A) Verapamil, phenylalkylamine; (B) desmethoxyverapamil (D888, phenylalkylamine); (C) isradipine, dihydropyridine; (D) nifedipine, dihydropyridine; (E) nonpeptide GVIA mimic, compound 1a.

3.5.2 Insect Ca$_V$

Early pharmacological studies have already revealed high-affinity binding sites for PA Ca$_V$ blockers (Fig. 5.4) in *D. melanogaster* head membrane preparations. Interestingly, the dissociation constants are remarkably low, ranging at picomolar concentrations. *Drosophila* PA receptors are insensitive to 1,4-dihydropyridines. However, low densities of DHP binding sites have also been detected in *Drosophila* membrane preparations (Pauron et al., 1987). Following these studies, photoaffinity labeling experiments revealed a single 135-kDa protein that was specifically labeled with a photoactivatable arylazido-PA in *Drosophila* head membranes (Pelzer et al., 1989). In addition, antibodies raised against a mammalian α2 subunit cross-reacted with a neuronal 156-kDa polypeptide in *Drosophila* neuronal membrane preparations. Similar to the mammalian α2δ subunits, the *Drosophila* 156-kDa protein was shown to be proteolytically cleaved into products of 136 and 26 kDa.

Reconstitution of *Drosophila* brain membrane fragments into artificial lipid bilayers resulted in the detection of Ca^{2+}/Ba^{2+}-permeable and Co^{2+}-sensitive single-channel activities (Gielow et al., 1995; Pelzer et al., 1989). These were highly sensitive to PA block, but insensitive to DHPs. On the other hand, also DHP-sensitive Ca^{2+}/Ba^{2+} currents were measured in *Drosophila* brain membranes, but at much lower densities. Finally, voltage-gated calcium currents were also reported in the body-wall muscles of *Drosophila* larvae. These were sensitive to L-type blockers like DHPs and to the T-type channel blocker amiloride.

The first cloning of insect Ca$_V$ genes and, in recent years, the insect genome sequencing projects, substantiated the notion that the general architecture of Ca$_V$ has been conserved during a long evolutionary history. A *Drosophila* Ca$_V$ α1 subunit was cloned from a head cDNA library by a PCR-based approach (Zheng et al., 1995). The cDNA sequence of about 8 kb showed homology to mammalian α1 subunits. Within the Ca$_V$ family, the highest sequence similarity (78.3%) and amino acid identity (64.2%) was found in the D α1 subunit of the rat, and in the insect Ca$_V$ α1 subunit, Mdlα that was cloned from body-wall muscles of housefly (*Musca domestica*) larvae (Grabner et al., 1994a). Consequently, the *Drosophila* α1 has been termed Dmca1D (*CG4894*). A housefly β-subunit β$_M$ shows 89% homology to its *Drosophila* ortholog. Diverse splice variants are expressed predominantly in the central nervous system of both larvae and adult flies (Grabner et al., 1994b).

Sequence analysis of a region in the *Drosophila* X chromosome, as defined by the courtship song mutant locus *cacophony* (*cac*) and by the *nightblind*-A mutation, led to the identification of a new α1 subunit gene, Dmca1A

(CG43368) that is expressed mainly in the nervous system. The sequence suggests that Dmca1A (Cac) constitutes a PA-sensitive, but DHP-insensitive calcium channel (Smith et al., 1996). This has been concluded, *inter alia*, from the presence of the canonical glutamate residues of the selectivity filter and from a putative PA binding site near the S6 helix of repeat IV.

Further Ca$_V$ subunit genes have been annotated in the *Drosophila* genome including a single α2-, an α/β-, and a T-type gene (www.flybase.org). In the meantime, genome sequencing projects on other insect species have greatly expanded the number of insect orthologs from the Ca$_V$ superfamily. Insect Ca$_V$ genes generally contain the characteristic sequence features which have been identified in vertebrate Ca$_V$ genes before. These include, for example, the putative S4 voltage-sensing helices, the glutamate residues conferring calcium selectivity, and homologous recognition sites for posttranslational modifications. In addition, C-terminal intracellular tail regions exhibit characteristic EF-hand (calcium-dependent inactivation) and CaM-binding motifs. Even the proposed binding sites of PAs and 1,4–dihydropyridines appear to share homology with their mammalian counterparts. Although the number of Ca$_V$ subunit genes is smaller in insect compared to mammalian genomes, there is substantial heterogeneity of calcium channel transcripts found in different tissues of the brain as detected by Northern analysis. This is the result of extensive alternative splicing and may also explain, together with mRNA-editing mechanisms, the diversity of physiological calcium currents measured in insect neuronal preparations (Peixoto et al., 1997).

In the long history of *Drosophila* as a genetic model organism, several mutant phenotypes were described that could be attributed to defective Ca$_V$ gene products. The altered male courtship song (*cacophony*) and changes in the visually controlled behavior (*nightblind*-A) result from allelic mutations of the Dmca1A (*cac*) gene. The *lethal(1)L13* allele of the *cac* gene results in late embryonic lethality in the homozygous state (Smith et al., 1996). A conditionally embryonic lethal (temperature-sensitive) *cac* phenotype results from the missense mutation of an invariant proline (P1385) to serine in the C-terminal tail next to the EF-hand motif. Embryonic lethality and paralysis at restrictive temperatures could be rescued by transgenic expression of the wild-type *cac* gene (Kawasaki et al., 2002).

3.5.3 Insect Ca$_V$ as potential insecticide target sites

Compelling physiological and genetic evidence suggests that Ca$_V$ have essential roles in signal transduction processes and development. Accordingly, it is plausible to regard insect Ca$_V$s as attractive molecular target sites for insecticides (Bloomquist, 1996; Hall et al., 1995). The first successful cloning of an insect Ca$_V$ has sparked the interest in potential application in insecticide discovery. Despite the fact that the predicted secondary structure and topology are similar between insect and mammalian Ca$_V$ genes, sequence and predicted secondary structure differences opened the possibility that insect-specific Ca$_V$ ligands may be identified (Zheng et al., 1995).

As mentioned before, the known vertebrate Ca$_V$ blockers, DHPs and PAs, also bind to insect channels. Interestingly, the PA verapamil (Fig. 5.4A) is toxic to *Drosophila* at low millimolar concentrations when applied orally (Hall et al., 1995). However, due to their physico-chemical properties, PAs and DHPs are not well suited to serve as agrochemical lead structures.

An interesting new avenue of research was opened by the finding that venoms of certain cone snails and spiders contain rich cocktails of peptide toxins with highly diverse pharmacological properties. Typically, cone snails and spiders use their venoms to rapidly paralyze or kill their prey, which usually includes a variety of invertebrate and lower vertebrate species. To fulfill these requirements peptide toxins need to be fast acting and highly potent. Therefore, neuronal and neuromuscular target sites, also including calcium channels, are affected by these toxins (King, 2007, King et al., 2008; Tsetlin and Hucho, 2004). Venom toxins have evolved to a unique diversity. It has been estimated that each of the 500–700 *Conus* species produces between 50 and 200 different peptide toxins (Pi et al., 2006), of which less than 0.1% have been characterized to an extent that will render them potential new leads for drug discovery (Lewis, 2009).

Conotoxins are peptides of 25–29 amino acids with a characteristic tertiary structure stabilized by three disulfide bonds. Among the 15 structural subfamilies of conotoxins, the ω-conotoxins block voltage-gated N-type calcium channels of presynaptic terminals (Escoubas and King, 2009; Olivera et al., 1991). Due to the role of N-type calcium channels in nociceptive neurons, ω-conotoxins have attracted attention as analgesic drugs. In fact, a synthetic analog of the *Conus magus* ω-conotoxin MVIIA, ziconotide, has been approved for the treatment of chronic pain (Lewis, 2009). However, the generally poor pharmacokinetic properties of these peptides, in particular, limited oral absorption and poor blood–brain barrier translocation, has stimulated the search for synthetic nonpeptide mimetics with better suited physico-chemical properties. As an example, substituted anthranilamides (Fig. 5.4E) mimicking the constrained conformation of the ω-conotoxin GVIA from *Conus geographus* compete with peptide binding in the low-micromolar range (Andersson et al., 2009).

Insect Ca_V are also sensitive to ω–conotoxins: both medium/low-voltage-activated calcium currents and HVA currents in the central nervous system of the American cockroach *Periplaneta americana*, are blocked by ω–conotoxins MVIIC and GVIA (Wicher and Penzlin, 1997). However, their affinity is substantially lower (micromolar range) compared to mammalian Ca_V (low nanonolar range). These results again indicate pharmacological differences between insect and mammalian Ca_V that may be exploited in the discovery of selective insecticides.

Venom peptide toxins of spiders are even more diverse than conotoxins: The web-based resource ArachnoServer (http://www.arachnoserver.org) currently contains more than 900 toxins from 85 species (Herzig et al., 2011). Considering the estimated 41,000 existing spider species, it is conceivable that several hundred thousand or even millions of biologically active peptides may represent an almost inexhaustible resource of bioactive molecules that could be used as tools to validate potential new target sites or even serve as potent leads for pharmaceutical drug as well as for pest control agents (Saez et al., 2010).

Agatoxins and atracotoxins isolated from funnel-web spider venoms are the best characterized blockers of insect Ca_V. The ω–agatoxin family consists of four subfamilies (I–IV) all targeting Ca_V of vertebrates and invertebrates. ω–Agatoxin IVA (ω–Aga-IVA, Fig. 5.5A) is a 48–amino acid residues peptide which inhibits two different types of HVA calcium

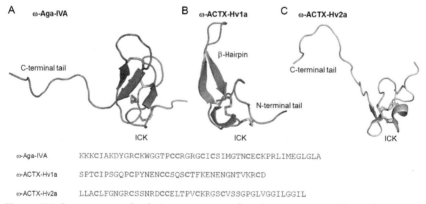

ω-Aga-IVA	KKKCIAKDYGRCKWGGTPCCRGRGCICSIMGTNCECKPRLIMEGLGLA
ω-ACTX-Hv1a	SPTCIPSGQPCPYNENCCSQSCTFKENENGNTVKRCD
ω-ACTX-Hv2a	LLACLFGNGRCSSNRDCCELTPVCKRGSCVSSGPGLVGGILGGIL

Figure 5.5 Sequences and solution structures of spider venom peptides acting on insect Ca_V. (A) ω-Agatoxin IVA, (B) ω-atracotoxin Hv1a, and (C) ω-atracotoxin Hv2a. ICK, inhibitor cystine knot. Coordinates of the solution structures by 1H-NMR were taken from PDB files: 1OAW (ω-agatoxin IVA), 1AXH (ω-atracotoxin Hv1a), and 1G9P (ω-atracotoxin Hv2a); graphs were generated using PyMOL v0.99 (DeLano Scientific). (See Color Plate 23 at the back of the book.)

currents in embryonic cockroach neurons. The first group of Ca_V channels are blocked by ω-agatoxin with high affinity ($IC_{50} = 9$ nM). In addition, these channels are insensitive to DHPs (Benquet et al., 1999). This is reminiscent of the findings with vertebrate channels: DHP-insensitive P/Q-type channels (Ca_V 2.1) are blocked by ω-agatoxin with nanomolar affinity. The second Ca_V population has a moderate affinity for the agatoxin ($IC_{50} = 900$ nM). The less-sensitive channels may represent other non-L-type channels that are observed in vertebrates as well. Further studies demonstrated that ω-Aga-IVA blocks not only the medium-voltage-activated calcium currents in bee neurons ($IC_{50} = 10$ nM) but also the Ca_V in mouse sensory neurons with about the same degree of affinity. In conclusion, ω-Aga-IVA does not discriminate between insect and mammalian channels (Wang et al., 2001). However, it should be noted that the pharmacology of insect Ca_V is much less elaborated and systematic pharmacological comparisons of insect and mammalian calcium channels are lacking (King, 2007).

Another family of insecticidal spider peptides, the ω-atracotoxins, was isolated from the venom of the Blue Mountain funnel-web spider *Hadronyche versuta* and from related species. ω-Atracotoxins of the ω-ACTX-2 family cause fast paralysis in several insect species after injection. The 45-residue peptide ω-ACTX-Hv2a (Fig. 5.5C) blocks calcium currents in bee neurons with an IC_{50} value of about 130 pM, but has little or any effect on mammalian neurons at concentrations up to 1.0 μM (Wang et al., 2001). ω-Atracotoxins of the ω-ACTX-1 family are active on a broad spectrum of species from diverse agronomically relevant insect orders such as Lepidoptera, Coleoptera, Hemiptera, Orthoptera, and Diptera. Injection of the toxins leads to fast and irreversible paralysis and lethality (Chong et al., 2007; Mukherjee et al., 2006; Vonarx et al., 2006).

A third interesting family of insecticidal spider toxin is represented by ω-plectotoxin-Pt1a (ω-PLTX-Pt1a) isolated from the venom of *Plectreurys tristis*. The toxin effectively blocks excitatory junctional potentials (ejp) at neuromuscular junctions of *Drosophila* larvae (Branton et al., 1987). Neurotransmitter release is inhibited without affecting transmission of action potentials, which is consistent with the hypothesis that the entry of calcium is blocked in the nerve terminals. In addition, blockade of both inactivating and noninactivating calcium currents in cultured embryonic neurons of *Drosophila* by plectotoxin has been shown. Apparently, the inhibition of the noninactivating calcium currents is apparently sufficient to block neurotransmitter release in *Drosophila* (Leung et al., 1989).

High-resolution structures of insecticidal spider toxins in solution have been solved from NMR-spectroscopic data (Fletcher et al., 1997; Wang et al., 1999; Yamaji et al., 2007). Like other ion channel-inhibiting peptide toxins, ω-agatoxins and ω-atracotoxins contain a characteristic three-dimensional structural motif known as "inhibitor cystine knot" (ICK) (Norton and McDonough, 2008; Norton and Pallaghy, 1998; Pringos et al., 2011). Essentially, a triple-strand, antiparallel β-sheet is forced into a highly constrained conformation by formation of three, in a few cases also four, disulfide bridges. The consensus sequence of the ICK motif is defined as $-CX_{3-7}CX_{3-6}CX_{0-5}CX_{1-4}CX_{4-13}C-$, in which X can be different amino acid residues. The structural rigidity of ICK peptides determines their stability at extreme pH values, high temperatures, and their resistance to proteolytic degradation. Head-to-tail cyclization further enhances the stability in some families of ICK peptides (Colgrave et al., 2008; Saez et al., 2010). In ω-agatoxin IVA (Fig. 5.5A), eight cysteine residues allow the formation of four disulfide bridges. The ICK core is linked to a long disordered C-terminus which is essential for receptor binding. Basic residues (Lys), which are primarily clustered at the N-terminus, confer a net positive charge to the toxin at physiological pH values. Therefore, it is conceivable that ω-Aga-IVA interacts with negatively charged residues in the binding site of the P-type Ca_V. Furthermore, mutational analysis indicates that ω-Aga-IVA binds to at least one of the voltage-sensing domains, probably in close proximity to the voltage-sensor S4 helix (Winterfield and Swartz, 2000). The mutational data corroborate the electrophysiological evidence that the toxin alters the gating properties of the channel after binding to the closed state with high affinity. This reduces the probability of the open state of the channel during excitation. Replacing a glutamate residue by lysine in the S3–S4 linker of repeat IV of the rat brain channel largely prevents inhibition by ω-Aga-IVA. This suggests that the S3–S4 linker of repeat IV forms at least part of the toxin binding site.

In *Drosophila* Dmca1, the S3–S4 linker sequence of repeat IV aligns well with the mammalian brain P/Q-type channel, while DmcaD and Ca-α1T contain sequence insertions in this region. Therefore, it is tempting to speculate that the Dmca1-type channels are the target sites of ω-Aga-IVA in insects (King, 2007).

The ω-ACTX-Hv2a has a core structure that closely resembles ω-Aga-IVA. The characteristic ICK motif (Fig. 5.5C) contains three disulfide bridges and a disordered C-terminal tail, which is essential for channel inhibition and insecticidal activity (Wang et al., 2001). As shown in

Fig. 5.5B, the ω–ACTX–Hv1a structure differs from ω–ACTX–Hv2a, for example, in the N-terminal region (Fletcher et al., 1997).

The exquisite specificity of certain spider toxins for invertebrate calcium channels has promoted the idea of using those toxins as novel insect control agents in different ways. Tedford et al. (2001) identified a highly conserved β-hairpin at the C-terminus of ω–ACTX–Hv1a which is essential for insecticidal activity as demonstrated by systematic mutational analysis of the hairpin sequence. Within this region, the suggested "insectophore" contains only three residues, Pro^{10}, Asn^{27}, and Arg^{35} that are likely to form a contact surface interacting with the channel (Fig. 5.6). These residues are conserved within the ω–ACTX–1 peptide family (Tedford et al., 2004). In principle, this structural and topological information could be applied to identify either simpler peptides or nonpeptide mimetics that bind to the ω–ACTX–1 site on insect Ca_V. The screening of compound/peptide libraries and rational structure-based design may be valuable approaches according to this strategy.

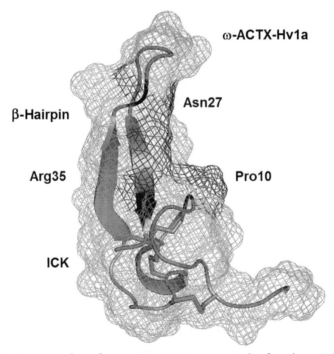

Figure 5.6 Contact surface of ω-atracotoxin Hv1a suggested to face the insect Ca_V. The contact surface is suggested to comprise only residues Pro10, Asn27, and Arg35 (labeled in blue). Coordinates of the solution structure from PDB file 1AXH; graph generated using PyMOL v0.99 (DeLano Scientific). (See Color Plate 24 at the back of the book.)

As mentioned earlier, the major obstacle of using peptides as pesticides is their limited bioavailability after contact or ingestion (Windley et al., 2012). Hopefully, better understanding the structural requirements of insect-specific Ca_V interaction may allow to translate this information into novel nonpeptide lead structures. Few examples in the field of medicinal chemistry suggest that peptide mimics of ICK toxins with reasonable target-site activity are not unrealistic (Andersson et al., 2009; Baell et al., 2006). Radiolabeled toxins have been successfully used as pharmacological tools in drug discovery for mammalian calcium channels (Dooley et al., 1988; Hirota and Lambert, 2006; Wada et al., 2005).

Following a different strategy, spider toxins could be used as insect control agents expressed in transgenic plants. As an example, the ω-ACTX-Hv1a toxin gene having the codon usage adapted for plant expression was introduced into tobacco plants (*Nicotiana tabacum*) and expressed under control of the cauliflower mosaic virus 35S promoter (Khan et al., 2006). Transgenic plants were protected against feeding by larvae of the cotton bollworm *Helicoverpa armigera* and of the cotton leafworm, *Spodoptera littoralis*. Intoxicated larvae exhibited characteristic symptoms starting from feeding cessation and progressing through uncontrolled movements up to paralysis. Finally, ω-atracotoxins have been regarded as potential biopesticides (Down et al., 2006; Fitches et al., 2004). Interestingly, it was demonstrated that fusion with other proteins such as lectins can improve oral uptake and biological activity.

In summary, several cystine knot toxins isolated from the venoms of cone snails and spiders specifically target insect Ca_V. The tremendous variability and complexity of invertebrate venoms represent a rich reservoir of pharmacological tools that has been developed during a long evolution. In particular, understanding the molecular basis of insect-specific blockers from the ω-atracotoxin family may be translated into insecticides with favorable toxicological profiles. In addition, there is also potential for insecticidal toxins as novel insect control agents in the field of plant biotechnology.

3.6. Ryanodine receptors
3.6.1 Mammalian ryanodine receptors
As the primary release channels of the ER/SR of excitable cells, ryanodine receptors (RyRs) mediate the massive, calcium-activated release of calcium into the cytosol. RyR are homotetrameric complexes of 565-kDa subunits (Fig. 5.7). Roughly, 85% of the protein mass is organized as a cytoplasmic assembly of protein domains which are involved in the regulation of channel

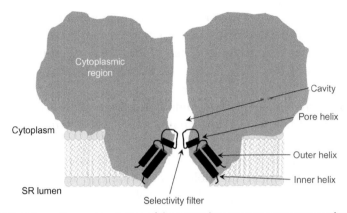

Figure 5.7 Schematic representation of the ryanodine receptor structure and topology. Two opposite subunits of the tetrameric quaternary structure are drawn. The pore region with inner and outer helices and the putative pore loop are assigned according to current structural models.

gating and in protein–protein interactions. The C-terminal transmembrane domain of about 700 amino acid residues forms the pore region. Cryo-electron microscopy combined with single-particle analysis has elucidated the overall architecture of the closed conformational state at about 10-Å resolution (Ludtke et al., 2005; Samso et al., 2005). Recently, high-resolution partial structures of N-terminal regions have been solved by X-ray crystallography to elucidate, for example, the structural effects of mutations involved in specific diseases (Lobo and Van Petegem, 2009; Tung et al., 2010).

It is far beyond the scope of this communication to discuss the structural biology of RyR in detail as it has been reviewed recently in a number of excellent papers (Hamilton and Serysheva, 2009; Kimlicka and Van Petegem, 2011; Wagenknecht and Liu, 2010). Consequently, only those architectural features with particular relevance for channel activity and modulation are briefly discussed in the following section.

The structural model of the transmembrane region displays canonical features of other ligand-gated and voltage-gated ion channels (Doyle et al., 1998; Dulhunty and Pouliquin, 2003; MacKinnon et al., 1998; Wagenknecht and Liu, 2010; Williams et al., 2001). Although the precise number of transmembrane helices has yet to be determined, there is now growing consensus that six helices likely form the channel pore. Secondary structure prediction and 3D-reconstruction of cryo-electron micrographs (cryo-EM) revealed that the transmembrane helices are arranged similarly to the inner, outer, and pore helices identified in the high-resolution structures of bacterial potassium channels. In addition, the apparent pore loop connecting the inner and outer helices contains a sequence motif, GGGIGD, which is reminiscent of the VGYG sequence of the selectivity filter in the structures of K^+ channels (Balshaw et al., 1999). In fact, alanine scanning of the proposed pore loop region of the rabbit cardiac RyR2 resulted in altered single-channel conductance (Du et al., 2001a). For example, the G4828A mutant had similar transport efficiency for monovalent and divalent cations.

In the bacterial KcsA potassium channel structure, the pore-forming region is organized around a fourfold axis of symmetry. Each RyR subunit contributes to the pore lining by the inner helices and to the selectivity filter of the pore loops. Disagreement in the literature exists about the proposed gating mechanism involving the kinking of pore-lining inner helices during channel opening (Carney et al., 2010; Ludtke et al., 2005; Samso et al., 2005). Nevertheless, there is now strong evidence supporting long-range conformational changes that accompany channel opening, which can be viewed like the movement of a camera lens aperture (Hamilton and Serysheva, 2009; Orlova et al., 1996).

The cytoplasmic assembly of the RyR homotetramer consists of 11 globular subregions which are interconnected. The four corners are termed "clamps" and comprise subregions 5–10 (Serysheva et al., 2008). It should be noted that at the relatively low resolution of reconstructed cryo-EM data (~ 10 Å), it is not possible to unequivocally assign secondary structural features such as α-helices and β-sheets.

The clamp structures are of particular importance as sites for protein–protein interactions (Jayaraman et al., 1992; Marks, 1996; Masumiya et al., 2003). The immunophilins FKBP12 (FK506-binding protein 12, calstabin1) and FKBP12.6 (FK506-binding protein 12.6, calsatbin2) are peptidyl-prolyl-*cis-trans* isomerases that bind with a stoichiometry of one calstabin per RyR monomer to the clamp regions. Depletion of calstabin by incubation with the immunosuppressant tacrolimus (FK506) increases the open probability and induces subconductance conformational states of RyR channels.

Structural as well as functional data favor the notion of a supramolecular organization of RyR complexes in ordered two-dimensional arrays (Liang et al., 2009). The physiological consequences integrating single-channel opening events into calcium transients generated by the concerted activation of many RyR channels are referred to as calcium waves or sparks. The dynamic coupling of RyR channels may provide the basis for the temporal and spatial initiation and termination of calcium signals in muscle cells to ensure tight control of muscle contraction and relaxation (Liang et al., 2007).

In mammals, three RyR isoforms can be distinguished: RyR1 is primarily expressed in skeletal muscles, whereas RyR2 has been found in the heart and in the smooth muscles of blood vessels and visceral organs. The third isoform, RyR3, is ubiquitously expressed in many tissues, but its functions are less understood (Hamilton, 2005; Lanner et al., 2010; Rossi and Sorrentino, 2002). In muscles, RyRs play a key role in the process of EC coupling. Depolarization of the plasma membrane activates DHP-sensitive L-type Ca_V (Ca_V 1.1). Ca_V then trigger calcium release from the SR by eliciting the opening of RyRs (Catterall, 1991; Ma et al., 2004). The increase in intracellular calcium induces muscle filament contraction by activating troponin C, a filament protein which contains two EF-hand motifs and is regarded as the calcium sensor component of the muscle thin filaments (Vinogradova et al., 2005). The mechanisms by which membrane depolarization and Ca_V activation trigger the RyR-mediated calcium mobilization are different in skeletal and cardiac muscles. The skeletal muscle RyR1 subtype physically interacts with Ca_V to constitute supramolecular structures frequently referred to as calcium-release units. In these structures, four Ca_V (tetrads) are in contact with one RyR1 of the SR membrane ("feet" structures). Additional proteins, for example, triadins, contribute for the formation of highly ordered triads, constituting junctions of plasma membrane T-tubules and SR membranes containing the calcium-release units (Paolini et al., 2004a; Protasi et al., 1998).

Binding of the ICK-type scorpion toxins imperatoxin and maurocalcine supports the finding of direct interaction of RyRs and Ca_V. These toxins are basic, 33-residue peptides that appear to mimic the structure of a peptide sequence in the connecting loop between domains II and III of $Ca_V1.1$ (Esteve et al., 2003; Mosbah et al., 2000). The toxins induce calcium release in skeletal muscles by stabilizing subconductance conformational states of RyR1. Radioligand binding studies have shown that maurocalcine and synthetic peptides matching the Ca_V loop II–III sequence share a common binding site on the RyR (Altafaj et al., 2005). Deletion of a sequence fragment (residues 3241–3661) proposed as part of the toxin docking site renders RyR insensitive to maurocalcine. However, it does not affect the efficacy of other pharmacological effectors such as ryanodine or caffeine.

In conclusion, activation of the $Ca_V1.1$ triggers RyR1 gating by direct conformational coupling without the necessity of calcium entry through the voltage-gated plasma membrane channel (Paolini et al., 2004b). In contrast to skeletal muscles, for RyR2-mediated calcium release in cardiac myocytes, a depolarization-induced limited calcium influx via Ca_V is required, which is known as calcium-induced calcium release (CICR) (Liu et al., 2010a).

Covalent protein modifications as well as protein–protein interactions control the activity of RyR channels. First, the small calcium-regulatory protein CaM binds to both RyR1 and RyR2 isoforms and regulates channel activity in a dual fashion (Rodney et al., 2001; Zhu et al., 2004). Calcium-free apo-CaM activates RyR1, whereas Ca^{2+}-bound CaM inhibits the channel. CaM-binding sites have been attributed to a region including residues 3614–3643 of the skeletal type RyR1. Synthetic peptides comprising these residues bind apo-CaM and Ca^{2+}-CaM with high affinity. In addition, they also affect the probability of channel opening in reconstituted RyR. Furthermore, RyRs are substrates of different protein kinases. The cardiac isoform RyR2 is phosphorylated by protein kinase A (PKA), a mechanism which has been discussed to play a critical role in the etiology of exercise-induced arrhythmias. In pathological situations, stress-induced hyper-phosphorylation of specific serine residues causes dissociation of calstabin 2 from the RyR2 complex, causing an increased, nonphysiological open probability of the channel (Wehrens et al., 2005a, 2006). Although the mechanistic details of this process are still controversial, several studies have substantiated, for example, the phosphorylation of RyR2 by PKA (Jones et al., 2008; Xiao et al., 2006). In addition, RyR2 is phosphorylated by Ca^{2+}/CaM-dependent protein kinase II (CaMKII) in response to fluctuations of the intracellular calcium concentrations. In some cases, the phosphorylation sites used by different protein kinases are overlapping: Ser^{2808}, for example, is phosphorylated by both PKA and CaMKII. The phosphorylation of other residues, such as Ser^{2030} and Ser^{2814}, is catalyzed by either PKA or CaMKII (Huke and Bers, 2008; Kockskaemper and Pieske, 2006; Lanner et al., 2010; Meng et al., 2007).

It has been known for some time that RyR activity is sensitive to changes of the cellular redox potential. Redox-sensitive thiols have been described in both skeletal and cardiac RyRs (Aracena-Parks et al., 2006; Gonzalez et al., 2010; Voss et al., 2004). As such, reactive thiols are potential sites for S-nitrosylation, a prototypic redox-dependent protein modification. Reversible reaction with nitric oxide in relation to the tissue oxygen tension is well documented for RyR channels (Ho et al., 2011; Niggli, 2011; Sun et al., 2008; Zissimopoulos and Lai, 2006).

The importance of intramolecular domain interactions for controlling channel gating has been implicated from studies of channel disease mutations

(Bannister et al., 2007). The RyR inhibitor dantrolene has been shown to stabilize these domain interactions thereby preventing activation (Kobayashi et al., 2004). In addition, synthetic peptide probes have been successfully applied to study subdomain interactions directly (Hamada et al., 2007). Based on these data, the proposed domain-switch concept postulates that reversible association and dissociation of N-terminal and central domains regulate channel activation (Liu et al., 2010b; Tatishi et al., 2009).

To summarize, ryanodine-sensitive calcium-release channels are the key components of supramolecular complexes functioning as calcium-signaling units. In skeletal muscles, RyRs are physically coupled to Ca$_V$. In contrast, calcium entry mediated by Ca$_V$ induces RyR opening in cardiac muscles (CICR). The interaction of RyRs with accessory proteins including CaM, calstabins, and triadins is essential both for structural integrity and function. The suggested mechanism of channel gating involves long-distance conformational changes and intramolecular protein domain interactions.

3.6.2 Modulation of RyR activity

Intracellular calcium transients need to be tightly controlled. As a self-reinforcing process, CICR requires counterregulatory mechanisms that lead to RyR channel closing. Finally, the efficient reuptake into the ER/SR stores terminates the intracellular calcium transients.

Genetic diseases are known that affect RyR function. Consequently, drugs able to restore the physiological channel regulation may be used in the treatment of RyR-associated channelopathies (Dulhunty et al., 2007; George and Lai, 2007; Meli et al., 2011). As an example, RyR2 has been implicated in the pathogenesis of arrhythmias and sudden cardiac death due to increased calcium leakiness of the channel (Lehnart et al., 2006). Compounds that prevent calcium leakage have been suggested for treatment of such diseases, although their precise mechanism of action is controversial (Hunt et al., 2007; Ye et al., 2012). In RyR1, mutations are implicated in genetically manifested myopathies such as malignant hyperthermia (MH) and central core disease (CCD). These diseases may be treated with appropriate channel-modulating drugs (Bannister et al., 2007; Boncompagni et al., 2010; Brini, 2004; Du et al., 2004). Taken together, these examples confirm that RyRs have great potential for drug discovery. Therefore, the pharmacology of RyRs will be discussed briefly.

Calcium is the most important physiological effector of RyR channel activity. Calcium-dependent RyR activation shows a characteristic bell-shaped function of the cytosolic Ca^{2+} concentration. At low micromolar concentrations, Ca^{2+} opens the channel following binding to high-affinity binding sites. At higher concentrations (~ 1.0 mM), the channel is inactivated after calcium binding to a second type of low-affinity binding sites (Meissner et al., 1997). In addition, the regulation by luminal Ca^{2+} has been reported. However, the relative importance of both cytoplasmic calcium-binding sites and sites accessible only from inside the SR are not clearly defined yet (Capes et al., 2011; Laver, 2007, 2009). Magnesium ions effectively inhibit RyR1 by reducing the open probability of the channel (Lanner et al., 2010; Voss et al., 2008). The inhibition can be reversed by high concentrations of calcium, indicating that both divalent cations bind to the same high-affinity and low-affinity binding sites. Generally, the cardiac RyR2 seems to be less sensitive to Mg^{2+} inhibition than the skeletal RyR1. ATP at low millimolar concentrations augments the activation of both RyR1 and RyR2 channels by calcium. In RyR1, activation by ATP has been described even at resting intracellular Ca^{2+} concentrations. RyR2 appears to be less sensitive to ATP (Laver et al., 2001; Meissner, 1984). A common binding site has been suggested for ATP, and other nucleotides such as adenosine diphosphate (ADP) and inositol monophosphate (IMP).

Cyclic ADP-ribose (cADPR) was originally identified in sea urchin eggs as a calcium-mobilizing second messenger acting on ER calcium stores. It is synthesized from nicotinamide adenine dinucleotide (NAD) by specific ADP-ribosyl cyclases (Guse, 1999, 2002). It has been shown that CICR is enhanced by nanomolar concentrations of cADPR suggesting a coactivator role for the cyclic nucleotide (Lee, 1993). Activity of cADPR on reconstituted RyRs has provided strong evidence that cADPR directly interacts with RyR (Sitsapesan and Williams, 1995; Sitsapesan et al., 1994). However, early labeling experiments with radioactive cADPR have failed to unequivocally identify RyR as the cADPR receptor (Thomas et al., 2001). Nevertheless, it is becoming more generally accepted that cyclic ADP-ribose activates RyR directly, although the mechanism remains to be clarified (Ogunbayo et al., 2011).

The alkaloid ryanodine (Fig. 5.8A), isolated from the tropical American shrub *Ryania speciosa* (Flacourtiaceae), has been known for a long time to affect muscle activity by interfering with calcium signaling (Jenden and Fairhurst, 1969). Ryanodine was subsequently demonstrated to bind to the SR/ER calcium-release channel with high affinity, which was thereafter named RyR. Ryanodine is an important pharmacological tool to study of

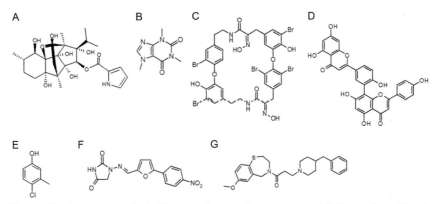

Figure 5.8 Pharmacological effectors of ryanodine receptors. (A) Ryanodine, (B) caffeine, (C) bastadin-5, (D) amentoflavone, (E) 4-Chloro-m-cresol, (F) dantrolene, and (G) JTV-519.

RyRs. Structural features essential for high-affinity binding were established with ryanoid analogs (Waterhouse et al., 1987; Welch et al., 1997). The concentration-dependent receptor modulation by ryanodine appears to be biphasic: at submicromolar and low micromolar concentrations, characteristic subconductance states with high open probabilities are induced (Du et al., 2001b; Pessah and Zimanyi, 1991). At higher micromolar concentrations, ryanodine behaves like a channel blocker. Consequently, high as well as low-affinity ryanodine binding sites have been postulated (Callaway et al., 1994; Pessah and Zimanyi, 1991). Mutagenesis experiments support the hypothesis that the ryanodine binds to the transmembrane part, probably in a large central cavity of the calcium channel close to the pore loop and the selectivity filter. Therefore, it is conceivable that the mode of action of ryanodine differs from that of classical ion-channel blockers as it does not simply plug the ion channel, but instead perturbs conformational movements involved in channel gating (Chen et al., 2002). In agreement with this suggestion, earlier biochemical studies showed that the isolated carboxy-terminal transmembrane part of RyR was sufficient to constitute functional channels which were sensitive to ryanodine modulation (Bhat et al., 1997). Furthermore, photoaffinity labeling with an azido-derivative of ryanodine followed by tryptic digestion revealed a 76-kDa proteolytic fragment assigned to the C-terminal part by immunochemical methods (Witcher et al., 1994). Since ryanodine binds preferentially to the activated calcium-conducting channel conformation(s), radiolabeled ryanodine is used as a conformation-sensitive, functional probe to experimentally assess the effects of mutations and the activity of pharmacological effectors on RyR activity (Herrmann-Frank et al., 1996; Lindsay and Williams, 1991; Meissner and el-Hashem, 1992; Suzuki et al., 1999).

The methylxanthine compound caffeine (Fig. 5.8B) is a well-established activator of RyR-mediated calcium release at millimolar concentrations (Pessah et al., 1987; Sitsapesan and Williams, 1990). It increases the frequency of channel openings cooperatively with calcium and ATP (Rousseau et al., 1988). As caffeine acts on all RyR subtypes and isoforms, it has become a widely adopted tool to test RyR functionality after recombinant expression and reconstitution.

Bastadins (Fig. 5.8C), which are macrocyclic natural products structurally derived from bromotyrosine, are produced by marine sponges (Mack et al., 1994). Although bastadin activity was originally identified in relation to subtype RyR1, recent evidence suggests that also RyR2 are targeted by synthetic bastadin derivatives (Zieminska et al., 2007). It has been proposed that bastadins stabilize the open-channel conformation by probably affecting the RyR–FKBP12 (calstabin 1) interaction. In addition, more complex mechanisms of action for bastadins have been discussed (Masuno et al., 2006). The flavonoid amentoflavone (Fig. 5.8D) elicits increased calcium release from the SR/ER by activating RyR similar to the methylxanthines (Suzuki et al., 1999). Binding of amentoflavone also to other ion channels has been described, as for example, its interaction with the benzodiazepine site of GABA-gated chloride channels (Hansen et al., 2005a).

Compounds including m-cresols and catechols have been reported to stimulate intracellular calcium release via RyRs (Herrmann-Frank et al., 1996; Lacava et al., 2012). 4-Chloro-m-cresol (4-CmC, Fig. 5.8E) is a relatively potent activator of RyRs. Structure–activity relationship data suggest the importance of the free 1-hydroxyl group as well as hydrophobic interactions for effector efficacy. Ther is no information on the binding site(s) of these aromatic RyR effectors (Jacobson et al., 2006).

Dantrolene (Fig. 5.8F) is the best studied RyR inhibitor. It is used as a muscle-relaxant in the treatment of MH. The skeletal RyR1 isoform and RyR3 are inhibited by micromolar dantrolene concentrations in an ATP-dependent fashion. However, the cardiac subtype seems to be unaffected (Kobayashi et al., 2009; Zhao et al., 2001). The dantrolene-binding peptide sequence (domain peptide 1 in the amino-terminal region of RyR1) is part of a region implicated in domain–domain interactions that presumably controls channel gating and activity (Kobayashi et al., 2004; Paul-Pletzer et al., 2002).

Medicinal chemistry research addressing RyR dysfunction in cardiovascular diseases led to the discovery of K201 (JTV-519, Fig. 5.8G), a tetrahydro-benzothiazepine suggested to prevent calcium leakiness of cardiac RyR2 channels (Blayney et al., 2010; Thomas and Williams, 2012; Wehrens et al., 2005b). Furthermore, RyRs are sensitive to general ion channel blockers such as ruthenium red, however, at fairly high concentrations (Anderson et al., 1989; DiJulio et al., 1997; Percival et al., 1994).

3.6.3 Insect RyRs
The role of RyRs in EC coupling seems to be similar in mammalian and insect muscles. Insect Ca_V and RyR are colocalized in contact areas of plasma membranes and the ER/SR (Takekura and Franzini-Armstrong, 2002). Electron-microscopy of flight muscles from flies and dragonflies revealed that although Ca_V are clustered close to the RyRs, they do not constitute ordered structures (tetrads) as in mammalian skeletal muscles by direct physical interaction Therefore, it has been concluded that the absence of tetrads in insect body muscles favors the mechanism of CICR similar to the situation in mammalian cardiac muscles (*vide supra*). Furthermore, it can be speculated that indirect EC coupling represents the more ancient mechanism of CICR, and that the physical Ca_V–RyR interaction appeared later in evolution after the divergence of vertebrates.

Radioligand binding data reported from John Casida´s laboratory revealed for the first time the pharmacology of high-affinity ryanodine binding sites in muscle and nervous tissues from different insect species including houseflies (*M. domestica*), cockroaches (*P. americana*), crickets (*Teleogryllus commodus*), and the corn earworm *Helicoverpa zea* (Lehmberg and Casida, 1994). The pharmacological characteristics of insect ryanodine binding sites exhibit clear similarities with the established mammalian pharmacology: First, the apparent dissociation constants (K_D) and receptor densities (B_{max}) are in the same range. Furthermore, ryanodine affinity is dependent on calcium and ATP. The role of calcium for ryanodine binding was further substantiated by studies with cockroach femoral and thoracic muscle preparations. The characteristic optimum curves showed a maximum at 100–500 µM calcium. Like mammalian RyR, ryanodine binding decreased at millimolar calcium concentrations (Schmitt et al., 1996, 1997).

Biochemical and cell-physiological studies with *Drosophila* embryos and adults identified the major players that control intracellular calcium homeostasis and signaling in the ER/SR including RyR, the IP_3R, and the SERCA (Vazquez-Martinez et al., 2003). High-affinity binding of ryanodine was stimulated by the nonhydrolyzable ATP analog adenylylmethylenediphosphonate (AMP-PCP), by 3-methylxanthine, and caffeine. As observed in earlier studies, millimolar calcium concentration reduced ryanodine binding, however, not completely. Furthermore, ryanodine binding was sensitive to inhibition by ruthenium red and dantrolene.

While mammalian genomes comprise three RyR genes, insects have only one RyR isoform. The *Drosophila* RyR gene codes for a protein of 5216 amino acid residues which exhibits only 45–47% sequence identity to the three mammalian genes (Takeshima et al., 1994). It has been localized to position 44F on chromosome 2 by *in situ* hybridization encompassing 25.7 kb of genomic sequence. The C-terminal region with the putative transmembrane segments is more conserved, whereas regions of the cytoplasmic part of the protein have diverged during evolution. Nevertheless, all RyR isoforms have likely evolved from a common ancestor (Kushnir et al., 2005).

Following successful functional expression, insect RyRs are becoming subjects of detailed molecular analysis. Expression in Chinese hamster ovary (CHO) cells revealed the biophysical characteristics of RyR-44F (Xu et al., 2000). The full-length channel exhibits fast gating kinetics and frequent occurrence of spontaneous subconductance states. Interestingly, a truncated RyR variant, which consists mainly of the C-terminal transmembrane part, constituted functional, ryanodine-sensitive channels. However, they were insensitive to inactivation by millimolar concentrations of calcium. It should be noted that the low level of channel activity yielded by the transient expression prevented a detailed biophysical analysis. This, however, is unlike mammalian channels which were functionally expressed in stably transformed cell lines (Pan et al., 2000).

P-element-mediated insertion mutagenesis showed that the *Drosophila* RyR-44F-coding gene is essential for viability and development (Sullivan et al., 2000). Interestingly, ubiquitous expression was found in muscles of the body wall, visceral muscles around the gut, and in the central nervous system. The contraction of the dorsal vessel driving the hemolymph circulation system is drastically impaired in *ryr* mutants. Similar symptoms in the mutant and in the wild-type larvae after ryanodine treatment were observed.

The first RyR sequence information of an agronomically important insect species was reported for *H. virescens* (Puente et al., 2000). Not surprisingly, the partial sequence including the carboxy-terminal 1172 amino acid residues is highly homologous (74% identity) to the *Drosophila* RyR-44F sequence, but only about 48% identical to the mammalian isoforms. Hydropathy plots predict six transmembrane helices that may form the channel pore. The proposed pore loop between helices M1 and M2 contains the consensus selectivity filter motif GGGIGD. Earlier, the *Heliothis* RyR was characterized pharmacologically in SR preparations from native flight muscles (Scott-Ward et al., 2001). Ryanodine binding was calcium dependent and cooperatively stimulated by caffeine. Unfortunately, single-channel analysis of SR membranes fused to planar lipid bilayers was impeded, probably by the low rate of successful fusion events or by insufficient stability of the membrane fusions. However, the characteristic biphasic regulation by calcium, and the occurrence of subconductance states induced by low ryanodine concentrations confirmed the radioligand binding results.

Several reports describing the recombinant expression of insect RyRs have been published recently (Cordova et al., 2006; Ebbinghaus-Kintscher et al., 2006; Kato et al., 2009). The major obstacles for successful functional expression are most likely associated with the size of the cDNAs ($\sim$15–16 kb), the need for proper folding and transmembrane localization, and the essential interactions with accessory proteins. Cordova et al. (2006) achieved functional expression of the *Drosophila ryr* cDNA in the fall armyworm *Spodoptera frugiperda* Sf9 cells in context of the mode of action elucidation of the new anthranilamide insecticides. Intracellular calcium transients were measured with the calcium-sensitive fluorescent dye Fura-2. Ryanodine (3 μM) had no effect on the basal calcium concentration, but caffeine-induced calcium transients could not be fully reverted to basal levels in the presence of ryanodine. A Sf9 clone stably expressing the *Drosophila* RyR was isolated under continuous selection pressure. In a subsequent report from the same group (Cordova et al., 2007), functional expression of *H. virescens* RyR in Sf9 cells was successfully demonstrated. Following a similar approach, Ebbinghaus-Kintscher et al. (2006) transiently expressed the *Drosophila* RyR in mammalian CHO cells. Intracellular calcium transients were evoked by caffeine that were similar to those measured in isolated *H. virescens* neurons. Calcium signals in both systems were sensitive to inhibition by high concentrations of ryanodine.

Recombinant expression of the full-length *B. mori* RyR-cDNA in human embryonic kidney (HEK293) cells has been shown recently by Y. Mori´s group working on the molecular mode of action of the phthalic acid diamide insecticide, flubendiamide (Kato et al., 2009). Although the cellular expression system lacks accessory proteins such as triadin, calsequestrin, and L-type calcium channels that potentially influence RyR expression and activity, functional RyR channels were formed. Recombinant silkworm RyRs were activated by caffeine and were sensitive to ryanodine. Finally, two recent publications describe the cloning and characterization of other lepidopteran RyR-coding genes. Wang et al. (2012b) reported the sequence of the full-length RyR-cDNA of the diamondback moth *Plutella xylostella*. It revealed six putative transmembrane helices localized in the carboxy-terminal region of the protein. The proposed pore loop including the characteristic GGGIGD motif is located between the C-terminal helices 5 and 6. Two EF-hand calcium-binding motifs were identified next to the transmembrane domain. As expected, the overall sequence identity is highest with insect RyRs ranging between 78% and 80%. As observed in other insect RyRs, homology to vertebrate RyR isoforms in generally only about 45%.

Expression of the *Plutella* RyR-mRNA in different body parts during development revealed deletion polymorphism that may result from alternative splicing. Functional expression and characterization of the *Plutella* RyR have not been shown yet. Alternative splicing of insect RyR-mRNA potentially generates receptor isoforms that can account for functional diversity in different tissues or in different developmental stages. This was shown for the first time with *Drosophila* RyR-44F (Takeshima et al., 1994) and has been confirmed in a recent publication on cloning and sequence analysis of the rice leaf-folder (*Cnaphalocrocis medinalis*) *ryr* gene (Wang et al., 2012a). Two putative splicing sites identified in this study are conserved also in *ryr* genes from other insect species. Remarkably, one of the splicing sites is located in the so-called SPRY2 site, which has been implicated in intramolecular protein–protein interaction. It can be speculated that changes in this region may alter the functional properties of the ion channels.

3.6.4 Insect RyRs as insecticide target sites
3.6.4.1 Ryanodine and ryanoids
For more than 50 years, extracts from roots and stems of *R. speciosa* had been used in the past as botanical insecticides (Ryania). The active ingredient was subsequently identified as an alkaloid termed ryanodine (Rogers et al., 1948). Later, ryanodine effects on insect muscle excitability, calcium mobilization, and the high-affinity binding of ryanodine to insect membrane preparations were discovered. Consequently, RyRs have been considered for a long time as an attractive molecular target for insecticides (Bloomquist, 1996; Lehmberg and Casida, 1994; Pessah, 1990; Usherwood, 1962; Waterhouse et al., 1987). A major drawback of ryanodine to be applied a pesticide is its

vertebrate toxicity. Consequently, approaches to find insect–selective ryanodine analogs have been followed, but with limited success. For example, ryanodol-type derivatives are less active on mammalian skeletal muscles. However, they have been shown to alter the ion permeability of insect potassium channels rather than affecting calcium homeostasis through RyRs (Usherwood and Vais, 1995; Vais and Usherwood, 1995). While the structural complexity of ryanodine makes it a largely unsuitable lead structure for pesticide chemistry, biochemical screening efforts have been undertaken to identify novel compounds interacting with the ryanodine binding site. The experimental difficulties to express insect RyRs for screening purposes motivated Dinsmore et al. (1998) to use an antibody retaining much of the specificity of the ryanodine binding site as a surrogate RyR. Binding of radiolabeled ryanodine to the monoclonal antibody was displaced by several ryanodine analogs with nanomolar IC_{50} values. However, despite these efforts, no pesticides addressing the ryanodine site have been developed so far.

3.6.4.2 Cyclodepsipeptides

The cyclic depsipeptide verticilide (Fig. 5.9A) with insecticidal and nematicidal activity was isolated from culture supernatants of the entomopathogenic fungus *Verticillium sp.* strain FKI-1033 (Monma et al., 2006). Verticilide inhibited ryanodine binding to cockroach microsomal membrane preparations with an IC_{50} value of 4.3 μM (Shiomi et al., 2010). Notably, the affinity to mouse RyRs was found to be roughly 10 times weaker. At present, the mechanism by which verticilide modifies invertebrate RyRs is unknown.

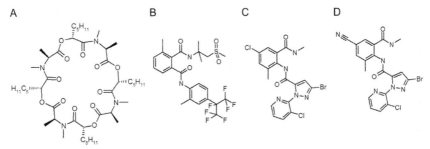

Figure 5.9 Insecticides reported to target insect ryanodine receptors. (A) Verticilide; (B) flubendiamide, 3-iodo-*N'*-(2-mesyl-1,1'-dimethylethyl)-*N*-{4-[1,2,2,2-tetrafluoro-1-(trifluoromethyl)ethyl]-*o*-tolyl}phthalamide; (C) chlorantraniliprole, 3-bromo-*N*-{4-chloro-2-methyl-6-[(methylamino)carbonyl]phenyl}-1-(3-chloropyridin-2-yl)-1H-pyrazole-5-carboxamide; and (D) cyantraniliprole, 3-bromo-1-(3-chloropyridin-2-yl)-*N*-{4-cyano-2-methyl-6-[(methylamino)carbonyl]phenyl}-1H-pyrazole-5-carboxamide.

3.6.4.3 Diamide insecticides

A few years ago, two new groups of chemical insecticides, the phthalic acid diamides and the anthranilamides, were reported to act on insect RyRs (Lahm et al., 2005; Tohnishi et al., 2005), and a summary of their modes of action has recently been reviewed by Sattelle et al. (2008). In 1993, the phthalic acid diamides originated from a herbicide research program at Nihon Nohyaku Co, Ltd (NNC). Flubendiamide (Fig. 5.9B), the first commercial phthalic diamide insecticide which is particularly active against lepidopteran pests, has been developed by NNC in collaboration with Bayer CropScience AG (Hamaguchi et al., 2012).

Lepidopteran larvae treated with flubendiamide exhibited characteristic symptoms of contraction paralysis suggesting that the disruption of insect muscle activity constituted the physiological basis for the insecticidal activity. Physiological investigations using single-cell calcium fluorescence measurements revealed that phthalic acid diamides elicited intracellular calcium transients in isolated *H. virescens* neurons in a concentration-dependent manner (Ebbinghaus-Kintscher et al., 2006). These calcium signals were independent of extracellular calcium and could be completely blocked by micromolar concentrations of ryanodine. Taken together, the results strongly suggest that phthalic acid diamides target insect RyRs. Finally, ryanodine-sensitive calcium transients evoked by flubendiamide and other phthalic diamides in CHO cells transformed with the *Drosophila* RyR-cDNA unequivocally proved that the new insecticides activated insect RyRs.

It has been shown before that the binding affinity of ryanodine to insect ER/SR membranes is regulated in a characteristic biphasic fashion by the calcium concentration (Lehmberg and Casida, 1994; Scott-Ward et al., 2001). Interestingly, the phthalic acid diamide compounds allosterically increased the ryanodine affinity at low calcium concentrations in ER/SR preparations obtained from *H. virescens* flight muscles. In addition, the decrease of ryanodine binding at millimolar calcium was prevented by nanomolar concentrations. Since ryanodine preferentially binds to the activated calcium-conducting conformational state, these results suggest that phthalic acid diamides seem, firstly, to sensitize the channel to calcium and, secondly, to stabilize an open-channel conformation leading to increased calcium flux. This probably results in emptying the intracellular calcium stores and in sustaining the increased intracellular calcium levels (Ebbinghaus-Kintscher et al., 2007). Interestingly, the activity of the SERCA was shown to be stimulated in the presence of phthalic acid diamides. This was interpreted as a compensatory indirect mechanism to counterbalance the permanent activation of RyRs (Masaki et al., 2007). Importantly, phthalic acid diamides were found to be almost ineffective on mammalian RyR subtypes expressed, for example, in neurons from the rat dorsal root ganglion, in mouse muscle cells (C2C12), and in rat heart cells (H9C2). Therefore, it is conceivable that the phthalic acid diamide binding site and the mechanism of RyR activation are insect specific and may explain the favorable toxicological profile of flubendiamide (Ebbinghaus-Kintscher et al., 2006, 2007; Masaki, 2008).

Radiolabeled flubendiamide binds to *H. virescens* SR membranes with high affinity (Lümmen et al., 2007). The apparent dissociation constant (K_D) of flubendiamide was determined as 3.4 nM. Remarkably, the high-affinity binding was dependent on calcium. The characteristic hyperbolic binding isotherm was observed at similar Ca^{2+} concentrations that were found to promote maximal ryanodine affinity. Lower calcium concentrations resulted in sigmoidal binding curves characteristic of a cooperative behavior between the binding sites. Complete removal of calcium prevented flubendiamide binding almost entirely.

In summary, phthalic acid diamides appear to bind with high affinity to the open channel induced by calcium. Due to the low nanomolar equilibrium dissociation constant of the receptor–ligand complex, the ion-conducting conformation is effectively stabilized (Fig. 5.10). Consequently, it can be predicted that RyR effectors which activate the channel should also increase its affinity for flubendiamide. Preliminary data confirm that low ryanodine concentrations ($\leq 1.0\ \mu M$) and millimolar concentrations of methylxanthines increase flubendiamide affinity at suboptimal calcium concentrations (Lümmen, P., unpublished). It should be mentioned that none of the established pharmacological RyR effectors such as ryanodine, cyclic ADP-ribose, methylxanthines, and dantrolene interfered with flubendiamide binding suggesting a new binding site for the phthalic diamide insecticides (Lümmen et al., 2007). Furthermore, Kato et al. (2009) carried out systematic deletion experiments on the carboxy-terminal region of the *B. mori* RyR, showing that the binding region of flubendiamide is located in the transmembrane part of the RyR channel comprising amino acid residues 4111–5084. In the same report, the RyR transmembrane region was specifically labeled with a photoreactive trifunctional flubendiamide derivative.

In summary, the results on the molecular mode of action of phthalic acid diamide insecticides indicate that the compounds interact with a new binding site likely to be located in the transmembrane region of insect RyRs. The site is distinct from, but allosterically coupled to, the ryanodine site

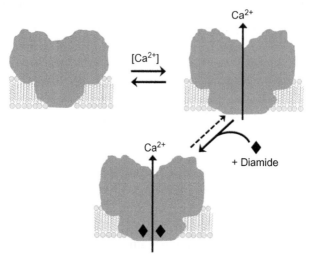

Figure 5.10 Hypothesis to explain the activity of diamide insecticides on insect ryanodine receptors. The equilibrium between calcium-conducting (open) and closed channels conformations is controlled by the calcium concentration. Diamide insecticides preferentially bind to the open channel with high affinity. Consequently, they shift the equilibrium to the calcium-conducting conformation(s) even at high calcium concentrations.

which is localized in the pore. Apparently, binding of phthalic diamides stabilizes the activated calcium-conduction channel conformation, thus causing elevation of intracellular calcium levels and depletion of calcium stores. Plausibly, these effects can explain the characteristic symptoms of muscle contraction and paralysis observed in treated lepidopteran larvae.

A second class of insecticides acting on insect RyRs was first discovered at DuPont in 1999. Soon thereafter, chlorantraniliprole (Fig. 5.9C) with excellent insecticidal activity on lepidopteran pests was developed and has recently been introduced into the market. Efforts to widen the biological target spectrum by adapting the physico-chemical properties of the compounds led to the discovery of the second commercial anthranilamide, cyantraniliprole (Fig. 5.9D). This compound controls, in addition to lepidopteran species, also other important insect pests from the Hemiptera, Thysanoptera, and Coleoptera orders (Lahm et al., 2005; Selby et al., 2008).

Physiological studies on cardiac muscles of the tobacco hornworm *Manduca sexta* revealed that anthranilamides affected EC coupling (Cordova et al., 2007). As application of anthranilamides on isolated *P. americana* neurons elicited intracellular calcium transients that were independent of calcium in the extracellular medium, it was concluded that calcium was rather released from intracellular stores. This explanation was further supported by treating the cells with the SERCA inhibitor CPA showing the importance of intracellular calcium stores. Application of ryanodine (1.0 μM) prolonged anthranilamide-induced calcium transients and inhibited further calcium mobilization by caffeine or anthranilamide. Measurements of anthranilamide-induced calcium transients in *S. frugiperda* Sf9 cells expressing the *Drosophila ryr* gene finally proved that the RyR is the anthranilamide target. In comparison with the short-lived caffeine-induced calcium signals, the calcium responses to anthranilamides were more sustained (Caspar et al., 2003; Cordova et al., 2006).

Radioligand binding experiments with ryanodine showed that an anthranilamide compound binds to a homogenous population of binding sites with nanomolar affinity (Cordova et al., 2007). The fact that anthranilamides did not inhibit ryanodine binding indicated that anthranilamides interact with a separate binding site (Cordova et al., 2007). However, athranilamide binding was enhanced by ryanodine. This could imply that although the binding sites are distinct, they are coupled allosterically. The anthranilamides may bind preferentially to a (partially) activated channel conformation induced by ryanodine. Similar to the phthalic acid diamides, the anthranilamides appear to be insect-specific RyR modulators: chlorantraniliprole was found to be >300-fold more potent on insect RyRs than on mammalian RyR1-type channels. In comparison with RyR2, selectivity factors were even higher (>2000-fold) which likely accounts for the highly favorable toxicological properties of the anthranilamides (Cordova et al., 2007; Sattelle et al., 2008).

It is tempting to speculate that the two classes of diamide insecticides share the same, or at least overlapping binding sites on insect RyRs, although the compounds are structurally different (Hamaguchi et al., 2012). Recently, radioligand displacement experiments using membrane preparations of housefly flight muscles suggest that chlorantraniliprole, flubendiamide, and ryanodine bind to distinct, but allosterically coupled sites (Isaacs et al., 2012). This is in agreement with earlier data from Kato et al. (2009) showing calcium mobilization mediated by the truncated *B. mori* RyR that the transmembrane region is critical for activity of both the experimental anthranilamide DP-23 (Lahm et al., 2005) and flubendiamide.

A recent publication describes field-collected strains of the diamondback moth (*Plutella xylostella*) that are highly resistant to diamide insecticides (Troczka et al., 2012). Interestingly, a point mutation G4946E located in the highly conserved connecting loop between the putative transmembrane helices 2 and 3 of the RyR has been identified in the resistant strains. Although it still has to be proven that the mutated RyR confers reduced sensitivity to diamides, it tempting to speculate that the mutation designates a region which is involved in diamide binding.

In summary, insect RyRs are specifically modulated by two new chemical insecticide families by a different mechanism than the established botanical insecticide ryanodine. Both phthalic acid diamides and anthranilamides evoke the rapid and sustained release of calcium from intracellular stores, probably by stabilizing a (partially) activated, calcium-conducting channel conformation.

It is noteworthy that in contrast to the wealth of structural and functional information that has been published on mammalian RyRs, insect RyRs appear to be an almost neglected field of research. The situation may hopefully change as the diamide insecticides offer new tools for fundamental research addressing insect RyR structure and function.

4. CONCLUDING REMARKS

Calcium plays a central role as a second messenger in controlling important physiological processes such as neurotransmission and muscle contraction. Consequently, calcium channels and pumps have been proposed as novel insecticide target sites.

Recent years have witnessed the introduction of two new classes of chemical insecticides, the diamides, which modulate the activity of insect ryanodine-sensitive ER/SR calcium-release channels by a mechanism which is distinct from that of established effectors including the plant alkaloid ryanodine. Since they are much less effective on mammalian RyR subtypes, the diamides represent an important step in the direction of developing safer pesticides.

Potentially, Ca_V may also constitute promising target sites for future pest control agents. Initial physiological and pharmacological data and the discovery of insect-specific venom peptides suggest that pharmacological differences exist between mammalian and insect Ca_V, which could be exploited for the development of pesticides with favorable toxicological properties. However, it should be mentioned that many problems in understanding the structure and function of insect calcium channel complexes still need to be resolved. Finally, the complexity of these proteins poses considerable experimental difficulties, like, for example, the functional expression of calcium channels in suitable cellular systems.

The existing commercial insect control solutions rely on a limited number of modes of action that are constantly compromised by resistance development in pest populations in the field. Furthermore, there is a permanent demand for toxicologically safer pesticides driving also the search for new modes of action. It has been demonstrated that calcium-signaling pathways comprise promising new pesticide targets to meet the requirements of future pest control.

ACKNOWLEDGMENTS

The author is indebted to Dr. Ulrich Ebbinghaus-Kintscher, Bayer CropScience AG, for helpful discussions on the mode of action of diamide insecticides and for general insights into the biophysical mechanisms of ion channels. Furthermore, the very valuable suggestions from Professor Ephraim Cohen, the Hebrew University of Jerusalem, are gratefully acknowledged. The excellent technical assistance by Claudia Wehr and Gabi Lachner in the author's work on RyRs is highly appreciated.

REFERENCES

Altafaj, X., Cheng, W., Esteve, E., Urbani, J., Grunwald, D., Sabatier, J.M., Coronado, R., De We, M., Ronjat, M., 2005. Maurocalcine and domain A of the II-III loop of the dihydropyridine receptor Cav 1.1 subunit share common binding sites on the skeletal ryanodine receptor. J. Biol. Chem. 280, 4013–4016.

Anderson, K., Lai, F.A., Liu, Q.Y., Rousseau, E., Erickson, H.P., Meissner, G., 1989. Structural and functional characterization of the purified cardiac ryanodine receptor-Ca^{2+} release channel complex. J. Biol. Chem. 264, 1329–1335.

Andersson, A., Baell, J.B., Duggan, P.J., Graham, J.E., Lewis, R.J., Lumsden, N.G., Tranberg, C.E., Tuck, K.L., Yang, A., 2009. Omega-conotoxin GVIA mimetics based on an anthranilamide core: effect of variation in ammonium side chain lengths and incorporation of fluorine. Bioorg. Med. Chem. 17, 6659–6670.

Aracena-Parks, P., Goonasekera, S.A., Gilman, C.P., Dirksen, R.T., Hidalgo, C., Hamilton, S.L., 2006. Identification of cysteines involved in S-nitrosylation, S-glutathionylation, and oxidation to disulfides in ryanodine receptor type 1. J. Biol. Chem. 281, 40354–40368.

Arnon, A., Cook, B.J., Montell, C., Selinger, Z., Minke, B., 1997. Calmodulin regulation of calcium stores in phototransduction of *Drosophila*. Science 275, 1119–1121.

Baell, J.B., Duggan, P.J., Forsyth, S.A., Lewis, R.J., Lok, Y.P., Schroeder, C.I., Shepherd, N.E., 2006. Synthesis and biological evaluation of anthranilamide-based non-peptide mimetics of ω-conotoxin GVIA. Tetrahedron 62, 7284–7292.

Bai, J., Binari, R., Ni, J.-Q., Vijayakanthan, M., Li, H.-S., Perrimon, N., 2008. RNA interference screening in *Drosophila* primary cells for genes involved in muscle assembly and maintenance. Development 135, 1439–1449.

Balshaw, D., Gao, L., Meissner, G., 1999. Luminal loop of the ryanodine receptor: a pore-forming segment? Proc. Natl. Acad. Sci. U.S.A. 96, 3345–3347.

Banerjee, S., Lee, J., Venkatesh, K., Wu, C.F., Hasan, G., 2004. Loss of flight and associated neuronal rhythmicity in inositol 1,4,5-trisphosphate receptor mutants of *Drosophila*. J. Neurosci. 24, 7869–7878.

Bannister, M.L., Hamada, T., Murayama, T., Harvey, P.J., Casarotto, M.G., Dulhunty, A.F., Ikemoto, N., 2007. Malignant hyperthermia mutation sites in the Leu2442-Pro2477

(DP4) region of RyR1 (ryanodine receptor 1) are clustered in a structurally and functionally definable area. Biochem. J. 401, 333–339.

Bass, C., Field, L.M., 2011. Gene amplification and insecticide resistance. Pest Manag. Sci. 67, 886–890.

Benquet, P., Guen, J.L., Dayanithi, G., Pichon, Y., Tiaho, F., 1999. Omega-AgaIVA-sensitive (P/Q-type) and -resistant (R-type) high-voltage-activated Ba^{2+} currents in embryonic cockroach brain neurons. J. Neurophysiol. 82, 2284–2293.

Bhat, M.B., Zhao, J., Takeshima, H., Ma, J., 1997. Functional calcium release channel formed by the carboxyl-terminal portion of ryanodine receptor. Biophys. J. 73, 1329–1336.

Blayney, L.M., Jones, J.L., Griffiths, J., Lai, F.A., 2010. A mechanism of ryanodine receptor modulation by FKBP12/12.6, protein kinase A, and K201. Cardiovasc. Res. 85, 68–78.

Bloomquist, J.R., 1996. Ion channels as targets for insecticides. Annu. Rev. Entomol. 41, 163–190.

Boncompagni, S., Loy, R.E., Dirksen, R.T., Franzini-Armstrong, C., 2010. The I4895T mutation in the type 1 ryanodine receptor induces fiber-type specific alterations in skeletal muscle that mimic premature aging. Aging Cell 9, 958–970.

Bootman, M.D., Collins, T.J., Peppiatt, C.M., Prothero, L.S., MacKenzie, L., De Smet, P., Travers, M., Tovey, S.C., Seo, J.T., Berridge, M.J., Ciccolini, F., Lipp, P., 2001. Calcium signalling—an overview. Semin. Cell Dev. Biol. 12, 3–10.

Branton, W.D., Kolton, L., Jan, Y.N., Jan, L.Y., 1987. Neurotoxins from *Plectreurys* spider venom are potent presynaptic blockers in *Drosophila*. J. Neurosci. 7, 4195–4200.

Brini, M., 2004. Ryanodine receptor defects in muscle genetic diseases. Biochem. Biophys. Res. Commun. 322, 1245–1255.

Brini, M., Carafoli, E., 2011. The plasma membrane Ca^{2+} ATPase and the plasma membrane sodium calcium exchanger cooperate in the regulation of cell calcium. Cold Spring Harb. Perspect. Biol. 3, 1–15.

Callaway, C., Seryshev, A., Wang, J.P., Slavik, K.J., Needleman, D.H., Cantu III, C., Wu, Y., Jayaraman, T., Marks, A.R., Hamilton, S.L., 1994. Localization of the high and low affinity [3H]ryanodine binding sites on the skeletal muscle Ca^{2+} release channel. J. Biol. Chem. 269, 15876–15884.

Cao, C.W., Zhang, J., Gao, X.W., Liang, P., Guo, H.L., 2008. Overexpression of carboxylesterase gene associated with organophosphorous insecticide resistance in cotton aphids, *Aphis gossypii* (Glover). Pestic. Biochem. Physiol. 90, 175–180.

Capes, E.M., Loaiza, R., Valdivia, H.H., 2011. Ryanodine receptors. Skeletal Muscle 1, 1–18.

Carafoli, E., Coletto, L., Brini, M., 2004. Plasma membrane calcium pumps. In: Futai, M., Wada, Y., Kaplan, J.H. (Eds.), Handbook of ATPases. Biochemistry, Cell Biology, Pathophysiology. Wiley-VCH, Weinheim.

Carney, J., Mason, S., Viero, C., Williams, A.J., 2010. The ryanodine receptor pore: is there a consensus view? Curr. Top. Membr. 66, 49–67.

Cartwright, E.I.J., Mohamed, T., Oceandy, D., Neyses, L., 2011. Calcium signaling dysfunction in heart disease. Biofactors 37, 175–181.

Casida, J.E., 2009. Pest toxicology: the primary mechanisms of pesticide action. Chem. Res. Toxicol. 22, 609–619.

Caspar, T., Cordova, D., Gutteridge, S., Rauh, J.J., Smith, R.M., Wu, L., Tao, Y., 2003. Isolation and use of ryanodine receptors. WO patent application number 2004027042.

Catterall, W.A., 1991. Excitation-contraction coupling in vertebrate skeletal muscle: a tale of two calcium channels. Cell 64, 871–874.

Catterall, W.A., 1993. Structure and modulation of Na+ and Ca^{2+} channels. Ann. N. Y. Acad. Sci. 707, 1–19.

Catterall, W.A., 2010. Signaling complexes of voltage-gated sodium and calcium channels. Neurosci. Lett. 486, 107–116.

Catterall, W.A., 2011. Voltage-gated calcium channels. Cold Spring Harb. Perspect. Biol. 3, a003947.

Catterall, W.A., Few, A.P., 2008. Calcium channel regulation and presynaptic plasticity. Neuron 59, 882–901.

Catterall, W.A., Yarov-Yarovoy, V., 2010. Helical motion of an S4 voltage sensor revealed by gating pore currents. Channels 4, 75–77.

Charalambous, K., Wallace, B.A., 2011. NaChBac: the long lost sodium channel ancestor. Biochemistry 50, 6742–6752.

Chen, S.R.W., Li, P., Zhao, M., Li, X., Zhang, L., 2002. Role of the proposed pore-forming segment of the Ca^{2+} release channel (ryanodine receptor) in ryanodine interaction. Biophys. J. 82, 2436–2447.

Chen, Y., Li, M., Zhang, Y., He, L., Yamada, Y., Fitzmaurice, A., Shen, Y., Zhang, H., Tong, L., Yang, J., 2004. Structural basis of the a1-b subunit interaction of voltage-gated Ca^{2+} channels. Nature 429, 675–680.

Chong, Y., Hayes, J.L., Sollod, B., Wen, S., Wilson, D.T., Hains, P.G., Hodgson, W.C., Broady, K.W., King, G.F., Nicholson, G.M., 2007. The omega-atracotoxins: selective blockers of insect M-LVA and HVA calcium channels. Biochem. Pharmacol. 74, 623–638.

Clapham, D.E., 1995. Calcium signaling. Cell 80, 259–268.

Colgrave, M.L., Kotze, A.C., Huang, Y.H., Ogrady, J., Simonsen, S.M., Craik, D.J., 2008. Cyclotides: natural, circular plant peptides that possess significant activity against gastrointestinal nematode parasites of sheep. Biochemistry 47, 5581–5589.

Cordova, D., Benner, E.A., Sacher, M.D., Rauh, J.J., Sopa, J.S., Lahm, G.P., Salby, T.P., Stevenson, T.M., Flexner, L., Gutteridge, S., Rhoades, D.F., Wu, L., Smith, R.M., Tao, Y., 2006. Anthranilic diamides: a new class of insecticides with a novel mode of action, ryanodine receptor activation. Pestic. Biochem. Physiol. 84, 196–214.

Cordova, D., Benner, E.A., Sacher, M.D., Rauh, J.J., Sopa, J.S., Lahm, G.P., Selby, T.P., Stevenson, T.M., Flexner, L., Gutteridge, S., Rhoades, D.F., WUu, L., Smith, R.M., Tao, Y., 2007. The novel mode of action of anthranilic diamide insecticides: ryanodine receptor activation. ACS Symp. Ser. 948, 223–234.

Damann, N., Voets, T., Nilius, B., 2008. TRPs in our senses. Curr. Biol. 18, R880–R889.

Davies, S.A., Terhzaz, S., 2009. Organellar calcium signalling mechanisms in *Drosophila* epithelial function. J. Exp. Biol. 212, 387–400.

Desai, S.A., Lnenicka, G.A., 2011. Characterization of postsynaptic Ca^{2+} signals at the *Drosophila* larval NMJ. J. Neurophysiol. 106, 710–721.

Desai-Shah, M., Cooper, R.L., 2009. Different mechanisms of Ca^{2+} regulation that influence synaptic transmission: comparison between crayfish and *Drosophila* neuromuscular junctions. Synapse 63, 1100–1121.

Desai-Shah, M., Papoy, A.R., Ward, M., Cooper, R.L., 2010. Roles of the sarcoplasmic/endoplasmic reticulum Ca^{2+}-ATPase, plasma membrane Ca^{2+}-ATPase and Na^{+}/Ca^{2+} exchanger in regulation of heart rate in larval *Drosophila*. Open Physiol. J. 3, 16–36.

Di Leva, F., Domi, T., Fedrizzi, L., Lim, D., Carafoli, E., 2008. The plasma membrane Ca^{2+} ATPase of animal cells: structure, function and regulation. Arch. Biochem. Biophys. 476, 65–74.

Dijulio, D.H., Watson, E.L., Pessah, I.N., Jacobson, K.L., Ott, S.M., Buck, E.D., Singh, J.C., 1997. Ryanodine receptor type III (Ry3R) identification in mouse parotid acini. Properties and modulation of [3H]ryanodine-binding sites. J. Biol. Chem. 272, 15687–15696.

Dinsmore, A.J., Rees-Blanchard, W., Bentley, P., Lewis, T., Kahl, S.D., McPherson, P.S., Mullinnix, M.J., Campbell, K.P., Windass, J.D., Earley, F.G.P., 1998. Characterization of antibody models of the ryanodine receptor for use in high-throughput screening. Pestic. Sci. 54, 345–352.

Dooley, D.J., Lickert, M., Lupp, A., Osswald, H., 1988. Distribution of [125I]ω-conotoxin GVIA and [3H]isradipine binding sites in the central nervous system of rats of different ages. Neurosci. Lett. 93, 318–823.

Down, R.E., Fitches, E.C., Wiles, D.P., Corti, P., Bell, H.A., Gatehouse, J.A., Edwards, J.P., 2006. Insecticidal spider venom toxin fused to snowdrop lectin is toxic to the peach-potato aphid, *Myzus persicae* (Hemiptera: *Aphididae*) and the rice brown planthopper, *Nilaparvata lugens* (Hemiptera: *Delphacidae*). Pest Manag. Sci. 62, 77–85.

Doyle, D.A., Morais, C.J., Pfuetzner, R.A., Kuo, A., Gulbis, J.M., Cohen, S.L., Chait, B.T., MacKinnon, R., 1998. The structure of the potassium channel: molecular basis of K^+ conduction and selectivity. Science 280, 69–77.

Du, Y., Dong, K., 2008. Molecular basis of knockdown resistance to pyrethroid insecticides. In: Liu, N. (Ed.), Recent Advances in Insect Physiology, Toxicology and Molecular Biology. Research Signpost, Kerala, India, pp. 29–40.

Du, G.G., Guo, X., Khanna, V.K., MacLennan, D.H., 2001a. Functional characterization of mutants in the predicted pore region of the rabbit cardiac muscle Ca^{2+} release channel (ryanodine receptor isoform 2). J. Biol. Chem. 276, 31760–31771.

Du, G.G., Guo, X., Khanna, V.K., MacLennan, D.H., 2001b. Ryanodine sensitizes the cardiac Ca^{2+} release channel (ryanodine receptor isoform 2) to Ca^{2+} activation and dissociates as the channel is closed by Ca^{2+} depletion. Proc. Natl. Acad. Sci. U.S.A. 98, 13625–13630.

Du, G.G., Khanna, V.K., Guo, X., MacLennan, D.H., 2004. Central core disease mutations R4892W, I4897T and G4898E in the ryanodine receptor isoform 1 reduce the Ca^{2+} sensitivity and amplitude of Ca^{2+}-dependent Ca^{2+} release. Biochem. J. 382, 557–564.

Dube, K., McDonald, D.G., O'Donnell, M.J., 2000. Calcium transport by isolated anterior and posterior Malpighian tubules of *Drosophila melanogaster*: roles of sequestration and secretion. J. Insect Physiol. 46, 1449–1460.

Dulhunty, A.F., Pouliquin, P., 2003. What we don't know about the structure of ryanodine receptor calcium release channels. Clin. Exp. Pharmacol. Physiol. 30, 713–723.

Dulhunty, A.F., Beard, N.A., Pouliquin, P., Casarotto, M.G., 2007. Agonists and antagonists of the cardiac ryanodine receptor: potential therapeutic agents? Pharmacol. Ther. 113, 247–263.

Ebbinghaus-Kintscher, U., Lümmen, P., Lobitz, N., Schulte, T., Funke, C., Fischer, R., Masaki, T., Yasokawa, N., Tohnishi, M., 2006. Phthalic acid diamides activate ryanodine-sensitive Ca^{2+} release channels in insects. Cell Calcium 39, 21–33.

Ebbinghaus-Kintscher, U., Lümmen, P., Raming, K., Masaki, T., Yasokawa, N., 2007. Flubendiamide, the first insecticide with a novel mode of action on insect ryanodine receptors. Pflanzenschutz-Nachr. Bayer (Engl. Ed.) 60, 117–140.

Enayati, A.A., Ranson, H., Hemingway, J., 2005. Insect glutathione transferases and insecticide resistance. Insect Mol. Biol. 14, 3–8.

Escoubas, P., King, G.F., 2009. Venomics as a drug discovery platform. Expert Rev. Proteomics 6, 221–224.

Esteve, E., Smida-Rezgui, S., Sarkozi, S., Szegedi, C., Regaya, I., Chen, L., Altafaj, X., Rochat, H., Allen, P., Pessah, I.N., Marty, I., Sabatier, J.M., Jona, I., De Waard, M., Ronjat, M., 2003. Critical amino acid residues determine the binding affinity and the Ca^{2+} release efficacy of maurocalcine in skeletal muscle cells. J. Biol. Chem. 278, 37822–37831.

Fahrner, M., Muik, M., Derler, I., Schindl, R., Fritsch, R., Frischauf, I., Romanin, C., 2009. Mechanistic view on domains mediating STIM1–Orai coupling. Immunol. Rev. 231, 99–112.

FAO, 2011. The State of the World's Land and Water Resources for Food and Agriculture (SOLAW)—managing Systems at Risk. Food and Agriculture Organization of the United Nations, Rome and Earthscan, London.

Federspiel, N., Stricker, J., Winberg, M., Tintrup, H., Yu, Y.-T., Zhou, L., Munagala, N., 2006a. Polynucleotides encoding insect plasma membrane Ca^{2+} ATPase. US patent application 20060269938.

Federspiel, N.D., Tintrup, H., Hanel, A., Breach, J.-C., 2006b. The SERCA calcium pump of Heliothis virescens and a cDNA encoding it and use of the ATPase as a target for insecticides. DE Patent application 2004–102004062248.

Feske, S., Gwack, Y., Prakriya, M., Srikanth, S., Puppel, S.H., Tanasa, B., Hogan, P.G., Lewis, R.S., Daly, M., Rao, A., 2006. A mutation in Orai1 causes immune deficiency by abrogating CRAC channel function. Nature 441, 179–185.

Feyereisen, R., 2006. Evolution of insect P450. Biochem. Soc. Trans. 34, 1252–1255.

Feyereisen, R., 2011. Arthropod CYPomes illustrate the tempo and mode in P450 evolution. Biochim. Biophys. Acta 1814, 19–28.

ffrench-Constant, R.H., 1999. Target site mediated insecticide resistance: what questions remain? Insect Biochem. Mol. Biol. 29, 397–403.

Findeisen, F., Minor Jr., D.L., 2010. Progress in the structural understanding of voltage-gated calcium channel (Cav) function and modulation. Channels 4, 459–474.

Fitches, E., Edwards, M.G., Mee, C., Grishin, E., Gatehouse, A.M., Edwards, J.P., Gatehouse, J.A., 2004. Fusion proteins containing insect-specific toxins as pest control agents: snowdrop lectin delivers fused insecticidal spider venom toxin to insect haemolymph following oral ingestion. J. Insect Physiol. 50, 61–71.

Fletcher, J.I., Smith, R., O'Donoghue, S.I., Nilges, M., Connor, M., Howden, M.E., Christie, M.J., King, G.F., 1997. The structure of a novel insecticidal neurotoxin, ω-atracotoxin-HV1, from the venom of an Australian funnel web spider. Nat. Struct. Biol. 4, 559–566.

George, C.H., Lai, F.A., 2007. Developing new anti-arrhythmics: clues from the molecular basis of cardiac ryanodine receptor (RyR2) Ca^{2+}-release channel dysfunction. Curr. Pharm. Des. 13, 3195–3211.

Gielow, M.L., Gu, G.G., Singh, S., 1995. Resolution and pharmacological analysis of the voltage-dependent calcium channels of Drosophila larval muscles. J. Neurosci. 15, 6085–6093.

Gonzalez, D.R., Treuer, A.V., Castellanos, J., Dulce, R.A., Hare, J.M., 2010. Impaired S-nitrosylation of the ryanodine receptor caused by xanthine oxidase activity contributes to calcium leak in heart failure. J. Biol. Chem. 285, 28938–28945.

Grabner, M., Bachmann, A., Rosenthal, F., Striessnig, J., Schultz, C., Tautz, D., Glossmann, H., 1994a. Insect calcium channels. Molecular cloning of an alpha 1-subunit from housefly (Musca domestica) muscle. FEBS Lett. 339, 189–194.

Grabner, M., Wang, Z., Mitterdorfer, J., Rosenthal, F., Charnet, P., Savchenko, A., Hering, S., Ren, D., Hall, L.M., Glossmann, H., 1994b. Cloning and functional expression of a neuronal calcium channel beta subunit from house fly (Musca domestica). J. Biol. Chem. 269, 23668–23674.

Guse, A.H., 1999. Cyclic ADP-ribose: a novel Ca^{2+} -mobilizing second messenger. Cell. Signal. 11, 309–316.

Guse, A.H., 2002. Cyclic ADP-ribose (cADPR) and nicotinic acid adenine dinucleotide phosphate (NAADP): novel regulators of Ca^{2+}-signaling and cell function. Curr. Mol. Med. 2, 273–282.

Gwack, Y., Srikanth, S., Feske, S., Cruz-Guilloty, F., Oh-Hora, M., Neems, D.S., Hogan, P.G., Rao, A., 2007. Biochemical and functional characterization of Orai proteins. J. Biol. Chem. 282, 16232–16243.

Hall, L.M., Ren, D., Feng, G., Eberl, D.F., Dubald, M., Yang, M., Hannan, F., Kousky, C.T., Zheng, W., 1995. Calcium channel as a new potential target for insecticides. In: Clark, J.M. (Ed.), Molecular Action of Insecticides on Ion Channels. ACS Symposium Series, Washington, DC, pp. 162–172.

Hamada, T., Bannister, M.L., Ikemoto, N., 2007. Peptide probe study of the role of interaction between the cytoplasmic and transmembrane domains of the ryanodine receptor in the channel regulation mechanism. Biochemistry 46, 4272–4279.

Hamaguchi, H., Hirooka, T., Masaki, T., Lahm, G.P., Cordova, D., Barry, J.D., Andaloro, J.T., Annan, I.B., Marcon, P.C., Portillo, H.E., Stevenson, T.M., Selby, T.P., 2012. Insecticides affecting calcium homeostasis. In: Krämer, W., Schirmer, U., Jeschke, P., Witschel, M. (Eds.), Modern Crop Protection Compounds. Wiley-VCH, Weinheim, pp. 1389–1425.

Hamilton, S.L., 2005. Ryanodine receptors. Cell Calcium 38, 253–260.

Hamilton, S.L., Serysheva, I.I., 2009. Ryanodine receptor structure: progress and challenges. J. Biol. Chem. 284, 4047–4051.

Hansen, R.S., Paulsen, I., Davies, M., 2005a. Determinants of amentoflavone interaction at the $GABA_A$ receptor. Eur. J. Pharmacol. 519, 199–207.

Hansen, S.B., Sulzenbacher, G., Huxford, T., Marchot, P., Taylor, P., Bourne, Y., 2005b. Structures of *Aplysia* AChBP complexes with nicotinic agonists and antagonists reveal distinctive binding interfaces and conformations. EMBO J. 24, 3635–3646.

Heinemann, S.H., Terlau, H., Stuhmer, W., Imoto, K., Numa, S., 1992. Calcium channel characteristics conferred on the sodium channel by single mutations. Nature 356, 441–443.

Hemingway, J., Field, L., Vontas, J., 2002. An overview of insecticide resistance. Science 298, 96–97.

Herrmann-Frank, A., Richter, M., Sarkozi, S., Mohr, U., Lehmann-Horn, F., 1996. 4-Chloro-m-cresol, a potent and specific activator of the skeletal muscle ryanodine receptor. Biochim. Biophys. Acta 1289, 31–40.

Herzig, V., Wood, D.L., Newell, F., Chaumeil, P.A., Kaas, Q., Binford, G.J., Nicholson, G.M., Gorse, D., King, G.F., 2011. ArachnoServer 2.0, an updated online resource for spider toxin sequences and structures. Nucleic Acids Res. 39, D653–D657.

Hirota, K., Lambert, D.G., 2006. Measurement of [3H]PN200-110 and [125I]omega-conotoxin MVIIA binding. Methods Mol. Biol. 312, 147–159.

Ho, H.-T., Stevens, S.C.W., Terentyeva, R., Carnes, C.A., Terentyev, D., Gyorke, S., 2011. Arrhythmogenic adverse effects of cardiac glycosides are mediated by redox modification of ryanodine receptors. J. Physiol. 589, 4697–4708.

Huke, S., Bers, D.M., 2008. Ryanodine receptor phosphorylation at Serine 2030, 2808 and 2814 in rat cardiomyocytes. Biochem. Biophys. Res. Commun. 376, 80–85.

Hunt, D.J., Jones, P.P., Wang, R., Chen, W., Bolstad, J., Chen, K., Shimoni, Y., Chen, S.R.W., 2007. K201 (JTV-519) suppresses spontaneous Ca^{2+} release and [3H] ryanodine binding to RyR2 irrespective of FKBP12.6 association. Biochem. J. 404, 431–438.

Ikura, M., 1996. Calcium binding and conformational response in EF-hand proteins. Trends Biochem. Sci. 21, 14–17.

Isaacs, A.K., Qi, S., Sarpong, S., Casida, J.E., 2012. Insect ryanodine receptor: distinct but coupled insecticide binding sites for [N-C^3H_3]chlorantraniliprole, flubendiamide, and [^{3}H]ryanodine. Chem. Res. Toxicol. 25, 1571–1573.

Iwamoto, T., 2007. Na^+/Ca^{2+} exchange as a drug target—insights from molecular pharmacology and genetic engineering. Ann. N. Y. Acad. Sci. 1099, 516–528.

Iwamoto, T., Watanabe, Y., Kita, S., Blaustein, M.P., 2007. Na^+/Ca^{2+} exchange inhibitors: a new class of calcium regulators. Cardiovasc. Hematol. Disord. Drug Targets 7, 188–198.

Jacobson, A.R., Moe, S.T., Allen, P.D., Fessenden, J.D., 2006. Structural determinants of 4-chloro-m-cresol required for activation of ryanodine receptor type 1. Mol. Pharmacol. 70, 259–266.

Jayaraman, T., Brillantes, A.M., Timerman, A.P., Fleischer, S., Erdjument-Bromage, H., Tempst, P., Marks, A.R., 1992. FK506 binding protein associated with the calcium release channel (ryanodine receptor). J. Biol. Chem. 267, 9474–9477.

Jenden, D.J., Fairhurst, A.S., 1969. The pharmacology of ryanodine. Pharmacol. Rev. 21, 1–25.

Jones, P.P., Meng, X., Xiao, B., Cai, S., Bolstad, J., Wagenknecht, T., Liu, Z., Chen, S.R.W., 2008. Localization of PKA phosphorylation site, Ser 2030, in the three-dimensional structure of cardiac ryanodine receptor. Biochem. J. 410, 261–270.

Joshi, R., Venkatesh, K., Srinivas, R., Nair, S., Hasan, G., 2004. Genetic dissection of *itpr* gene function reveals a vital requirement in aminergic cells of *Drosophila* larvae. Genetics 166, 225–236.

Kato, K., Kiyonaka, S., Sawaguchi, Y., Tohnishi, M., Masaki, T., Yasokawa, N., Mizuno, Y., Mori, E., Inoue, K., Hamachi, I., Takeshima, H., Mori, Y., 2009. Molecular characterization of flubendiamide sensitivity in lepidopterous ryanodine receptor Ca^{2+} release channel. Biochemistry 48, 10342–10352.

Kawasaki, F., Collins, S.C., Ordway, R.W., 2002. Synaptic calcium-channel function in *Drosophila*: analysis and transformation rescue of temperature-sensitive paralytic and lethal mutations of cacophony. J. Neurosci. 22, 5856–5864.

Ketterman, A.J., Saisawang, C., Wongsantichon, J., 2011. Insect glutathione transferases. Drug Metab. Rev. 43, 253–265.

Khan, S.A., Zafar, Y., Briddon, R.W., Malik, K.A., Mukhtar, Z., 2006. Spider venom toxin protects plants from insect attack. Transgenic Res. 15, 349–357.

Kimlicka, L., Van Petegem, F., 2011. The structural biology of ryanodine receptors. Sci. China Life Sci. 54, 712–724.

King, G.F., 2007. Modulation of insect Ca_V channels by peptidic spider toxins. Toxicon 49, 513–530.

King, G.F., Escoubas, P., Nicholson, G.M., 2008. Peptide toxins that selectively target insect Na_V and Ca_V channels. Channels 2, 100–116.

Kissin, I., Szallasi, A., 2011. Therapeutic targeting of TRPV1 by resiniferatoxin, from preclinical studies to clinical trials. Curr. Top. Med. Chem. 11, 2159–2170.

Kobayashi, S., Bannister, M.L., Gangopadhyay, J.P., Hamada, T., Parness, J., Ikemoto, N., 2004. Dantrolene stabilizes domain interactions within the ryanodine receptor. J. Biol. Chem. 280, 6580–6587.

Kobayashi, S., Yano, M., Suetomi, T., Ono, M., Tateishi, H., MochiZzuki, M., Xu, X.J., Uchinoumi, H., Okuda, S., Yamamoto, T., Koseki, N., Kyushiki, H., Ikemoto, N., Matsuzaki, M., 2009. Dantrolene, a therapeutic agent for malignant hyperthermia, markedly improves the function of failing cardiomyocytes by stabilizing interdomain interactions within the ryanodine receptor. J. Am. Coll. Cardiol. 53, 1993–2005.

Kockskaemper, J., Pieske, B., 2006. Phosphorylation of the cardiac ryanodine receptor by Ca^{2+}/calmodulin-dependent protein kinase II. Circ. Res. 99, 333–335.

Kono, Y., Tomita, T., 2006. Amino acid substitutions conferring insecticide insensitivity in Ace-paralogous acetylcholinesterase. Pestic. Biochem. Physiol. 85, 123–132.

Kushnir, A., Mollah, A.K.M.M., Wehrens, X.H.T., 2005. Evolution of the ryanodine receptor gene family. In: Wehrens, X.H.T., Marks, A.R. (Eds.), Ryanodine Receptors. Structure, Function and Dysfunction in Clinical Disease. Springer-Verlag, Berlin, pp. 1–8.

Lacava, C., Sgaragli, G., Fusi, F., 2012. 3,5-di-t-butylcatechol as a ryanodine receptor agonist in rat intact skeletal muscle fibers. Drug Dev. Res. 73, 138–145.

Lahm, G.P., Selby, T.P., Freudenberger, J.H., Stevenson, T.M., Myers, B.J., Seburyamo, G., Smith, B.K., Flexner, L., Clark, C.E., Cordova, D., 2005. Insecticidal anthranilic diamides: a new class of potent ryanodine receptor activators. Bioorg. Med. Chem. Lett. 15, 4898–4906.

Lanner, J.T., Georgiou, D.K., Joshi, A.D., Hamilton, S.L., 2010. Ryanodine receptors: structure, expression, molecular details, and function in calcium release. Cold Spring Harb. Perspect. Biol. 2, a003996.

Lanzafame, A.A., Christopoulos, A., Mitchelson, F., 2003. Cellular signaling mechanisms for muscarinic acetylcholine receptors. Receptors Channels 9, 241–260.

Laver, D.R., 2007. Ca^{2+} stores regulate ryanodine receptor Ca^{2+} release channels via luminal and cytosolic Ca^{2+} sites. Clin. Exp. Pharmacol. Physiol. 34, 889–896.

Laver, D.R., 2009. Luminal Ca^{2+} activation of cardiac ryanodine receptors by luminal and cytoplasmic domains. Eur. Biophys. J. 39, 19–26.

Laver, D.R., Lenz, G.K.E., Lamb, G.D., 2001. Regulation of the calcium release channel from rabbit skeletal muscle by the nucleotides ATP, AMP, IMP and adenosine. J. Physiol. 537, 763–778.

Lee, H.C., 1993. Potentiation of calcium- and caffeine-induced calcium release by cyclic ADP-ribose. J. Biol. Chem. 268, 293–299.

Lee, S.Y., Lee, A., Chen, J., MacKinnon, R., 2005. Structure of the KvAP voltage-dependent K^+ channel and its dependence on the lipid membrane. Proc. Natl. Acad. Sci. U.S.A. 102, 15441–15446.

Lehmberg, E., Casida, J.E., 1994. Similarity of insect and mammalian ryanodine binding sites. Pestic. Biochem. Physiol. 48, 145–152.

Lehnart, S.E., Terrenoire, C., Reiken, S., Wehrens, X.H.T., Song, L.S., Tillman, E.J., Mancarella, S., Coromilas, J., Lederer, W.J., Kass, R.S., Marks, A.R., 2006. Stabilization of cardiac ryanodine receptor prevents intracellular calcium leak and arrhythmias. Proc. Natl. Acad. Sci. U.S.A. 103, 7906–7910.

Leung, H.T., Branton, W.D., Phillips, H.S., Jan, L., Byerly, L., 1989. Spider toxins selectively block calcium currents in Drosophila. Neuron 3, 767–772.

Leung, F.P., Yung, L.M., Yao, X., Laher, I., Huang, Y., 2008. Store-operated calcium entry in vascular smooth muscle. Br. J. Pharmacol. 153, 846–857.

Lewis, R.J., 2009. Conotoxin venom peptide therapeutics. Adv. Exp. Med. Biol. 655, 44–48.

Liang, X., Hu, X., Hu, J., 2007. Dynamic inter-receptor coupling: a novel working mechanism of 2-D ryanodine receptor array. Biophys. J. 92, 1215–1223.

Liang, X., Hu, X.F., Hu, J., 2009. Dynamic interreceptor coupling contributes to the consistent open duration of ryanodine receptors. Biophys. J. 96, 4826–4833.

Liao, J., Li, H., Zeng, W., Sauer, D.B., Belmares, R., Jiang, Y., 2012. Structural insight into the ion-exchange mechanism of the sodium/calcium exchanger. Science 335, 686–690.

Lindsay, A.R., Williams, A.J., 1991. Functional characterisation of the ryanodine receptor purified from sheep cardiac muscle sarcoplasmic reticulum. Biochim. Biophys. Acta 1064, 89–102.

Littleton, J.T., Ganetzky, B., 2000. Ion channels and synaptic organization: analysis of the Drosophila genome. Neuron 26, 35–43.

Liu, Y., Porta, M., Qin, J., Ramos, J., Nani, A., Shannon, T.R., Fill, M., 2010a. Flux regulation of cardiac ryanodine receptor channels. J. Gen. Physiol. 135, 15–27.

Liu, Z., Wang, R., Tian, X., Zhong, X., Gangopadhyay, J., Cole, R., Ikemoto, N., Chen, S.R.W., Wagenknecht, T., 2010b. Dynamic, inter-subunit interactions between the N-terminal and central mutation regions of cardiac ryanodine receptor. J. Cell Sci. 123, 1775–1784.

Lobo, P.A., Van Petegem, F., 2009. Crystal structures of the N-terminal domains of cardiac and skeletal muscle ryanodine receptors: insights into disease mutations. Structure 17, 1505–1514.

Lockyer, P.J., Puente, E., Windass, J., Earley, F., East, J.M., Lee, A.G., 1998. Cloning and expression of an insect Ca^{2+}-ATPase from Heliothis virescens. Biochim. Biophys. Acta 1369, 14–18.

Ludtke, S.J., Serysheva, I.I., Hamilton, S.L., Chiu, W., 2005. The pore structure of the closed RyR1 channel. Structure 13, 1203–1211.

Lümmen, P., Ebbinghaus-Kintscher, U., Funke, C., Fischer, R., Masaki, T., Yasokawa, N., Tohnishi, M., 2007. Phthalic acid diamides activate insect ryanodine receptors. ACS Symp. Ser. 948, 235–248.

Lytton, J., 2007. Na^+/Ca^{2+} exchangers: three mammalian gene families control Ca^{2+} transport. Biochem. J. 406, 365–382.

Ma, J., Hayek, S.M., Bhat, M.B., 2004. Membrane topology and membrane retention of the ryanodine receptor calcium release channel. Cell Biochem. Biophys. 40, 207–224.

Mack, M.M., Molinski, T.F., Buck, E.D., Pessah, I.N., 1994. Novel modulators of skeletal muscle FKBP12/calcium channel complex from Lanthella basta. Role of FKBP12 in channel gating. J. Biol. Chem. 269, 23236–23249.

MacKinnon, R., Cohen, S.L., Kuo, A., Lee, A., Chait, B.T., 1998. Structural conservation in prokaryotic and eukaryotic potassium channels. Science 280, 106–109.

Marks, A.R., 1996. Immunophilin modulation of calcium channel gating. Methods 9, 177–187.

Marom, M., Sebag, A., Atlas, D., 2007. Cations residing at the selectivity filter of the voltage-gated Ca^{2+}-channel modify fusion-pore kinetics. Channels 1, 377–386.

Masaki, T., 2008. Study on the mechanism of insecticidal activity through disruption of intracellular calcium homeostasis. J. Pestic. Sci. 33, 271–277.

Masaki, T., Yasokawa, N., Ebbinghaus-Kintscher, U., Lümmen, P., 2007. Flubendiamide stimulates Ca^{2+} pump activity coupled to RyR-mediated calcium release in lepidopterous insects. In: Ohkawa, H., Miyagawa, H., Lee, P.W. (Eds.), Pesticide Chemistry. Wiley-VCH, Weinheim, pp. 137–140.

Masumiya, H., Wang, R., Zhang, J., Xiao, B., Chen, S.R., 2003. Localization of the 12.6-kDa FK506-binding protein (FKBP12.6) binding site to the NH_2-terminal domain of the cardiac Ca^{2+} release channel (ryanodine receptor). J. Biol. Chem. 278, 3786–3792.

Masuno, M.N., Pessah, I.N., Olmstead, M.M., Molinski, T.F., 2006. Simplified cyclic analogues of bastadin-5. structure-activity relationships for modulation of the RyR1/FKBP12 Ca^{2+} channel complex. J. Med. Chem. 49, 4497–4511.

McLennan, D.H., Kranias, E.G., 2004. Regulation of the sarco(endo)plasmic reticulum Ca^{2+}-ATPase by phospholamban and sarcolipin. In: Futai, M., Wada, Y., Kaplan, J. H. (Eds.), Handbook of ATPases. Wiley-VCH, Weinheim, pp. 25–62.

Meissner, G., 1984. Adenine nucleotide stimulation of Ca^{2+}-induced Ca^{2+} release in sarcoplasmic reticulum. J. Biol. Chem. 259, 2365–2374.

Meissner, G., El-Hashem, A., 1992. Ryanodine as a functional probe of the skeletal muscle sarcoplasmic reticulum Ca^{2+} release channel. Mol. Cell. Biochem. 114, 119–123.

Meissner, G., Rios, E., Tripathy, A., Pasek, D.A., 1997. Regulation of skeletal muscle Ca^{2+} release channel (ryanodine receptor) by Ca^{2+} and monovalent cations and anions. J. Biol. Chem. 272, 1628–1638.

Meli, A.C., Refaat, M.M., Dura, M., Reiken, S., Wronska, A., Wojciak, J., Carroll, J., Scheinman, M.M., Marks, A.R., 2011. A novel ryanodine receptor mutation linked to sudden death increases sensitivity to cytosolic calcium. Circ. Res. 109, 281–290.

Meng, X., Xiao, B., Cai, S., Huang, X., Li, F., Bolstad, J., Trujillo, R., Airey, J., Chen, S.R.W., Wagenknecht, T., Liu, Z., 2007. Three-dimensional localization of serine 2808, a phosphorylation site in cardiac ryanodine receptor. J. Biol. Chem. 282, 25929–25939.

Messutat, S., Heine, M., Wicher, D., 2001. Calcium-induced calcium release in neurosecretory insect neurons: fast and slow responses. Cell Calcium 30, 199–211.

Michiels, J., XI, C., Verhaert, J., Vanderleyden, J., 2002. The functions of Ca^{2+} in bacteria: a role for EF-hand proteins? Trends Microbiol. 10, 87–93.

Mikhaylova, M., Hradsky, J., Kreutz, M.R., 2011. Between promiscuity and specificity: novel roles of EF-hand calcium sensors in neuronal Ca^{2+} signalling. J. Neurochem. 118, 695–713.

Minke, B., 1982. Light-induced reduction in excitation efficiency in the trp mutant of Drosophila. J. Gen. Physiol. 79, 361–385.

Minke, B., 2010. The history of the Drosophila TRP channel: the birth of a new channel superfamily. J. Neurogenet. 24, 216–233.

Møller, J.V., Nissen, P., Sørensen, T.L.M., Maire, M.L., 2005. Transport mechanism of the sarcoplasmic reticulum Ca^{2+}-ATPase pump. Curr. Opin. Struct. Biol. 15, 387–393.

Monma, S., Sunazuka, T., Nagai, K., Arai, T., Shiomi, K., Matsui, R., Omura, S., 2006. Verticilide: elucidation of absolute configuration and total Synthesis. Org. Lett. 8, 5601–5604.

Montell, C., Rubin, G.M., 1989. Molecular characterization of the *Drosophila trp* locus: a putative integral membrane protein required for phototransduction. Neuron 2, 1313–1323.

Mosbah, A., Kharrat, R., Fajloun, Z., Renisio, J.G., Blanc, E., Sabatier, J.M., El Ayeb, M., Darbon, H., 2000. A new fold in the scorpion toxin family, associated with an activity on a ryanodine-sensitive calcium channel. Proteins: Struct. Funct. Bioinformat. 40, 436–442.

Mukherjee, A.K., Sollod, B.L., Wikel, S.K., King, G.F., 2006. Orally active acaricidal peptide toxins from spider venom. Toxicon 47, 182–187.

Nagata, K., 2007. TRP channels as target sites for insecticides: physiology, pharmacology and toxicology. Invert. Neurosci. 7, 31–37.

Nash, M.S., Young, K.W., Willars, G.B., Challiss, R.A., Nahorski, S.R., 2001. Single-cell imaging of graded Ins(1,4,5)P$_3$ production following G- protein-coupled-receptor activation. Biochem. J. 356, 137–142.

Nauen, R., 2006. Insecticide mode of action: return of the ryanodine receptor. Pest Manag. Sci. 62, 690–692.

Niggli, E., 2011. Ryanodine receptors: waking up from refractoriness. Cardiovasc. Res. 91, 563–564.

Noble, D., Herchuelz, A., 2007. Role of Na/Ca exchange and the plasma membrane Ca^{2+}-ATPase in cell function. EMBO Rep. 8, 228–232.

Norton, R.S., McDonough, S.I., 2008. Peptides targeting voltage-gated calcium channels. Curr. Pharm. Des. 14, 2480–2491.

Norton, R.S., Pallaghy, P.K., 1998. The cystine knot structure of ion channel toxins and related polypeptides. Toxicon 36, 1573–1583.

Oberwinkler, J., 2002. Calcium homeostasis in fly photoreceptor cells. Adv. Exp. Med. Biol. 514, 539–583.

Oerke, E.-C., 2006. Crop losses to pests. J. Agric. Sci. 144, 31–43.

Oerke, E.C., Dehne, H.W., 2004. Safeguarding production—losses in major crops and the role of crop protection. Crop Protect. 23, 275–285.

Ogunbayo, O.A., Zhu, Y., Rossi, D., Sorrentino, V., Ma, J., Zhu, M.X., Evans, A.M., 2011. Cyclic adenosine diphosphate ribose activates ryanodine receptors, whereas NAADP activates two-pore domain channels. J. Biol. Chem. 286, 9136–9140.

Olivera, B.M., Imperial, J.S., Cruz, L.J., Bindokas, V.P., Venema, V.J., Adams, M.E., 1991. Calcium channel-targeted polypeptide toxins. Ann. N. Y. Acad. Sci. 635, 114–122.

Orlova, E.V., Serysheva, I.I., Van Heel, M., Hamilton, S.L., Chiu, W., 1996. Two structural configurations of the skeletal muscle calcium release channel. Nat. Struct. Biol. 3, 547–552.

Ortega, C., Ortolando, S., Carafoli, E., 2007. The plasma membrane calcium pump. New Compr. Biochem. 41, 179–197.

Pan, Z., Damron, D., Nieminen, A.L., Bhat, M.B., Ma, J., 2000. Depletion of intracellular Ca^{2+} by caffeine and ryanodine induces apoptosis of chinese hamster ovary cells transfected with ryanodine receptor. J. Biol. Chem. 275, 19978–19984.

Paolini, C., Protasi, F., Franzini-Armstrong, C., 2004a. The relative position of RyR feet and DHPR tetrads in skeletal muscle. J. Mol. Biol. 342, 145–153.

Paolini, C., Fessenden, J.D., Pessah, I.N., Franzini-Armstrong, C., 2004b. Evidence for conformational coupling between two calcium channels. Proc. Natl. Acad. Sci. U.S.A. 101, 12748–12752.

Paul-Pletzer, K., Yamamoto, T., Bhat, M.B., Ma, J., Ikemoto, N., Jimenez, L.S., Morimoto, H., Williams, P.G., Parness, J., 2002. Identification of a dantrolene-binding sequence on the skeletal muscle ryanodine receptor. J. Biol. Chem. 277, 34918–34923.

Pauron, D., Qar, J., Barhanin, J., Fournier, D., Cuany, A., Pralavorio, M., Berge, J.B., Lazdunski, M., 1987. Identification and affinity labeling of very high affinity binding sites for the phenylalkylamine series of Ca^{2+} channel blockers in the *Drosophila* nervous system. Biochemistry 26, 6311–6315.

Payandeh, J., Scheuer, T., Zheng, N., Catterall, W.A., 2011. The crystal structure of a voltage-gated sodium channel. Nature 475, 353–358.

Peixoto, A.A., Smith, L.A., Hall, J.C., 1997. Genomic organization and evolution of alternative exons in a *Drosophila* calcium channel gene. Genetics 145, 1003–1013.

Pelzer, S., Barhanin, J., Pauron, D., Trautwein, W., Lazdunski, M., Pelzer, D., 1989. Diversity and novel pharmacological properties of Ca^{2+} channels in *Drosophila* brain membranes. EMBO J. 8, 2365–2371.

Peng, I.F., Wu, C.F., 2007. *Drosophila cacophony* channels: a major mediator of neuronal Ca^{2+} currents and a trigger for K^+ channel homeostatic regulation. J. Neurosci. 27, 1072–1081.

Percival, A.L., Williams, A.J., Kenyon, J.L., Grinsell, M.M., Airey, J.A., Sutko, J.L., 1994. Chicken skeletal muscle ryanodine receptor isoforms: ion channel properties. Biophys. J. 67, 1834–1850.

Perry, T., Batterham, P., Daborn, P.J., 2011. The biology of insecticidal activity and resistance. Insect Biochem. Mol. Biol. 41, 411–422.

Pessah, I.N., 1990. Chemical effectors of the ryanodine receptor: a novel strategy for insect control. Spec. Publ. Roy. Soc. Chem. 79, 278–296.

Pessah, I.N., Zimanyi, I., 1991. Characterization of multiple [3H]ryanodine binding sites on the Ca^{2+} release channel of sarcoplasmic reticulum from skeletal and cardiac muscle: evidence for a sequential mechanism in ryanodine action. Mol. Pharmacol. 39, 679–689.

Pessah, I.N., Stambuk, R.A., Casida, J.E., 1987. Ca^{2+}-activated ryanodine binding: mechanisms of sensitivity and intensity modulation by Mg^{2+}, caffeine, and adenine nucleotides. Mol. Pharmacol. 31, 232–238.

Pi, C., Liu, J., Peng, C.C., Liu, Y., Jiang, X., Zhao, Y., Tang, S., Wang, L., Dong, M., Chen, S., Xu, A., 2006. Diversity and evolution of conotoxins based on gene expression profiling of *Conus litteratus*. Genomics 88, 809–819.

Powl, A.M., O'Reilly, A.O., Miles, A.J., Wallace, B.A., 2010. Synchrotron radiation circular dichroism spectroscopy-defined structure of the C-terminal domain of NaChBac and its role in channel assembly. Proc. Natl. Acad. Sci. U.S.A. 107, 14064–14069.

Pringos, E., Vignes, M., Martinez, J., Rolland, V., 2011. Peptide neurotoxins that affect voltage-gated calcium channels: a close-up on omega-agatoxins. Toxins 3, 17–42.

Protasi, F., Franzini-Armstrong, C., Allen, P.D., 1998. Role of ryanodine receptors in the assembly of calcium release units in skeletal muscle. J. Cell Biol. 140, 831–842.

Puente, E., Suner, M., Evans, A.D., McCaffery, A.R., Windass, J.D., 2000. Identification of a polymorphic ryanodine receptor gene from *Heliothis virescens* (Lepidoptera: *Noctuidae*). Insect Biochem. Mol. Biol. 30, 335–347.

Qazi, S., Trimmer, B.A., 1999. The role of inositol 1,4,5-trisphosphate 5-phosphatase in inositol signaling in the CNS of larval *Manduca sexta*. Insect Biochem. Mol. Biol. 29, 161–175.

Randall, A.D., Tsien, R.W., 1997. Contrasting biophysical and pharmacological properties of T-type and R-type calcium channels. Neuropharmacology 36, 879–893.

Randall, A., Tsien, R.W., 1995. Pharmacological dissection of multiple types of Ca^{2+} channel currents in rat cerebellar granule neurons. J. Neurosci. 15, 2995–3012.

Ranson, H., Hemingway, J., 2005. Mosquito glutathione transferases. Methods Enzymol. 401, 226–241.

Ren, D., Navarro, B., Xu, H., Yue, L., Shi, Q., Clapham, D.E., 2001. A prokaryotic voltage-gated sodium channel. Science 294, 2372–2375.

Richards, M.W., Butcher, A.J., Dolphin, A.C., 2004. Ca^{2+} channel ß-subunits: structural insights AID our understanding. Trends Pharmacol. Sci. 25, 626–632.

Rodney, G.G., Moore, C.P., Williams, B.Y., Zhang, J.-Z., Krol, J., Pedersen, S.E., Hamilton, S.L., 2001. Calcium binding to calmodulin leads to an N-terminal shift in its binding site on the ryanodine receptor. J. Biol. Chem. 276, 2069–2074.

Rogers, E.F., Koniuszy, F.R., Shavel, J., Folkers, K., 1948. Plant Insecticides. I. Ryanodine, a new alkaloid from *Ryania speciosa* Vahl. J. Am. Chem. Soc. 70, 3086–3088.

Rosay, P., Davies, S.A., Yu, Y., Sozen, A., Kaiser, K., Dow, J.A., 1997. Cell-type specific calcium signalling in a *Drosophila* epithelium. J. Cell Sci. 110, 1683–1692.

Rossi, D., Sorrentino, V., 2002. Molecular genetics of ryanodine receptors Ca^{2+}-release channels. Cell Calcium 32, 307–319.

Rousseau, E., Ladine, J., Liu, Q.Y., Meissner, G., 1988. Activation of the Ca^{2+} release channel of skeletal muscle sarcoplasmic reticulum by caffeine and related compounds. Arch. Biochem. Biophys. 267, 75–86.

Ruiz-Meana, M., Fernandez-Sanz, C., Garcia-Dorado, D., 2010. The SR–mitochondria interaction: a new player in cardiac pathophysiology. Cardiovasc. Res. 88, 30–39.

Ryglewski, S., Lance, K., Levine, R.B., Duch, C., 2012. $Ca_V 2$ channels mediate low and high voltage-activated calcium currents in *Drosophila* motoneurons. J. Physiol. 590, 809–825.

Saez, N.J., Senff, S., Jensen, J.E., Er, S.Y., Herzig, V., Rash, L.D., King, G.F., 2010. Spider-venom peptides as therapeutics. Toxins 2, 2851–2871.

Sagara, Y., Wade, J.B., Inesi, G., 1992. A conformational mechanism for formation of a dead-end complex by the sarcoplasmic reticulum ATPase with thapsigargin. J. Biol. Chem. 267, 1286–1292.

Samso, M., Wagenknecht, T., Allen, P.D., 2005. Internal structure and visualization of transmembrane domains of the RyR1 calcium release channel by cryo-EM. Nat. Struct. Biol. 12, 539–544.

Sander, T., Bruun, A.T., Balle, T., 2010. Docking to flexible nicotinic acetylcholine receptors: a validation study using the acetylcholine binding protein. J. Mol. Graph. Model. 29, 415–424.

Sanyal, S., Consoulas, C., Kuromi, H., Basole, A., Mukai, L., Kidokoro, Y., Krishnan, K.S., Ramaswami, M., 2005. Analysis of conditional paralytic mutants in *Drosophila* sarco-endoplasmic reticulum calcium ATPase reveals novel mechanisms for regulating membrane excitability. Genetics 169, 737–750.

Sattelle, D.B., Cordova, D., Cheek, T.R., 2008. Insect ryanodine receptors: molecular targets for novel pest control chemicals. Invert. Neurosci. 8, 107–119.

Schmitt, M., Turberg, A., Londershausen, M., Dorn, A., 1996. Binding sites for Ca^{2+}-channel effectors and ryanodine in *Periplaneta americana*—possible targets for new insecticides. Pestic. Sci. 48, 375–385.

Schmitt, M., Turberg, A., Londershausen, M., 1997. Characterization of a ryanodine receptor in *Periplaneta americana*. J. Recept. Signal Transduct. Res. 17, 185–197.

Schwarz, E.M., Benzer, S., 1997. Calx, a Na–Ca exchanger gene of *Drosophila melanogaster*. Proc. Natl. Acad. Sci. U.S.A. 94, 10249–10254.

Scott-Ward, T.S., Dunbar, S.J., Windass, J.D., Williams, A.J., 2001. Characterization of the ryanodine receptor-Ca^{2+} release channel from the thoracic tissues of the lepidopteran insect *Heliothis virescens*. J. Membr. Biol. 179, 127–141.

Selby, T.P., Lahm, G.P., Stevenson, T.M., Hughes, K.A., Annan, I.B., Cordova, D., Bellin, C.A., Benner, E.A., Wing, K.D., Barry, J.D., Currie, M.J., Pahutski, T.F., 2008. Discovery of Cyazypyr[TM]: a new cross-spectrum insecticide from the anthranilic diamide class of ryanodine receptor activators. In: Abstracts of Papers, 236th ACS National Meeting, Philadelphia, PA, United States, August 17–21, 2008, AGRO-039.

Seo, M.-D., Velamakanni, S., Ishiyama, N., Stathopulos, P.B., Rossi, A.M., Khan, S.A., Dale, P., Li, C., Ames, J.B., Ikura, M., Taylor, C.W., 2012. Structural and functional conservation of key domains in $InsP_3$ and ryanodine receptors. Nature 483, 108–112.

Serysheva, I.I., Ludtke, S.J., Baker, M.R., Chiu, W., Hamilton, S.L., 2002. Structure of the voltage-gated L-type Ca 2 + channel by electron cryomicroscopy. Proc. Natl. Acad. Sci. U.S.A. 99, 10370–10375.

Serysheva, I.I., Ludtke, S.J., Baker, M.L., Cong, Y., Topf, M., Eramian, D., Sali, A., Hamilton, S.L., Chiu, W., 2008. Subnanometer-resolution electron cryomicroscopy-based domain models for the cytoplasmic region of skeletal muscle RyR channel. Proc. Natl. Acad. Sci. U.S.A. 105, 9610–9615.

Shafrir, Y., Durell, S.R., Guy, H.R., 2008. Models of voltage-dependent conformational changes in NaChBac channels. Biophys. J. 95, 3663–3676.

Shiomi, K., Matsui, R., Kakei, A., Yamaguchi, Y., Masuma, R., Hatano, H., Arai, N., Isozaki, M., Tanaka, H., Kobayashi, S., Turberg, A., Omura, S., 2010. Verticilide, a new ryanodine-binding inhibitor, produced by Verticillium sp. FKI-1033. J. Antibiot. 63, 77–82.

Sinkins, W.G., Estacion, M., Prasad, V., Goel, M., Shull, G.E., Kunze, D.L., Schilling, W.P., 2009. Maitotoxin converts the plasmalemmal Ca^{2+} pump into a Ca^{2+}-permeable non-selective cation channel. Am. J. Physiol. 297, C1533–C1543.

Sitsapesan, R., Williams, A.J., 1990. Mechanisms of caffeine activation of single calcium-release channels of sheep cardiac sarcoplasmic reticulum. J. Physiol. 423, 425–439.

Sitsapesan, R., Williams, A.J., 1995. Cyclic ADP-ribose and related compounds activate sheep skeletal sarcoplasmic reticulum Ca^{2+} release channel. Am. J. Physiol. 268, C1235–C1240.

Sitsapesan, R., McGarry, S.J., Williams, A.J., 1994. Cyclic ADP-ribose competes with ATP for the adenine nucleotide binding site on the cardiac ryanodine receptor Ca^{2+}-release channel. Circ. Res. 75, 596–600.

Smith, L.A., Wang, X., Peixoto, A.A., Neumann, E.K., Hall, L.M., Hall, J.C., 1996. A Drosophila calcium channel α1 subunit gene maps to a genetic locus associated with behavioral and visual defects. J. Neurosci. 16, 7868–7879.

Spassova, M.A., Soboloff, J., He, L.P., Hewavitharana, T., Xu, W., Venkatachalam, K., van Rossum, D.B., Patterson, R.L., Gill, D.L., 2004. Calcium entry mediated by SOCs and TRP channels: variations and enigma. Biochim. Biophys. Acta 1742, 9–20.

Srikanth, S., Wang, Z., Hasan, G., Bezprozvanny, I., 2004a. Functional properties of a pore mutant in the Drosophila melanogaster inositol 1,4,5-trisphosphate receptor. FEBS Lett. 575, 95–98.

Srikanth, S., Wang, Z., Tu, H., Nair, S., Mathew, M.K., Hasan, G., Bezprozvanny, I., 2004b. Functional properties of the Drosophila melanogaster inositol 1,4,5-trisphosphate receptor mutants. Biophys. J. 86, 3634–3646.

Steinbeck, J.A., Henke, N., Opatz, J., Gruszczynska-Biegala, J., Schneider, L., Theiss, S., Hamacher, N., Steinfarz, B., Golz, S., Brüstle, O., Kuznicki, J., Methner, A., 2011. Store-operated calcium entry modulates neuronal network activity in a model of chronic epilepsy. Exp. Neurol. 232, 185–194.

Strehler, E.E., 2004. Plasma-membrane calcium pump: structure and function. Encycl. Biol. Chem. 3, 385–389.

Sullivan, K.M., Scott, K., Zuker, C.S., Rubin, G.M., 2000. The ryanodine receptor is essential for larval development in Drosophila melanogaster. Proc. Natl. Acad. Sci. U.S.A. 97, 5942–5947.

Sun, J., Yamaguchi, N., Xu, L., Eu, J.P., Stamler, J.S., Meissner, G., 2008. Regulation of the cardiac muscle ryanodine receptor by O_2 tension and S-nitrosoglutathione. Biochemistry 47, 13985–13990.

Suzuki, A., Matsunaga, K., Mimaki, Y., Sashida, Y., Ohizumi, Y., 1999. Properties of amentoflavone, a potent caffeine-like Ca^{2+} releaser in skeletal muscle sarcoplasmic reticulum. Eur. J. Pharmacol. 372, 97–102.

Takekura, H., Franzini-Armstrong, C., 2002. The structure of Ca^{2+} release units in arthropod body muscle indicates an indirect mechanism for excitation-contraction coupling. Biophys. J. 83, 2742–2753.

Takeshima, H., Nishi, M., Iwabe, N., Miyata, T., Hosoya, T., Masai, I., Hotta, Y., 1994. Isolation and characterization of a gene for a ryanodine receptor/calcium release channel in Drosophila melanogaster. FEBS Lett. 337, 81–87.

Tatishi, H., Yan, M., Mochizuki, M., Suetomi, T., Ono, M., Xu, X., Uchinoumi, H., Okuda, S., Oda, T., Kobyashi, S., Yamamoto, T., Ikeda, Y., Ohkusa, T., Ikemoto, N., Matsuzaki, M., 2009. Defective domain-domain interactions within the ryanodine receptor as a critical cause of diastolic Ca^{2+} leak in failing hearts. Cardiovasc. Res. 81, 536–545.

Taylor, D.A., Sack, J.S., Maune, J.F., Beckingham, K., Quiocho, F.A., 1991. Structure of a recombinant calmodulin from Drosophila melanogaster refined at 2.2-A resolution. J. Biol. Chem. 266, 21375–21380.

Tedford, H.W., Fletcher, J.I., King, G.F., 2001. Functional significance of the beta hairpin in the insecticidal neurotoxin ω-atracotoxin-Hv1a. J. Biol. Chem. 276, 26568–26576.

Tedford, H.W., Gille, N., Ménez, A., Doehring, C.J., Zamponi, G.W., King, G.F., 2004. Scanning mutagenesis of ω-atracotoxin-Hv1a reveals a spatially restricted epitope that confers selective activity against insect calcium channels. J. Biol. Chem. 279, 44133–44140.

Thastrup, O., Cullen, P.J., Drøbak, B.K., Hanley, M.R., Dawson, A.P., 1990. Thapsigargin, a tumor promoter, discharges intracellular Ca^{2+} stores by specific inhibition of the endoplasmic reticulum Ca^{2+}-ATPase. Proc. Natl. Acad. Sci. U.S.A. 87, 2466–2470.

Thomas, N.L., Williams, A.J., 2012. Pharmacology of ryanodine receptors and Ca^{2+}-induced Ca^{2+} release. Wiley Interdisc. Rev. Membr. Transp. Signal. 1, 383–397. http://dx.doi.org/10.1002/wmts.34.

Thomas, J.M., Masgrau, R., Churchill, G.C., Galione, A., 2001. Pharmacological characterization of the putative cADP-ribose receptor. Biochem. J. 359, 451–457.

Tohnishi, M., Nako, H., Furuja, T., Seo, A., Kodama, H., Tsubata, K., Fujioka, S., Hirooka, T., Nishimatsu, T., 2005. Flubendiamide, a novel class insecticide with high lepidopterous activity. J. Pestic. Sci. 30, 354–360.

Tomizawa, M., Casida, J.E., 2009. Molecular recognition of neonicotinoid insecticides: the determinants of life or death. Acc. Chem. Res. 42, 260–269.

Tomizawa, M., Casida, J.E., 2011. Unique neonicotinoid binding conformations conferring selective receptor interactions. J. Agric. Food Chem. 59, 2825–2828.

Toyoshima, C., Nomura, H., 2002. Structural changes in the calcium pump accompanying the dissociation of calcium. Nature 418, 605–611.

Toyshima, C., Naksako, M., Nomura, H., Ogawa, H., 2000. Crystal structure of the calcium pump of sarcoplasmic reticulum at 2.6 angstrom resolution. Nature 405, 647–655.

Troczka, B., Zimmer, C.T., Elias, J., Schorn, C., Bass, C., Davies, T.G.E., Field, L.M., Williamson, M.S., Slater, R., Nauen, R., 2012. Resistance to diamide insecticides in diamondback moth, Plutella xylostella (Lepidoptera: Plutellidae) is associated with a mutation in the membrane-spanning domain of the ryanodine receptor. Insect Biochem. Mol. Biol. 42, 873–880.

Tsetlin, V.I., Hucho, F., 2004. Snake and snail toxins acting on nicotinic acetylcholine receptors: fundamental aspects and medical applications. FEBS Lett. 557, 9–13.

Tung, C.C., Lobo, P.A., Kimlicka, L., van Petegem, F., 2010. The amino-terminal disease hotspot of ryanodine receptors forms a cytoplasmic vestibule. Nature 468, 585–588.

Usherwood, P.N.R., 1962. The action of the alkaloid ryanodine on insect skeletal muscle. Comp. Biochem. Physiol. 6, 181–199.

Usherwood, P.N., Vais, H., 1995. Towards the development of ryanoid insecticides with low mammalian toxicity. Toxicol. Lett. 82–83, 247–254.

Vais, H., Usherwood, P.N., 1995. Novel actions of ryanodine and analogues—perturbers of potassium channels. Biosci. Rep. 15, 515–530.

Van Leeuwen, T., Vontas, J., Tsagkarakou, A., Dermauw, W., Tirry, L., 2010. Acaricide resistance mechanisms in the two-spotted spider mite Tetranychus urticae and other important Acari: a review. Insect Biochem. Mol. Biol. 40, 563–572.

Vazquez-Martinez, O., Canedo-Merino, R., Diaz-Munoz, M., Riesgo-Escovar, J.R., 2003. Biochemical characterization, distribution and phylogenetic analysis of Drosophila *melanogaster* ryanodine and IP$_3$ receptors, and thapsigargin-sensitive Ca^{2+} -ATPase. J. Cell Sci. 116, 2483–2494.

Venkatesh, K., Hasan, G., 1997. Disruption of the IP3 receptor gene of *Drosophila* affects larval metamorphosis and ecdysone release. Curr. Biol. 7, 500–509.

Venkiteswaran, G., Hasan, G., 2009. Intracellular Ca^{2+} signaling and store-operated Ca^{2+} entry are required in *Drosophila* neurons for flight. Proc. Natl. Acad. Sci. U.S.A. 106, 10326–10331.

Verkhratsky, A., 2002. The endoplasmic reticulum and neuronal calcium signalling. Cell Calcium 32, 393–404.

Vinogradova, M.V., Stone, D.B., Malanina, G.G., Kartzaferi, C., Cooke, R., Mendelson, R.A., Fletterick, R.J., 2005. Ca^{2+}-regulated structural changes in troponin. Proc. Natl. Acad. Sci. U.S.A. 102, 5038–5043.

Vonarx, E.J., Tyler, M.I., Atkinson, R.K., Howden, M.E.H., 2006. Characterization of insecticidal peptides from venom of Australian funnel-web spiders. J. Venom. Anim. Toxins 12, 215–233.

Voss, A.A., Lango, J., Ernst-Russell, M., Morin, D., Pessah, I.N., 2004. Identification of hyperreactive cysteines within ryanodine receptor type 1 by mass spectrometry. J. Biol. Chem. 279, 34514–34520.

Voss, A.A., Allen, P.D., Pessah, I.N., Perez, C.F., 2008. Allosterically coupled calcium and magnesium binding sites are unmasked by ryanodine receptor chimeras. Biochem. Biophys. Res. Commun. 366, 988–993.

Wada, T., Imanishi, T., Kawaguchi, A., Mori, M.X., Mori, Y., Imoto, K., Ichida, S., 2005. Effects of calmodulin and Ca^{2+} channel blockers on ω-conotoxin GVIA binding to crude membranes from α1B subunit (Ca$_V$2.2) expressed BHK cells and mice brain lacking the α1B subunits. Neurochem. Res. 30, 1045–1054.

Wagenknecht, T.C., Liu, Z., 2010. Electron microscopy of ryanodine receptors. Curr. Top. Membr. 66, 27–47.

Wang, X.-H., Smith, R., Fletcher, J.I., Wilson, H.I., Wood, C.J., Howden, M.E.H., King, G.F., 1999. Structure-function studies of ω-atracotoxin, a potent antagonist of insect voltage-gated calcium channels. Eur. J. Biochem. 264, 488–494.

Wang, X.-H., Connor, M., Wilson, D.T., Wilson, H.I., Nicholson, G.M., Smith, R., Shaw, D., Mackay, J.P., Alewood, P.F., Christie, M.J., King, G.F., 2001. Discovery and structure of a potent and highly specific blocker of insect calcium channels. J. Biol. Chem. 276, 40306–40312.

Wang, M.C., Velarde, G., Ford, R.C., Berrow, N.S., Dolphin, A.C., Kitmitto, A., 2002. 3D structure of the skeletal muscle dihydropyridine receptor. J. Mol. Biol. 323, 85–98.

Wang, S.-Q., Zhu, M., Carafoli, E., 2011. Ca^{2+}: a versatile master key for intracellular signaling cascades. Sci. China Life Sci. 54, 683–685.

Wang, J., Li, Y., Han, Z., Zhu, Y., Xie, Z., Wang, J., Liu, Y., Li, X., 2012a. Molecular characterization of a ryanodine receptor gene in the rice leaffolder, *Cnaphalocrocis medinalis* (Guenee). PLoS One 7, e36623.

Wang, X., Wu, S., Yang, Y., Wu, Y., 2012b. Molecular cloning, characterization and mRNA expression of a ryanodine receptor gene from diamondback moth, *Plutella xylostella*. Pestic. Biochem. Physiol. 102, 204–212.

Waterhouse, A.L., Pessah, I.N., Francini, A.O., Casida, J.E., 1987. Structural aspects of ryanodine action and selectivity. J. Med. Chem. 30, 710–716.

Wehrens, X.H.T., Lehnart, S.E., Marks, A.R., 2005a. Intracellular calcium release and cardiac disease. Annu. Rev. Physiol. 67, 69–98.

Wehrens, X.H.T., Lehnart, S.E., Reiken, S., v. d. Nagel, R., Morales, R., Sun, J., Cheng, Z., Deng, S.X., de Windt, L.J., Landry, D.W., Marks, A.R., 2005b. Enhancing calstabin binding to ryanodine receptors improves cardiac and skeletal muscle function in heart failure. Proc. Natl. Acad. Sci. U.S.A. 102, 9607–9612.

Wehrens, X.H.T., Lehnart, S.E., Reiken, S., Vest, J.A., Wronska, A., Marks, A.R., 2006. Ryanodine receptor/calcium release channel PKA phosphorylation: a critical mediator of heart failure progression. Proc. Natl. Acad. Sci. U.S.A. 103, 511–518.

Wei, X.Y., Perez-Reyes, E., Lacerda, A.E., Schuster, G., Brown, A.M., Birnbaumer, L., 1991. Heterologous regulation of the cardiac Ca^{2+} channel $\alpha 1$ subunit by skeletal muscle beta and gamma subunits. Implications for the structure of cardiac L-type Ca^{2+} channels. J. Biol. Chem. 266, 21943–21947.

Welch, W., Williams, A.J., Tinker, A., Mitchell, K.E., Deslongchamps, P., Lamothe, J., Gerzon, K., Bidasee, K.R., Besch Jr., H.R., Airey, J.A., Sutko, J.L., Ruest, L., 1997. Structural components of ryanodine responsible for modulation of sarcoplasmic reticulum calcium channel function. Biochemistry 36, 2939–2950.

Wicher, D., Penzlin, H., 1997. Ca^{2+} currents in central insect neurons: electrophysiological and pharmacological properties. J. Neurophysiol. 77, 186–199.

Williams, R.J., 2006. The evolution of calcium biochemistry. Biochim. Biophys. Acta 1763, 1139–1146.

Williams, A.J., West, D.J., Sitsapesan, R., 2001. Light at the end of the Ca^{2+} -release channel tunnel: structures and mechanisms involved in ion translocation in ryanodine receptor channels. Q. Rev. Biophys. 34, 61–104.

Windley, M.J., Herzig, V., Dziemborowicz, S.A., Hardy, M.C., King, G.F., Nicholson, G.M., 2012. Spider-venom peptides as bioinsecticides. Toxins 4, 191–227.

Winterfield, J.R., Swartz, K.J., 2000. A hot spot for the interaction of gating modifier toxins with voltage-dependent in channels. J. Gen. Physiol. 116, 637–644.

Witcher, D.R., McPherson, P.S., Kahl, S.D., Lewis, T., Bentley, P., Mullinnix, M.J., Windass, J.D., Campbell, K.P., 1994. Photoaffinity labeling of the ryanodine receptor/Ca^{2+} release channel with an azido derivative of ryanodine. J. Biol. Chem. 269, 13076–13079.

Wu, L.G., Westenbroek, R.E., Borst, J.G., Catterall, W.A., Sakmann, B., 1999. Calcium channel types with distinct presynaptic localization couple differentially to transmitter release in single calyx-type synapses. J. Neurosci. 19, 726–736.

Wu, M., Wang, M., Nix, J., Hryshko, L.V., Zheng, L., 2009. Crystal structure of CBD2 from the Drosophila Na^+/Ca^{2+} exchanger: diversity of Ca^{2+} regulation and its alternative splicing modification. J. Mol. Biol. 387, 104–112.

Wu, M., Le Hoa, D., Wang, M., Yurkov, V., Omelchenko, A., Hnatowich, M., Nix, J., Hryshko, L.V., Zheng, L., 2010. Crystal structures of progressive Ca^{2+} binding states of the Ca^{2+} sensor Ca^{2+} binding domain 1 (CBD1) from the CALX Na^+/Ca^{2+} exchanger reveal incremental conformational transitions. J. Biol. Chem. 285, 2554–2561.

Xiao, B., Zhong, G., Obayashi, M., Yang, D., Chen, K., Walsh, M.P., Shimoni, Y., Cheng, H., ter Keurs, H., Chen, S.R.W., 2006. Ser-2030, but not Ser-2808, is the major phosphorylation site in cardiac ryanodine receptors responding to protein kinase A activation upon b-adrenergic stimulation in normal and failing hearts. Biochem. J. 396, 7–16.

Xu, X., Bhat, M.B., Nishi, M., Takeshima, H., Ma, J., 2000. Molecular cloning of cDNA encoding a Drosophila ryanodine receptor and functional studies of the carboxyl-terminal calcium release channel. Biophys. J. 78, 1270–1281.

Xu, C., Ma, H., Inesi, G., Al Shawi, M.K., Toyoshima, C., 2004. Specific structural requirements for the inhibitory effect of thapsigargin on the Ca^{2+} ATPase SERCA. J. Biol. Chem. 279, 17973–17979.

Yamaji, N., Sugase, K., Nakajima, T., Miki, T., Wakamori, M., Mori, Y., Iwashita, T., 2007. Solution structure of agelenin, an insecticidal peptide isolated from the spider Agelena opulenta, and its structural similarities to insect-specific calcium channel inhibitors. FEBS Lett. 581, 3789–3794.

Yan, S., Cui, F., Qiao, C., 2009. Structure, function and applications of carboxylesterases from insects for insecticide resistance. Protein Pept. Lett. 16, 1181–1188.

Yáñez, M., Gil-Longo, J., Campos-Toimil, M., 2012. Calcium binding proteins. In: Islam, S. (Ed.), Calcium Signaling. Advances in Experimental Medicine and Biology. Springer-Berlag, Berlin, pp. 461–482.

Yao, S., Gallenkamp, D., Wölfel, K., Lüke, B., Schindler, M., Scherkenbeck, J., 2011. Synthesis and SERCA activities of structurally simplified cyclopiazonic acid analogues. Bioorg. Med. Chem. 19, 4669–4678.

Yarov-Yarovoy, V., Baker, D., Catterall, W.A., 2006. Voltage sensor conformations in the open and closed states in ROSETTA structural models of K(+) channels. Proc. Natl. Acad. Sci. U.S.A. 103, 7292–7297.

Yarov-Yarovoy, V., DeCaen, P.G., Westenbroek, R.E., Pan, C.-Y., Scheuer, T., Baker, D., Catterall, W.A., 2012. Structural basis for gating charge movement in the voltage sensor of a sodium channel. Proc. Natl. Acad. Sci. U.S.A. 109, E93–E102.

Yatime, L., Buch-Pedersen, M.J., Musgaard, M., Morth, J.P., Lund Winther, A.M., Pedersen, B.P., Olesen, C., Andersen, J.P., Vilsen, B., Schiott, B., Palmgren, M.G., Møller, J.V., Nissen, P., Fedosova, N., 2009. P-type ATPases as drug targets: tools for medicine and science. Biochim. Biophys. Acta 1787, 207–220.

Ye, Y., Yaeger, D., Owen, L.J., Escobedo, J.O., Wang, J., Singer, J.D., Strongin, R.M., Abramson, J.J., 2012. Designing calcium release channel inhibitors with enhanced electron donor properties: stabilizing the closed state of ryanodine receptor type 1. Mol. Pharmacol. 81, 53–62.

Yu, F.H., Catterall, W.A., 2004. The VGL-chanome: a protein superfamily specialized for electrical signaling and ionic homeostasis. Sci. STKE 2004, re15.

Yuchi, Z., van Petegem, F., 2011. Common allosteric mechanisms between ryanodine and inositol-1,4,5-triphosphate receptors. Channels 5, 120–123.

Yue, L., Navarro, B., Ren, D., Ramos, A., Clapham, D.E., 2002. The cation selectivity filter of the bacterial sodium channel, NaChBac. J. Gen. Physiol. 120, 845–853.

Zhao, F., Li, P., Chen, S.R.W., Louis, C.F., Fruen, B.R., 2001. Dantrolene inhibition of the ryanodine receptor Ca2+ release channels. Molecular mechanism and isoform selectivity. J. Biol. Chem. 276, 13810–13816.

Zheng, W., Feng, G., Ren, D., Eberl, D.F., Hannan, F., Dubald, M., Hall, L.M., 1995. Cloning and characterization of a calcium channel α1 subunit from *Drosophila melanogaster* with similarity to the rat brain type D isoform. J. Neurosci. 15, 1132–1143.

Zhou, Y., Yang, W., Kirberger, M., Lee, H.W., Ayalasomayajula, G., Yang, J.J., 2006. Prediction of EF-hand calcium-binding proteins and analysis of bacterial EF-hand proteins. Proteins 65, 643–655.

Zhu, X., Ghanta, J., Walker, J.W., Allen, P.D., Valdivia, H.H., 2004. The calmodulin binding region of the skeletal ryanodine receptor acts as a self-modulatory domain. Cell Calcium 35, 165–177.

Zieminska, E., Stafiej, A., Pitsinos, E.N., Couladouros, E.A., Moutsos, V., Kozlowska, H., Toczylowska, B., Lazarewicz, J.W., 2007. Synthetic bastadins modify the activity of ryanodine receptors in cultured cerebellar granule cells. Neurosignals 15, 283–292.

Zissimopoulos, S., Lai, F.A., 2006. Redox regulation of the ryanodine receptor/calcium release channel. Biochem. Soc. Trans. 34, 919–921.

Zuccotti, A., Clementi, S., Reinbothe, T., Torrente, A., Vandael, D.H., Pirone, A., 2011. Structural and functional differences between L-type calcium channels: crucial issues for future selective targeting. Trends Pharmacol. Sci. 32, 366–375.

Zwaal, R., Kaletta, T., van den Craen, M., Logghe, M., Smits, E., van Creikinge, W., Bogaert, T., 2002. Methods for identifying pesticidal compounds using gene sca-1 for sarco-endoplasmic reticulum Ca^{2+} ATPase cloned from C. elegans. WO patent application number 2002033405.

INDEX

Note: Page numbers followed by "*f*" indicate figures, and "*t*" indicate tables.

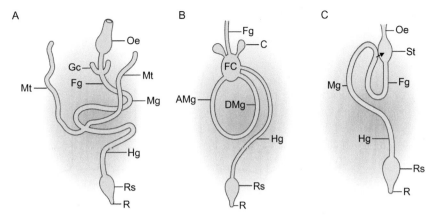

PLATE 1 Schemes of insect digestive tracts: A, typical insect; B, whitefly; C, aphid. AMg, ascending midgut; C, coecum; DMg, descending midgut; Fc, filter chamber; Fg, foregut; Gc, gastric coecum; Hg, hindgut; Mg, midgut; Mt, Malpighian tubule; Oe, oesophagus; R, rectum; Rc, rectal sac; St, stomach. Anatomical gut structures associated with water-shunting in whiteflies and aphids are depicted in B and C, respectively. Cohen, E and OMICS Publishing Group (ISSN-2161-0983) are credited for this figure. (For b/w version, see page 4 in the volume.)

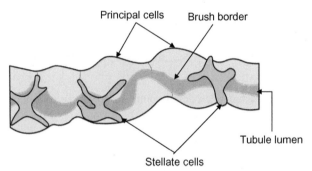

PLATE 2 The scheme depicts the two different cell types of nearly all Malpighian tubules. Principal cells are abundant, large and densely brush-bordered. Stellate cells are smaller, star-shaped, scattered among principal cells and have fewer and smaller apical microvilli. (For b/w version, see page 5 in the volume.)

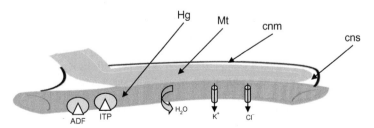

Cryptonephridial complex

PLATE 3 A generalized scheme of a cryptonephridial complex. Hg, hindgut; Mt, Malpighian tubule attached to the hindgut; cnm, cryptonephridial membrane; cns, cryptonephridial space. The putative receptors for ADF (antidiuretic factor) and ITP (ion transport peptide) neurohormones are depicted in circles. The cylinders represent channels for K^+ and Cl^- ions. (For b/w version, see page 6 in the volume.)

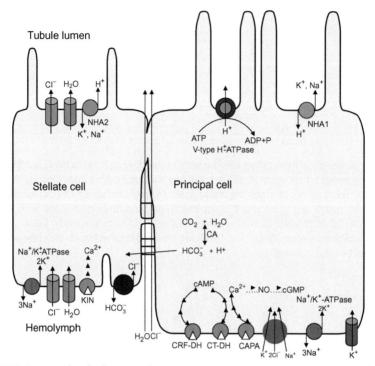

PLATE 4 A generalized schematic diagram summarizing various current models for ion and water transport as well as receptor signalling pathways in principal and stellate cells of Malpighian tubules. Cylinders in basal and apical domains of the stellate cell represent transcellular chloride and water channels (aquaporins), whereas the transmembrane basolateral-rectified potassium ion channel is depicted in the principal cell. In addition, the paracellular pathway for water molecules and chloride ions is included. The apical ion exchangers NHA1 and NHA2 are represented in the principal cell and in the stellate cell, respectively. The V-type H^+-ATPase that transports protons across apical cell membranes is illustrated in the principal cell. Basolateral G-protein-coupled receptors (GPCRs) for kinins (KIN), corticotrophin-releasing factor-like DHs (CRF-DH), cardioacceleratory peptide (CAP_{2b}), and calcitonin-like DHs (CT-DH) are depicted. Intracellular second messengers such as cAMP and Ca^{2+}, which are triggered by CRF-DH and CT-DH and by KIN, correspondingly, are presented. Generation of cAMP by interaction of serotonin with its GPCR is not included in the diagram. The cascade pathway generating cGMP following interaction of CAP_{2b} with its receptor starts with Ca^{2+} activating NOS, and the produced NO interact with guanylate cyclase (GC) to form cGMP. The K^+, Na^+, $2Cl^-$ co-transporter (red circle) in the principal cell and the Na^+/K^+-ATPase (purple circle) in both cell types are represented. The bicarbonate anion (HCO_3^-) produced in principal cell by carbonic anhydrase (CA) is transported via septate junctions to the stellate cell where the basal bicarbonate/chloride ions exchanger functions. (For b/w version, see page 7 in the volume.)

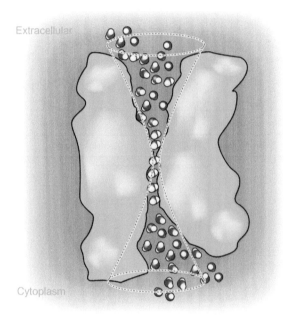

PLATE 5 "Hourglass" model of AQP (side view). Circles represent water molecules inside the narrow channel and in the extracellular and cytoplasmic vestibules. Cohen, E. and OMICS Publishing Group (ISSN-2161-0983) are credited for this figure. (For b/w version, see page 13 in the volume.)

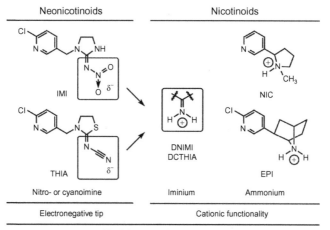

Neonicotinoids   Nicotinoids

IMI

THIA

Nitro- or cyanoimine

DNIMI
DCTHIA

NIC

EPI

Iminium   Ammonium

Electronegative tip   Cationic functionality

Nicotinic agonist pharmacophores

ESP

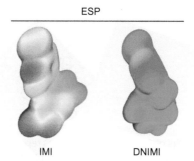

IMI   DNIMI

PLATE 6 Neonicotinoid and nicotinoid chemotypes of nicotinic agonists with an electronegative pharmacophore and cationic functionality, respectively (*upper*). Electrostatic potential (ESP) mapping on the molecular surfaces of representative neonicotinoid IMI and nicotinoid DNIMI (protonated at physiological pH) was obtained by B3LYP/6-311G** (Tomizawa et al., 2003) (*lower*). ESP surfaces are exhibited in red for negative and are graded through orange, yellow, and green to blue for positive with an overall energy range of −60 to 160 kcal/mol. (For b/w version, see page 65 in the volume.)

Heteropentameric nAChR

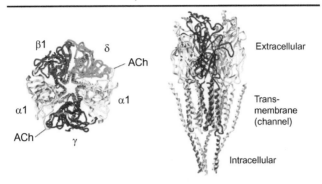

Homopentameric AChBP

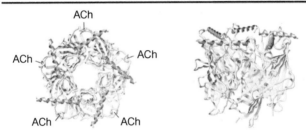

An interfacial ligand-binding pocket of AChBP

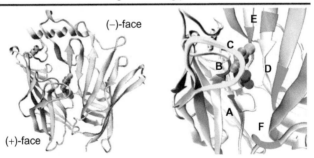

PLATE 7 Protein structures of *Torpedo marmorata* heteropentameric nAChR (top) and *A. californica* homopentameric AChBP (*middle*) (views from top (*left*) and side (*right*) in each protein) based on refined electron microscopy and crystallography data (PDB ID codes 2BG9 and 2BYN, respectively) (Hansen et al., 2005; Unwin, 2005). AChBP lacks the transmembrane domain (ion channel pore) and cytoplasmic end of the nAChR. An interfacial ligand-binding pocket of AChBP (extracted from 3C79 pentamer (Talley et al., 2008)) (overview, *bottom left*) consisting of six regions (loops A–F as highlighted by colours) (zoomed-in, *bottom right*). Loops A–C domains are from the (+)-face principal subunit, while loops D– F areas are from the (−)-face partnering subunit. IMI molecule is also displayed as a landmark. (For b/w version, see page 70 in the volume.)

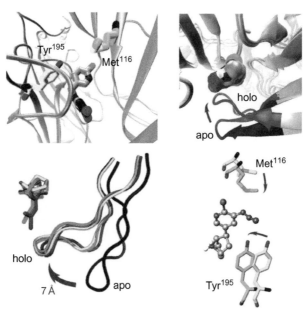

PLATE 8 Agonist-binding site at the *Aplysia* AChBP as a suitable insect nAChR surrogate. Representative insecticide IMI is nestled in the interfacial agonist-binding pocket between the primary or (+)-face (silver) and complementary or (−)-face (pink) subunits (*upper left*). Sites for photoderivatization (Tyr[195] and Met[116] on the principal and partnering subunits, respectively) are featured. Conformational rearrangement of the loop C domain induced by agonist occupation based on crystal structures apo- and agonist-bound (holo) *Aplysia* AChBPs. Typical image for the interfacial binding pocket embracing the IMI molecule (PDB ID code 3C79) which is superimposed onto the apo form (2BYN) (*upper right*). The loop C region of the ligand vacant apo form (blue, 2BYN) is overlaid with those of the IMI- (red, 3C79), THIA- (green, 3C84), and EPI-bound (yellow, 2BYQ) forms (Hansen et al., 2005; Talley et al., 2008) (*lower left*). Comparisons of geometries for Tyr[195] and Met[116] between the azidoepibatidine–AChBP complex (docked onto 2BYQ) and the apo form (2BYN) (*lower right*). Photoderivatization sites Tyr[195] and Met[116] in the photoprobe-bound conformation (aquamarine) are superimposed onto those in the apo structure (yellow) (Tomizawa et al., 2007b). (For b/w version, see page 74 in the volume.)

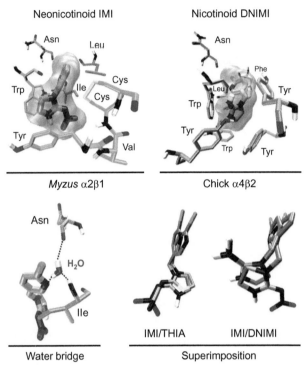

Neonicotinoid IMI — Nicotinoid DNIMI

Myzus α2β1 — Chick α4β2

Water bridge — Superimposition

IMI/THIA — IMI/DNIMI

PLATE 9 Structural models for binding site interactions of IMI and DNIMI with the α–β subunit interfacial agonist-binding pocket of nicotinic receptors based on chemical and structural biology investigations. Representative neonicotinoid IMI and nicotinoid DNIMI are embedded in the aphid (*Myzus persicae*) α2β1 and chick α4β2 interfaces, respectively (*upper*). Amino acids in lime-green or pink are from aphid α2 or chick α4 subunit, and those in orange or cyan are from aphid β1 or chick β2 subunit, respectively. A water molecule near the pyridine nitrogen atom, captured in AChBP-IMI or AChBP-THIA crystal structure (PDB ID code 3C79 or 3C84, respectively) (Talley et al., 2008), is superimposed onto this IMI-bound structure (*lower left*). Consistently, a water or solvent bridge is also observed in the AChBP crystals liganded with nicotinoids NIC and EPI (Celie et al., 2004; Hansen et al., 2005). Binding conformations of IMI, THIA, and DNIMI as observed in the agonist-binding pocket are compared (lower right). IMI and THIA are nicely superimposable, whereas IMI and DNIMI (or THIA and DCTHIA (not shown)) pharmacophores are reversed relative to each other. DNIMI- and EPI-bound conformations are suitably overlaid (not shown) (Tomizawa et al., 2007a, 2008a). (For b/w version, see page 76 in the volume.)

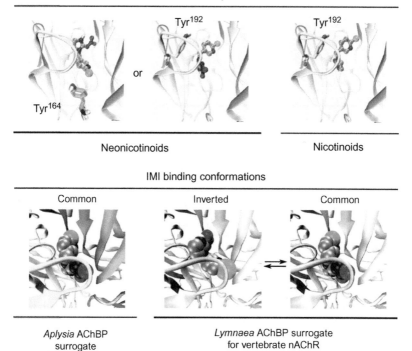

Modification sites on *Lymnaea* AChBP

Tyr^{192} Tyr^{192}

Tyr^{164}

Neonicotinoids Nicotinoids

IMI binding conformations

Common Inverted Common

Aplysia AChBP
surrogate
for insect nAChR

Lymnaea AChBP surrogate
for vertebrate nAChR

PLATE 10 Molecular recognition conferring selective receptor interaction. Identified photoderivatization sites on the *Lymnaea* AChBP as a structural homologue of the vertebrate nAChR (*upper*). Neonicotinoid photoprobe specifically modifies either Tyr^{192} on the primary face or Tyr^{164} on the complementary face, whereas nicotinoid photoprobe exclusively derivatizes Tyr^{192} (Tomizawa et al., 2008a). Comparative binding conformations of IMI embraced by *Aplysia* AChBP as an insect nAChR homologue and by *Lymnaea* AChBP as a vertebrate nAChR surrogate (*lower*). IMI binds in two conformations designated "common" and "inverted". (For b/w version, see page 79 in the volume.)

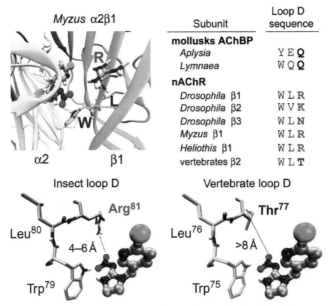

Subunit	Loop D sequence
mollusks AChBP	
Aplysia	Y E **Q**
Lymnaea	W Q **Q**
nAChR	
Drosophila β1	W L R
Drosophila β2	W V K
Drosophila β3	W L N
Myzus β1	W L R
Heliothis β1	W L R
vertebrates β2	W L **T**

PLATE 11 Insect nAChR structural model in IMI-bound state reveals a unique cavity extending from the bound IMI nitro oxygen tip towards the loop D region. IMI is docked into the aphid (*Myzus persicae*) α2β1 interfacial agonist-binding pocket (Tomizawa et al., 2008b) (*upper left*). Three amino acid residues (W (Trp), L (Leu), and R (Arg)) on the β1 subunit make up the loop D cavity. Alignment of the loop D amino acid sequences from two mollusk AChBP subtypes, five insect β subunits (represented by fruit fly *Drosophila melanogaster*, green peach aphid *Myzus persicae*, and tobacco budworm *Heliothis virescens*), and the vertebrate (human, rodent, and chick) β2 subunit of the nAChR (*upper right*). The loop D Arg on the *Myzus* β1 subunit is spatially equivalent to Thr (T) with a shorter side chain on the β2 subunit of the vertebrate neuronal nAChR, suggesting a difference between insect and vertebrate loop D cavities in their depths and functional residues (*lower*). (For b/w version, see page 83 in the volume.)

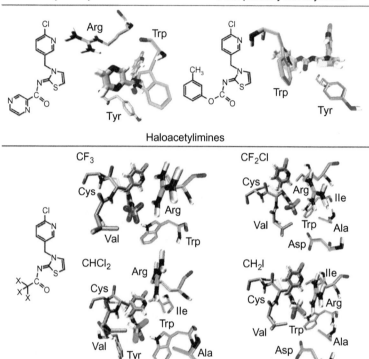

PLATE 12 Structural models for the binding site interactions of candidate nicotinic insecticides with the α–β interfacial agonist-binding pocket of the insect nAChR (*Myzus* α2β1), featuring the functions of novel pharmacophores in the regional binding niche (particularly, loops C–D and their adjacent areas). The relevant amino acids in pink are from α2 subunit and in lime-green are from β1 subunit. The binding interactions of the 2-imino-3-(6-chloropyridin-3-ylmethyl)-thiazoline moiety are identical to those of 2-imino-3-(6-chloropyridin-3-ylmethyl)-imidazolidine moiety of IMI (see Fig. 2.6) (Ohno et al., 2009b, 2010b; Tomizawa et al., 2008b, 2011b). (For b/w version, see page 86 in the volume.)

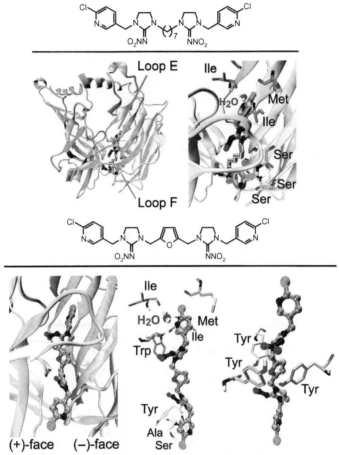

PLATE 13 Predicted binding site interactions of bis-IMI insecticides bearing a heptamethylene linker (*upper*) and a furan-2,5-dimethylene fulcrum (*lower*) with the *Aplysia* AChBP (PDB ID code 3C79 (Talley et al., 2008)) as an insect nAChR homologue (Kagabu et al., 2010; Ohno et al., 2009c). The bis-IMIs are docked in the subunit interfacial binding pocket between the (+)-face (lime-green) and (−)-face (orange) subunits, wherein the chloropyridine moieties contact loops E and F domains. One of five interfacial binding pockets is extracted and shown from the homopentameric AChBP structure. The water molecule (Ohno et al., 2009a; Talley et al., 2008), bridging between pyridine nitrogen and related amino acids, is displayed. The heptamethylene spacer optimally bridges these two subsites, yet the linker itself binds in a relatively non-specific manner. In contrast, the furan ring is specifically embraced by a hydrophobic pocket, consisting of three aromatic amino acids, and is stabilized via H-bonding. (For b/w version, see page 91 in the volume.)

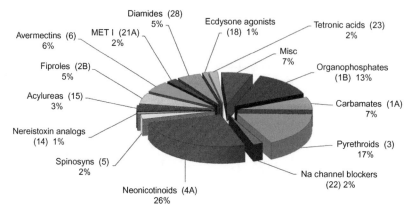

PLATE 14 Percentage of total global insecticide sales 2010 [total = $12.6 billion end user dollars]. Numbers in ( ) are IRAC mode of action group classification. *Source: Agranova (2012).* (For b/w version, see page 103 in the volume.)

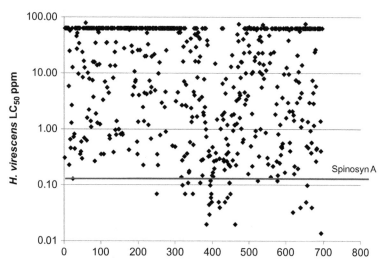

PLATE 15 Progression of spinosyn analogue efficacy. (For b/w version, see page 132 in the volume.)

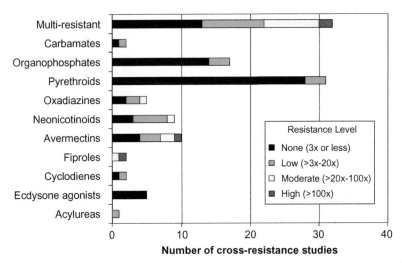

PLATE 16 Distribution of examples of cross-resistance to the spinosyns in insect species resistant to other insecticides. Level of cross-resistance to the spinosyns—none (3 × or less); low (> 3×–20 ×); moderate (> 20–100); high (>100). Multiresistant = likely presence of more than resistance mechanism. Total number of studies = 116. (For b/w version, see page 199 in the volume.)

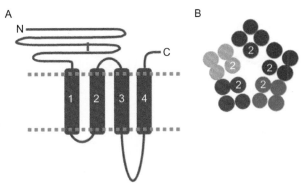

PLATE 17 Schematic presentation of an LGIC subunit. (A) Side view. (B) Top view showing arrangement of subunits and transmembrane domains. (For b/w version, see page 216 in the volume.)

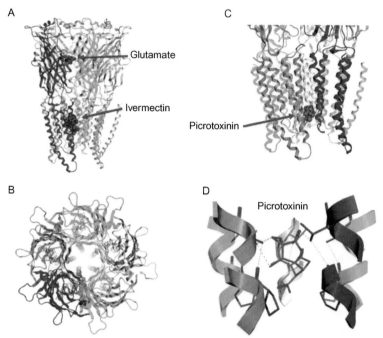

PLATE 18 Structures of *C. elegans* GluCl. (A) Side view of GluCl bound with glutamate and ivermectin. Illustration is based on X-ray crystal structure (PDB code 3RHW). (B) Top view. (C) Side view of GluCl transmembrane domain bound with picrotoxinin. Illustration is based on X-ray crystal structure (PDB code 3RI5). (D) Close-up of the picrotoxinin-binding site. (For b/w version, see page 241 in the volume.)

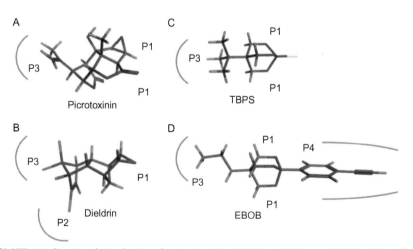

PLATE 19 Structural similarity of picrotoxinin, dieldrin, TBPS, and EBOB. (For b/w version, see page 251 in the volume.)

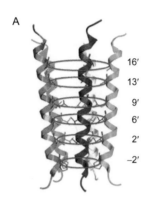

A

16'
13'
9'
6'
2'
−2'

B

```
                 -2'   2'    6'  9'  13'  16'
DmRDL        TPARVALGVTTVLTMTTLMSSTNA
DmGluClα     VPARVSLGVTTLLTMATQTSGINA
CeGluClα     IPARVTLGVTTLLTMTAQSAGINS
RatGABARα1   VPARTVFGVTTVLTMTTLSISARN
RatGABARβ2   SAARVALGITTVLTMTTINTHLRE
RatGABARγ2   VPARTSLGITTVLTMTTLSTIARK
```

PLATE 20 Channel pore of *C. elegans* GluCl. (A) Channel-lining amino acid residues of the TM2 region of the *C. elegans* GluCl α subunit. Illustration is based on X-ray crystal structure (PDB code 3RI5). (B) Alignment of the TM2 sequences of GABACl (GABAR) and GluCl subunits. Dm, *Drosophila melanogaster*; Ce, *Caenorhabditis elegans*. (For b/w version, see page 254 in the volume.)

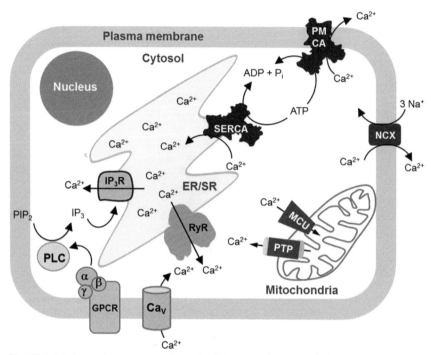

PLATE 21 Schematic representation of calcium regulation and the roles of calcium-handling proteins. The internal calcium stores are represented by the sarco-/endoplasmatic reticulum and the mitochondria. Voltage-gated calcium channels (Ca_V) mediate a limited calcium entry after depolarization of the cell membrane. This is followed by a massive calcium release from the internal stores mediated by the ryanodine receptor (RyR). The second intracellular calcium-release channel is the IP_3 receptor (IP_3R). IP_3 is produced by phospholipase C (PLC) through hydrolysis of phosphatidylinositol 4,5-bisphosphate. PLC activity is stimulated through certain G-protein-coupled receptors. The sarcoplasmic Ca^{2+}-ATPase (SERCA) refills the internal calcium stores, whereas the plasma membrane Ca^{2+}-ATPase (PMCA) and the Na^+/Ca^{2+} exchanger (NCX) transport calcium to the extracellular space. In mitochondria, the calcium uniporter (MCU) and the permeability transition pore (PTP) carry out calcium uptake and release in/from the mitochondrial stores. (For b/w version, see page 291 in the volume.)

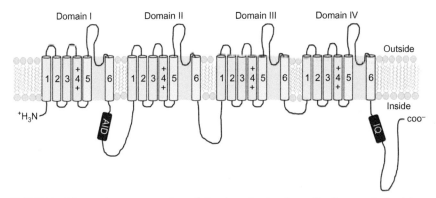

PLATE 22 Schematic representation of the domain structure of voltage-gated calcium channels. Highlighted are helices 5 and 6 of each domain. The inner pore is formed by helices 6 of domain I–IV. The pore loops connect helices 5 and 6 of each domain. The positive charges depict the voltage-sensor helix 4. The β-subunit interacts with the AID site located in the intracellular loop between domain I and II. The carboxy-terminal IQ-motif binds calmodulin. (For b/w version, see page 302 in the volume.)

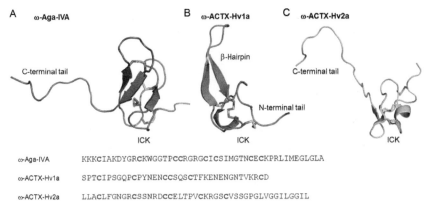

ω-Aga-IVA	KKKCIAKDYGRCKWGGTPCCRGRGCICSIMGTNCECKPRLIMEGLGLA
ω-ACTX-Hv1a	SPTCIPSGQPCPYNENCCSQSCTFKENENGNTVKRCD
ω-ACTX-Hv2a	LLACLFGNGRCSSNRDCCELTPVCKRGSCVSSGPGLVGGILGGIL

PLATE 23 Sequences and solution structures of spider venom peptides acting on insect Ca$_V$. (A) ω-Agatoxin IVA, (B) ω-atracotoxin Hv1a, and (C) ω-atracotoxin Hv2a. ICK, inhibitor cystine knot. Coordinates of the solution structures by 1H-NMR were taken from PDB files: 1OAW (ω-agatoxin IVA), 1AXH (ω-atracotoxin Hv1a), and 1G9P (ω-atracotoxin Hv2a); graphs were generated using PyMOL v0.99 (DeLano Scientific). (For b/w version, see page 308 in the volume.)

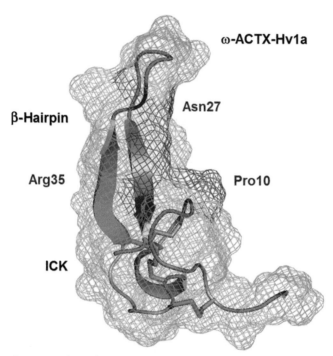

PLATE 24 Contact surface of ω-atracotoxin Hv1a suggested to face the insect Ca$_V$. The contact surface is suggested to comprise only residues Pro10, Asn27, and Arg35 (labeled in blue). Coordinates of the solution structure from PDB file 1AXH; graph generated using PyMOL v0.99 (DeLano Scientific). (For b/w version, see page 311 in the volume.)

Printed and bound by CPI Group (UK) Ltd, Croydon, CR0 4YY

08/05/2025

01864955-0002